HYBRID

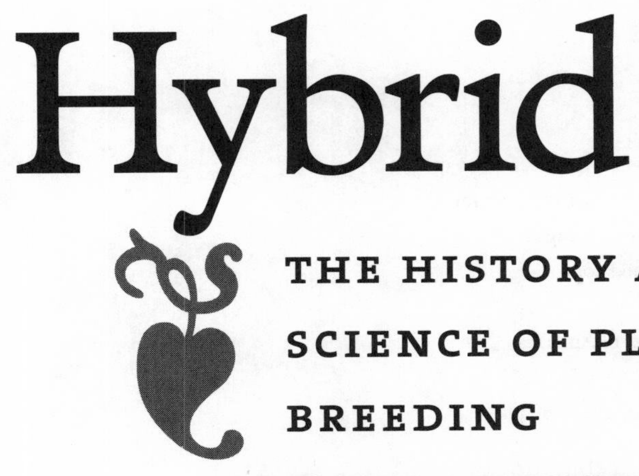

Hybrid

THE HISTORY AND SCIENCE OF PLANT BREEDING

NOEL KINGSBURY

The University of Chicago Press

Chicago and London

The University of Chicago Press, Chicago 60637
The University of Chicago Press, Ltd., London

Paperback edition 2011
Printed in the United States of America

20 19 18 17 16 15 14 13 12 11   2 3 4 5 6

ISBN-13: 978-0-226-43704-0 (cloth)
ISBN-10: 0-226-43704-3 (cloth)
ISBN-13: 978-0-226-43713-2 (paper)
ISBN-10: 0-226-43713-2 (paper)

Library of Congress Cataloging-in-Publication Data
Kingsbury, Noel.
Hybrid : the history and science of plant breeding / Noel Kingsbury.
p. cm.
Includes bibliographical references and index.
ISBN-13: 978-0-226-43704-0 (cloth : alk. paper)
ISBN-10: 0-226-43704-3 (cloth : alk. paper)
1. Plant breeding—History. 2. Hybridization. I. Title.
SB123.K554 2009
631.5′2—dc22

2008054051

⊗ This paper meets the requirements of ANSI/NISO Z39.48-1992 (Permanence of Paper).

CONTENTS

The greatest service which can be rendered any country is to add a useful plant to its culture; especially a bread grain.—Thomas Jefferson

❦

If you want to feed someone, give them some rice, if you want to give them a reason for living, give them a flower.—Chinese proverb

❦

History . . . celebrates the battlefields whereon we meet our death, but scorns to speak of the ploughed fields whereby we thrive; it knows the names of the king's bastards, but cannot tell us of the origin of wheat. That is the way of human folly. —J. Henri Fabre

ACKNOWLEDGMENTS

First and foremost I am very grateful to David Shaw, of Bangor, Wales, for his very generously agreeing to read through the manuscript and for his advice on the presentation of the technical information in the text regarding genetics. I'm also very grateful to Ethne Clarke, for reading through the manuscript and reassuring me that what I write is not full of "deeply ingrained, suck it up with mother's milk Briticisms, untranslatable to an American readership." Fundamentally though, I owe a debt of gratitude to Christie Henry at University of Chicago Press for seeing the potential in the idea of this book in the first place. Thanks also to Mary Gehl, the readers, and those others at the press who have played their part in bringing the idea to fruition.

In writing this history, I have depended very heavily on libraries and their invariably helpful staff. I would like to express my appreciation of the facilities and the staff of the following: the Swedish Agricultural University Library at Alnarp; Iowa State University; Cornell University; the university libraries of Sheffield, Bristol, Birmingham, Oxford, and Cambridge; and the Lindley Library of the Royal Horticultural Society in London. Also to Jayne at my local library at Hay-on-Wye in Wales and the staff of the British Inter-Library Loans system—one of the truly marvelous institutions of British public life. I am particularly grateful to Sukey Haney of the library of the Tower Hill Botanical Gardens, Massachusetts, and to the staff of the Kerala Agricultural University, Mannuthy, India, for their assistance beyond the call of duty.

It has been very useful talking to active breeders, or former breeders, or others in the worlds of agriculture and horticulture, both to gain specific information and also to get a sense of what a breeder's life is like: Tremar Menendez and Andrew Ormerod in Cornwall; Colin Leakey of Lincoln, England; Dr. K. Arivandakshan, Dr. C. R. Elsy, Dr. Presanna Kumari, Dr. V. K. Malika, Professor Sreevalsan Menon, and Dr. V. V. Radhakrishnan, at Kerala Agricultural University; Professor Ronnie Coffman, Dr. Kathleen Gale, Dr. Lee Kass, and Professor Royse Murphy at Cornell; Dr. P. Gypmanteseri at Chiang Mai University, Thailand; Dawn Buell, Dale Flowerday, and Stan Jensen in Nebraska; Professor Jerald Pataky at the University of Illinois; and others in Britain, including Claire Austin, Charles Chesshire, Raymond Evison, Penny Maplestone, Professor Denis Murphy, Sarah Ritterhausen, and Jennifer Trehane, also Bev Larson of Northrup King.

I'd also like to say thanks to my indefatigable agent Fiona Lindsay and to my dear partner Jo Eliot, for her encouragement and support whilst writing this book and for her assistance in reading the manuscript and indexing. Then there are those friends who have so kindly put me up whilst using library facilities, in particular Helen and Jonathan Barnes of London and Ethne and Don Clarke of Des Moines, Iowa.

A NOTE ON NAMES

Names cause endless confusion, partly because scientific ("Latin") names are so often different to common names, the extent to which they are used varies, and because of differences between different variants of English. Anyone unfamiliar with the basics of scientific names—the *binomial* system—should refer to the technical notes at the end of the book now.

As a general rule, all crop plants are referred to by their English names; where there is confusion over British and American English variants, this is noted. *Zea mays* is referred to throughout as "corn"—rather than "maize"—although "corn" in the British English vernacular is understood to mean *wheat*.

The first time a crop is mentioned, its scientific name is given—in many cases this may only be a genus (see technical note 1: Binomial Names, Selections, and Hybrids), as many crops are derived from, and include, several species.

In the case of ornamentals, most of those discussed have scientific names that have entered into the vernacular—we all know what rhododendrons and camellias are. Where this is not the case, a common name is used only if it is widely accepted (and on first use is followed by the scientific name). Individual species are often given the scientific name—as English common names are often not widely used or recognized.

Names of specific plant varieties in this book will be given in single quotation marks, for example, 'Little Joss.' This is common practice in horticulture but not necessarily in other fields, such as agriculture. The advantage of doing so is that it stresses that the variety in question is either a cultivar (i.e., all individuals are genetically identical) or that all individuals are very similar. Where a name is given that is not in single quotes, for example, Sea Island cotton, then the reference is to a race or grouping where all individuals share certain characteristics but are by no means genetically identical—and where a number of distinct identities can be recognized.

INTRODUCTION

Shopping for food: we all do it, whether at the supermarket, or from traditional neighborhood shops, or in a market. It's the modern equivalent of what our ancestors would have done in long-gone hunter-gatherer days. The hunters (nearly always men) have perhaps earned a bit too much of the limelight in our popular reconstructions of life in these times; in most such societies, the bulk of the food would have been gathered rather than hunted, mostly by groups of women. Digging up roots and picking berries is a rather more unglamorous activity than chasing after and spearing mammoths, but it undoubtedly brought in more calories and fed more mouths. Shopping is more like gathering than hunting, although those who like to track down obscure wines and rare cheeses might disagree.

Our ancestors would have had an immediate and very direct experience of what they were gathering as they clawed tubers out of the ground, got pricked by thorny stems as they gathered berries, or trekked miles to find a particularly rich source of mushrooms. By contrast we know so little about where our food actually comes from. I do not mean its country of origin, although we often know little enough about that, but its historical origins. As you trundle your shopping cart around the aisles and gather up tomatoes (*Solanum lycopersicum*), carrots (*Daucus carota*), and apples (*Malus domestica*), think about this: is there an Eden somewhere in the world where tribal people pick big, juicy tomatoes—just like the supermarket ones, but from vines clambering over bushes; ease nice fat orange carrots from the earth with a digging stick, or gather perfectly ripe apples from trees in large woven baskets? A moment's reflection and we realize this is too rosy a vision of Eden to be true.

Having been shopping for food, let's go and have a look around the garden center and ponder the same sort of questions. Roses are perhaps the obvious place to start: the favorite flower of both the Christian and the Muslim civilizations, laden with literary, symbolic, and spiritual meanings. They are also a dramatic illustration of how different a wild plant and a cultivated one can be. Wild roses have relatively small single flowers with five petals and are nearly always pink; also they do not flower for very long—only a few weeks—so fail to take a country walk for a while and you will miss them.

Garden roses are, for the most part, so different that if they did not have the same name and very similar foliage, many would not believe they could possibly be the same plant. Their flowers are large, with a great many petals; they flower for months on end and come in every color except, famously, *blue*. It is hard to imagine that there can be a corner of the globe where wild roses grow that look even remotely like the cultivated ones; such a place could only be a fantasy, a *Through the Looking-Glass* world where a Red Queen rules among trees and shrubs covered in perpetually flowering multicolor roses.

This book is the story of how the plants that feed us and clothe us came to be the way they are. It is the story of how our fruit, vegetables, and grain crops came to be far more appetizing than their wild ancestors, but also more productive, easier to grow, and very often more nutritious. We will also look, but more briefly, at the other side of plant breeding—that of gardens, flowers and ornamental horticulture.

MEET THE ANCESTORS—WILD CROPS

For a better idea of what our ancestors had as plant food resources take a trip to the seaside—I'm thinking specifically of the coastline of Britain and northwest Europe. Whether it is just coincidence or not might be interesting to speculate, but it is possible to find the ancestors of several of our most familiar vegetables growing on the cliff tops. Meeting them and realizing their connections with the plastic and Styrofoam-wrapped contents of our supermarket cart (or even our vegetable patch in the garden—if we have one) brings home very forcefully just how far we have come. Let's start with wild cabbage (*Brassica oleracea*). It grows usually on the very cliff edge itself—tough, almost woody stems jutting out into the wind, with large grey succulent-looking leaves, and sometimes, heads of yellow flowers. Many who walk by will miss it, for it looks nothing like cabbage, although home vegetable growers may recognize more than a passing resemblance to a broccoli plant. Rubbing a leaf and smelling it however will trigger instant recognition. Pick some and take it home or back to the self-catering cottage you may have rented for your coastal break, chop it up, cook it, and see what you think.[1] Most will find it tough and very strongly flavored, but recogniz-

1. It should be pointed out that there are two provisos here: one is that you should never pick and eat wild food without being sure that you haven't picked anything poisonous—check in a wildflower identification guide first. The second is that if

ably cabbage-like. Having met wild cabbage, we inevitably ask how the wild plant got to look like the neat, tightly wrapped balls of leaves we buy in the shops. And what is the connection with broccoli? And of those other vegetables that are grouped under the heading of 'brassicas': cauliflower, brussels sprouts, kale, collard greens?

There is another wild vegetable that can sometimes be found along the coast—the wild carrot, with ferny foliage and circular heads made up of many hundreds of tiny flowers. Dig one up—the root smells of carrot, but there the resemblance ends, for not only are the roots very narrow but also they are also not orange, but dirty white, and acrid in taste. So why are garden carrots tasty, plump, and juicy, but also orange?

NATURE AND NURTURE—
EVOLUTION, NATURAL AND HUMAN-DIRECTED

The gap between wild plants, which would have surrounded our tribal ancestors and those with which we are familiar, is often huge—or it may not be. It is widest and most dramatic in many cereal, fruit, and vegetable crops, where the ancestors may be almost unrecognizable, and least so in herbs and in some ornamental plants, where differences can be very slight. The gap between the wild and the cultivated is all about the difference between nature's requirements and ours. At first glance, life in the wild—the forest, marsh, or prairie—and life in cultivation—the field, the orchard, the garden, or the window box—may seem very different. However, at root, they are the same, as both are subject to evolutionary pressures. On the one hand it is nature that provides this pressure, on the other it is us, the human race.

Plants growing in the wild are subject to a struggle for survival, "the survival of the fittest," "fittest" here meaning the best adapted for coping with a given set of conditions. It was Charles Darwin writing in the nineteenth century who first formulated the idea of evolution, the idea that plants and animals are in constant competition, which results in the fittest or the best adapted surviving and passing on their genes to the next generation, those less well-adapted dying or failing to reproduce and so not passing on their genetic code. Life, then, is a constant winnowing and sifting of genetic material.

Plants in cultivation are protected from many of the vagaries of nature. They are watered if they get too dry, fed with fertilizers if the soil is too poor

---

everyone started harvesting wild food, there would soon be none left! Wild plants should only be collected for food or other purposes if they are locally abundant.

and weeded—that is, competing plants removed. Life may seem like luxury compared with the wild. But is it? Think about weeding for a start. What is a weed? "A plant in the wrong place" is a common witty answer, but it is actually very accurate, for we define weeds as plants we do not want that compete for resources with those we do want. Clearly we have criteria about which plants we want and those that fail those criteria. If we think about our attitudes about weeding in evolutionary terms, it is the cultivated plants that are "fitter" than the weeds, as they have characteristics which we want, and since in the field and the garden we have largely substituted ourselves for nature, it is *we* who control the evolutionary process.

In thinking about the great gap between wild plants and modern agricultural and garden plants, I find it is helpful to think like this: we are exerting an evolutionary pressure on plants. This is where plant breeding and selection comes in. A good example may be a peasant farmer in India who keeps some of the crop from their highest-yielding lentil plants for sowing next year. The farmer is exerting an evolutionary pressure on the lentil (*Lens culinaris*)—perhaps towards larger numbers of lentils per pod, or more pods per plant. Such plants are "fitter" in evolutionary terms. Scientists working with plants do the same, except that they understand much more of the process of heredity and are able to take direct action to get particular targeted results. "Selection" is about picking the best from each generation; "breeding" is about a more active process of crossing different strains of plant for particular results.

It is plant breeders who stand between ancestral crops or wild plants and modern varieties. In the case of wheat (*Triticum* species), barley (*Hordeum vulgare*), and rice (*Oryza* species[2]), which were first cultivated thousands of years ago, there must be a long chain reaching through history, as generation passed on seed to generation, making small and incremental changes along the way. Each picked the best of its crop, with occasional big leaps as sudden (and accidental) genetic changes happened along the way. For most of this time, plant breeding (or more accurately, selection) has been a matter of traditional knowledge and techniques, of seed being stored in woven baskets and perhaps traded at annual markets, good seed from high-yielding plants being bartered for metal tools or cloth. It is only over the last century or so that plant selection and breeding has become a systematic science.

2. There are two species of rice in cultivation: the African *O. glaberrima* and the Asian *O. sativa.* The latter is highly complex, but there are two distinct varieties, the *indica* types south of the Himalayas and the *japonica* north of them.

Nowadays test tubes and computers have replaced the baskets, and huge research budgets have replaced the barter.

Whether traditional or modern, it is humans selecting plants and, later, plant breeders who have transformed wild plants into cultivated; it is they and their stories who explain the dramatic gap between the thin white roots of wild carrot and the plump orange ones on the supermarket shelf.

THRUST UNWILLINGLY TO THE STAGE—PLANT BREEDING ENTERS POLITICS

The turn of the twentieth century into the twenty-first has seen the most acrimonious debate ever in the history of plant breeding, indeed, it is perhaps the only time when plant breeding has become the topic of frenzied debate, newspaper headlines, demonstrations, and television discussion programs. "Genetic engineering" (hereafter referred to as GM—genetic modification) has been criticized from a variety of standpoints and yet its progress as a tool of plant breeding appears only occasionally checked. There are other issues too, which have brought plant breeding into an obviously political arena.

- The use of traditional crops from developing countries to breed new commercial crops in the industrialized world—the *north-south gene drain*.
- Patenting of genes by corporations so restricting their usage by other breeders, or even traditional users—allegations of *bio-piracy*.
- Public institutions involved in plant breeding being sold off to multinational corporations, mostly agrochemical or pharmaceutical companies, so that they have a near monopoly of food production—*the corporate food chain*.
- Modern crops replacing traditional crops, so endangering biodiversity and the ability of traditional communities to manage their resources the way they see fit—*loss of crop biodiversity*.
- A loss of control by farmers (particularly traditional farmers in poor countries) over their sources of seed, with laboratory-based corporate researchers taking over the business of breeding new varieties—*loss of sovereignty*.

These issues have been widely discussed and written about. There are enough books on genetic engineering in particular to fill a barn. Here the aim is not to go over this well-plowed ground but to look at the background to the GM debate and the issues raised above—the history of plant breeding, from the origins of agriculture to the dawn of laboratory-based breeding. With the exception of the origins of our crops, surprisingly little has been written about the history of plant breeding, perhaps because the very nature of breeding attracts highly focused people rather than "big picture" people. This book is aimed at filling that gap, looking at how plant breeding developed within its historical context and that of the science and technology of its day. Plant breeding took place alongside social, economic, cultural, and philosophical developments and so became embroiled in politics and conflict. These are very much part of the story.

Understanding the history of plant breeding will not only help provide a background to debates over GM and gene sovereignty but will also help explain where we have reached as a species. Most of us are alive not just because of historically recent advances in medical science and an awareness of basic hygiene but also because of advances in agriculture: improved plant nutrition, crop protection, and . . . plant breeding. Our access to food is so much more secure than that of our ancestors, our choices are incomparably wider, and our food is safer than ever before. Much of this security and choice are down to plant breeding. With the varied diet we have today it is difficult to imagine what it must have been like for our ancestors; their diet was often not just poor but incredibly monotonous. How many of us could imagine eating nothing but potatoes or rice for every meal for our entire lives, with vegetables dependent entirely on the season and meat or fish a rare luxury? The fact that many in the world still live like this, and suffer malnutrition as a result, is a powerful illustration of why we need plant breeding to continue to make advances. What kind of advances and where they will take us is in our hands. But to make decisions and so to control our destiny we need to understand how we got where we are.

From feeding the world to breeding stripy petunias (*Petunia x hybrida*) may seem like turning from the sublime to the ridiculous, but it seems to me that the Chinese proverb that heads this book encapsulates much of our attitude toward the plant world. Once people have enough to live on and their survival is assured, they turn almost at once to the improvement of their surroundings and the creation of objects of beauty. Archaeologists will tell us that no sooner was the pot invented than it was decorated. The story of plant breeding is the story not just of food but of ornamental plants, too. Luxury always follows necessity, so in terms of developments, ornamental plant breeding has tended to lag behind that concerned with edible crops. Instead of being "the main story," the interest of ornamental plant breeding history is more in the nature of a sideshow, fascinating for the light it shines on the social and cultural development of humanity, in particular, attitudes to the relationship between nature and culture.

Improving upon nature is the very essence of plant breeding, and so it goes to the heart of one of the central debates of the human condition: the relationship between humanity and nature and the degree to which the human race has a right (or indeed a responsibility) to change plant life for its own ends. Long before the current furor about genetic engineering, various aspects of plant breeding were controversial, and it is quite clear that particular ideas about society and politics, even religion, are implicated in people's attitudes regarding plant breeding.

As plant breeding has become more and more sophisticated, the further from their wild ancestors our crop plants, fruits and vegetables, and garden plants have become. Over time, our domestic plants have been bent more and more to our will—nature is being molded to our tastes, desires, and needs. Modern advances in genetic engineering bring the "designer plant" much closer, a prospect that fills some people with alarm. And yet . . . as many plant breeders will agree, our ancestors achieved more significant changes to many plants than a whole century of scientific breeding has produced. Maize, or corn (*Zea mays*), is perhaps the most dramatic example. If we did not have corn, and a scientist proposed engineering its existence—a species utterly unable to function in the wild and completely built to serve human interests—those opposed to genetic engineering would launch a storm of protest. Yet the crop's transformation from spindly grass to supercrop was managed by traditional Native American farmers not through intention but through a series of natural changes whose significance they recognized and cherished. Corn is the ultimate construction of nature and culture working hand in hand.

With the continued unraveling of the genetic code of life, the possibilities for plant breeding become ever greater. Ensuring that the results serve all humanity is a deeply moral as well as an intensely political issue. Perhaps never before has there been such a need for a history of plant breeding. Plants are the bedrock of life on this planet, so to more deeply understand this aspect of our relationship with them is a vital task.

THE STORY OF PLANT BREEDING

This book is aimed at those in the plant sciences or those just interested in plants, agriculture, or the environment who want to know more about the history behind today's high technology and headlines. It is also aimed at those interested in history and human culture who are perhaps unfamiliar with agriculture, gardening, or genetics. Some understanding of genetics does help with the story, so some technical explanation is included in a series of technical notes at the end. We finish just short of laboratory-based plant breeding in the form of genetic engineering—to have included this would probably have doubled the size of the book, and much of this history is readily available elsewhere.

The book is divided into two parts. Part One deals with what is almost a linear historical narrative, from the domestication of the first crops to the birth of Mendelian genetics. Mendel's theories of how heredity works were not universally accepted at first by any means, and a discussion of this process of their acceptance, rapid in some countries, slow in others, is an important part of the story. Part Two looks at what happened when Mendelian genetics became accepted unequivocally. It, and a host of other scientific discoveries, were then able to transform plant breeding. From here on a clear narrative thread disappears, and the science soon leaves nonspecialists behind. Consequently, in Part Two a different approach is taken, looking more broadly at trends in different plant breeding technologies and their social and political implications.

Chapter one deals with the domestication of crops; chapter two, with traditional agriculture, which is dependent upon crop varieties that have evolved for a particular locale, are very particular to it, and are hence incredibly diverse. The future of traditional crops, in these days of hi-tech crops, may seem bleak, but reasons are suggested as to why this may not necessarily be the case. These two chapters are essentially about the prehistory of plant breeding; we have no names or dates, but instead the evidence of archaeology and genetics and observations made by anthropologists and others of people farming in traditional ways today.

We may all depend on agriculture, but growing crops has never been a high-status activity. Chapter three describes how the wealthy and educated of Europe began to become more interested in agriculture around the seventeenth century—one sign among many that a great shift in attitudes was happening. Europeans at this time were colonizing the Americas, and the role of English-speaking American colonists in early plant breeding and in making discoveries in plant sciences are a key part of this European story. That it was not only Europeans who at particular times in history became more interested in farming is recognized; there were similar movements in China and Japan. The age of "improvement" led to a more self-consciously scientific age, described in chapter four, with a more academic interest in actively trying to improve plants. It was during the eighteenth and nineteenth centuries that people started to make a more determined effort to cross plants to produce novelties, either for purposes of research or to produce more or better crops—*breeding*, in other words, not just selecting.

The age of agricultural improvement was part of a complex interrelated series of changes in European society that resulted in the continent bursting out and thrusting its influence over the rest of globe, to explore, to trade, to rule, to plunder, and often to overwhelm. European world domination and what this meant for crop plants and their breeding is explored in chapter five. Imperial expansion led to unprecedented exchanges of plant material between continents, leading to a sharp increase in breeding, all of which was conducted without a clear scientific understanding of genetics. It was Johann Mendel (1822–1884), generally known by his monastic name of Gregor, whose discovery of the basic principles of genetics in the late nineteenth century changed plant breeding from a technology to an applied science.[3] The background to the acceptance of Mendel and the discovery of his work is discussed in chapter six, and its varied reception and influence in different countries is covered in chapter seven.

Perhaps one reason plant breeding has attracted so little interest from historians is that most plant breeders are dedicated to their work rather than publicizing it or playing politics. One character is head and shoulders above all others, however—Luther Burbank, a legend in his lifetime—but did he deserve his reputation? He certainly deserves a chapter all of his own—chapter eight. Burbank was an admirer of Darwin but not of Mendel—arguably making him the last great prescientific plant breeder.

That the river of history sometimes follows markedly different routes in different places was illustrated by the importance of Marxist-Leninist political regimes in twentieth-century history; chapter nine follows the story of how a remarkably strong start in plant breeding was made by Soviet Communists—and then shattered. Intrigue and ideology ensured that Mendelian genetics was discarded, with devastating results.

Part Two, the era of a genetics dominated by Mendelian principles, starts off by looking at one of the most decisive breakthroughs of all, the use of hybridization to produce high-yielding and consistent crops, with chap-

3. The distinction between technology and science is often not understood; Wolpert (1992) provides a clear discussion of the issues.

ter ten. As the twentieth century advanced, plant breeding flowered into a myriad of techniques that bent more and more plants to humanity's will. This period of immense productivity is covered in chapter eleven. Particular developments in plant breeding were then applied to a problem that had always affected humanity, but that was now finally being recognized as not just a moral outrage but a political threat—famine. That mass famine has been averted thanks largely to the role of plant breeding is explored in chapter twelve.

The story of garden plants, discussed in chapter thirteen, is really another story entirely, even though it parallels that of crops in many ways. It engages much more with the cultural history of humanity. Increasingly, however, many of the more political aspects of the ownership of genetic resources are as relevant for ornamental plant breeding as they are for crops. Finally, it is time to look at this increasingly politicized arena. Ownership of genetic resources and related issues are widely covered elsewhere, and are therefore looked at somewhat discursively in chapter fourteen.

In the final chapter, it is time to look at the grand narrative and to extract basic themes about how plant breeding has developed and its relationship with the rest of the human story. This is also a useful place to stand back and look at the GM controversy in the light of plant breeding history. Controversy inevitably means engaging with politics and ethics—aspects of life which once would have seemed completely alien to a field which appeared to deal only with benefiting humanity. The fact that plant breeding is now a politically contested area, even to the extent of there being opposition to the whole concept, has come as a shock to many in the profession—and we close with a look at the post-modern revolt against science.

THE TECHNICAL NOTES

In order to make this book as accessible as possible to a wide readership, it has been written using as few technical terms as possible. Some, however, and a basic knowledge of plant genetics, are essential in order to understand the story, including some of its economic and political aspects. When a new technical term is introduced there is a footnote reference to the Technical Notes section on page 421. For those outside the plant sciences, this section acts as a very basic primer in plant genetics.

ABBREVIATIONS

A number of abbreviations are used extensively throughout the text:

FAO	Food and Agriculture Organization of the United Nations
HYV	high-yielding variety
NGO	nongovernmental organization (i.e., a nonprofit organization)
RHS	Royal Horticultural Society (of Great Britain)
USDA	United States Department of Agriculture

PART ONE ❦ FROM THE BIRTH OF AGRICULTURE TO THE BIRTH OF GENETICS

ONE ❦ ORIGINS ❦

THE DOMESTICATION OF PLANTS

Every spring we go into the woods to collect wild garlic (*Allium ursinum*)[1]—usually to make soup, which despite the strong smell of the fresh leaves makes a deliciously mild broth. Also in our collecting bags and baskets are a few other abundant wild plants as well: stinging nettles (*Urtica diocia*), also for soup, and hedge garlic (*Alliaria petiolata*) for salads. Not as many people in Britain do this as elsewhere in Europe, where it is still quite common practice, especially in more recently industrialized Eastern Europe. If there wasn't so much wild garlic around, we'd want to make the most of what there was; so if brambles were threatening to overwhelm a particularly good clump, we might be tempted to get out the clippers to cut them back. This is hardly farming, but it is a beginning—it is managing a natural plant community to the benefit of a particular species we have an interest in.

The blackberry brambles (*Rubus fruticosus*) that often overwhelm the low-growing wild garlic tell a different story. A spiny, aggressively spreading scrambling plant thriving in the mild climate of northwest Europe (and to parts of the United States and Canadian Pacific Northwest where it has been introduced), it produces a generous crop of juicy and tasty black fruit, the blackberry, which is the most popular wild-gathered food in Britain. Wild garlic collecting may be limited to a few aficionados of wild food, but whole families descend on bramble clumps in September, bearing plastic boxes and buckets, which they fill with the shiny fruit, nearly always getting stabbed and scratched by thorns in the process. Curses and children's tears are almost inevitable, and a good afternoon's picking usually ends with a feeling of wounded satisfaction. Large amounts can be collected quickly to make a delicious filling for pies and a very good jam.

The story of the British wild garlic and bramble illustrate in a small way the origins of farming and gardening. Of the vast number of wild plants which our hunter-gatherer ancestors lived off, few have been actively cultivated. The bramble eventually joined the ranks of the cultivated (in the

1. A European wild plant that, in fact, is not the ancestor of cultivated garlic, *A. sativum*.

nineteenth century), although it is not a major commercial crop; the name of the most common variety illustrates its most important characteristic: 'Merton Thornless.' *Allium ursinum* never became a crop plant.

Theories about how and why our Neolithic hunting and gathering ancestors became farmers are many: population pressure, a reduction in wild food sources because of climate change or the overexploitation of the environment by grazing animals, and the development of settlements with large and increasingly complex societies that led to increased localized pressure on food sources. The general consensus now is that it began a bit like our hypothetical management of the garlic patch in the woods, with attempts by tribal hunter-gatherers to manage nature for their benefit—sometimes dubbed "protoculture." Many so-called natural environments across the world are the result of ancient human intervention, particularly through the use of fire; Native Americans burned prairie to ensure good grazing for buffalo, Australian Aborigines burned woodland to open it up to make hunting game easier, Amazonian Indians planted favored trees.

Agricultural origins have been widely researched—but what happened to plants during this period of major transformation? How did they change as a result of the human interface with nature, which became agriculture, and how did this change evolve into plant breeding? The seeds of change involved the development of a mutual dependency, which grew as our ancestors moved from foraging to farming; Neolithic people became dependent on plants for their survival, but the plants in turn coevolved and became dependent on people for their survival, assuming genetic identities distinct from their wild ancestors.

"A GATHERING WE WILL GO"—FROM FORAGING TO FARMING

Visitors to the uplands of Britain are often surprised to see fires sweeping the landscape. Is this a cause for alarm? An occasion to ring emergency services? No. This is land management for the sake of a game bird, the grouse. Burning kills young trees, scrub, and old heather plants, and stimulates younger heather (*Calluna vulgaris*) to produce lots of new shoots which are food for the grouse. When the shooting season starts on August 12, wealthy Russian, American, and Japanese businessmen and Arab sheiks join aristocratic landowners to "bag" as many grouse as they can. "The glorious twelfth," as it is known, is renowned as a key date in the calendar of the wealthy. The idea of burning vegetation in order to maintain a predominance of particular plant species is analogous to our clearing away brambles from the wild garlic. It is a form of land management thousands of years old. The irony is that the management technique that lies behind the sport of some of the world's richest people is remarkably similar to that of our Neolithic ancestors.

Farming emerged during a protracted phase between hunting-gathering and settled agriculture. During this time, there was a continuum of activities resulting in the domestication of some plant species. Burning appears to have been a key part of early land management and is still used in many cultures. It tends to encourage herbaceous plants, many of which may have been attractive as food for their leaves and plentifully produced, and often large, seed; slower to recover after burns are the less edible shrubs and trees. Protective tending would probably have come next, along with other practices such as draining wet ground and watering the dry. At some stage our ancestors would have started propagating: distributing seed, dividing and planting tubers and suckers. With propagation comes the real beginning of cultivation, of the transformation from the wild to the domesticated. Once humans started to sow seed or collect and plant tubers, they began to change the genetic composition of plant populations, sprouting a gradual divergence between wild and cultivated members of the species.

Much of the literature on the origins of agriculture is focused on the Fertile Crescent—The Middle East, in particular, parts of current-day Turkey, Iraq, Israel/Palestine, and Lebanon—and on the crops that were domesticated there, notably wheat and barley. Partly this is a consequence of a long history of archaeological work in a region whose climate preserves the remains of human culture well. But there is surely also a bias—this is, after all, the home of Western civilization and of Christianity—more ancient and more fundamental agricultural origins may have had less attention.

Early agriculture in the Middle East was based on plants, which were reproduced from seed. Older, however, than seed-based agriculture might have been "vegiculture," the propagation of food plants by transplanting tubers. Australian Aboriginal hunter-gatherers are known to dig up yams for food but then transplant a proportion. This is a bit like potatoes in the vegetable patch: the majority are harvested and eaten, but a few can be kept back to start next year's crop, which will be genetically identical. The ease with which such vegetatively propagated crops may be cultivated and transported suggests that this may have preceded seed-based farming.[2] It is easy to imagine that tastier, faster-growing, or larger yams would be favored, re-

2. As suggested by Harlan 1992.

sulting in cultivated *clones*.[3] The origins of yams[4] are known to be multicentered, the knowledge of their exploitation spread by contacts between tribal peoples.[5] In contrast, certain seed-propagated crops, such as emmer wheat, pea (*Pisum sativum*), and lentil, are known (through DNA analysis) to have been brought into cultivation only once rather than a number of times in different locations. This suggests that growing from seed was inherently more difficult than vegiculture—both practically and/or conceptually.[6]

Early settlers in California reported that there were oak groves planted in straight lines—the legacy of now-vanished Native American tribes who relied on carbohydrate-rich acorns for food. This is an example of one intermediary stage between hunting gathering and agriculture—"casual cultivation," whereby nomadic or partly nomadic peoples plant crops but do not live alongside them. The oaks were obviously a long-term proposition; other Native American groups in semiarid areas of the American southwest are known to have planted fields of crops, undertaken basic watering and weeding, and then left them until it was time to return to the area for harvest.[7] Casual cultivation could involve many species—elders of the Kumeyaay tribe interviewed in the early twentieth century described how they would undertake experimental plantings of a large number of medicinal and food plants in a variety of habitats.[8]

The enormous global importance of grains (primarily wheat, barley, rice, and corn) has influenced the attention of researchers. Yet in many parts of the world people undertake "gardening" rather than farming, where small plots support a limited number of staple crops and a much larger number of fruits, vegetables, herbs, and medicinal plants. In many cases these are part of swidden agriculture, where temporary fields are made on newly cleared ground, and desired crops grow amongst a wide variety of other more or less weedy species. Many of the latter may contribute to the diet of the household as potherbs or be forage for animals. The diversity of garden agriculture can be

3. See technical note 2: Clones and Cultivars.

4. "Yam" can mean several things: 1) in British and Caribbean English it is generally used for species of *Dioscorea*; 2) in American English it is used colloquially for *Ipomaea batatas*, always known as "sweet potato" in British and Caribbean English; 3) since there are many starchy tuberous rooted perennial plant species used for food in the tropics, it is sometimes used as a general term for all such species.

5. Alexander and Coursey 1969.

6. Zohary 1996.

7. Minnis 1992.

8. Shipek 1989.

quite extraordinary, with hundreds of species growing in cultivated patches.[9] Some of the species unintentionally grown may end up by being domesticated—it is possible that this is how some salad crops may have originally been taken into cultivation. Even in completely settled farming systems, diverse mixtures of annuals or biennials are grown. In the Hunza Valley, high in the Himalayas, such mixtures are usually grown near the house to be used as potherbs or chicken feed. This particular mixture seems to have become a self-sustaining edible weed flora; very few undesirables seem to germinate, and the mixture comes up with little active encouragement.[10] Prominent is a large-flowered form of the wild mallow *Malva sylvestris,* used as a vegetable but occasionally grown as an ornamental in gardens.

The origins of cultivated crops have long fascinated researchers, and the inevitable controversy around rival theories is worthy of another book. One name stands above all others, and although today's evidence does not always support all his theories, the work and career of the Russian Nikolai Vavilov (1887–1943) provided the first coherent unifying account of crop plant origins. His full story will be discussed later (see chapter nine); here we are concerned with his theory of "centers." Based on exhaustive research and a great many expeditions, often into very remote areas, he concluded that the vast majority of crops had originated independently in a limited number of centers: the Middle East, northern China, and Central America. Later work, particularly by the American geneticist, plant breeder, and "archaeobotanist" Jack R. Harlan (1917–1998) supported the idea of these three basic centers but also suggested that there were other crops that originated over a far more diffuse area.

Given that Western civilization had its origins in the Fertile Crescent of the Middle East, the intense interest in the region's history is understandable—but the phases it went through were not necessarily mirrored in other centers of origin. The earliest period (11,000 BCE–8300 BCE) has been described as the "agrotechnical phase," with finds of sickle blades, pestles, mortars, and storage pits, as well as some permanent settlements. It is thought that this technology was developed to exploit wild crop plants, which may well have increased greatly in quantity owing to land management practices such as the clearing of woodland. The next phase (8300 BCE–5500 BCE) was that of the "domestication revolution," with the cultivation of wild forms of cereals followed by their gradual domestication: emmer wheat (*Triticum di-*

9. D. R. Harris 1989; De Tapia 1992.
10. Shebir 2004.

coccon) and barley in the southern part of Fertile Crescent corridor, einkorn wheat (*Triticum monococcum* var. *boeticum*), rye (*Secale cereale*), and barley, plus legumes and flax (*Linum usitatissimum*) in the northern part.[11]

In vegiculture, the key domestication event was the transplanting of selected tubers (or suckers in the case of above ground crops like bananas [*Musa*] species). This did not involve any genetic changes in the individuals selected—so what was in cultivation was a selection of the wild clones. In seed-based agriculture, the very first act of saving seed is an act of unconscious breeding rather than selection, as the seed is the result of sexual reproduction, that is, the shuffling of genes. Seed saving is inevitably and intrinsically the product of nature *and* culture, of plant genes being acted upon by human agency. The woman (for it probably *was* a woman) who saved the very first head of seed for resowing next season made a decision; she picked *this* rather than *that*—maybe it had ripened earlier or was bigger. With this act, the genetic makeup of the crop that she sowed would be different than that from which it was chosen. Plant breeding had begun.

MAIZE MYSTERIES

Most breeding work is slow and incremental. Occasionally dramatic advances are made and provide the quantum leaps of history. Today such advances attract attention, sometimes bringing honors to the breeders; they make good newspaper copy, briefly bringing plant breeding into the limelight, such as with the new semidwarf wheat varieties of the Green Revolution. In the past, such leaps—the discovery of a productive mutation or accidental crossing in a peasant field leading to an improved seed strain—would not necessarily have been recorded; or if they were, the origins would be given a mythological status, as a gift of the gods, or credited to the wisdom of a powerful ruler. The contrast between gradual and dramatic progress is one familiar to students of history as the contrast between slow reform and revolution. This contrast will be familiar too to students of evolution; it was once assumed that evolution proceeded slowly and methodically, but recent work has suggested that it too might make most of its progress in sudden spurts.

Quantum leaps inevitably bring us to the story of corn, or maize. This is the crop upon which the great civilizations of Mesoamerica were built, the staple of many Native American tribes, and now one of the world's most

11. Feldman 2001.

important foodstuffs, both directly through human consumption and indirectly through being fed to animals reared for meat or milk; it is also an important biofuel crop. More than with any other crop, the story of corn is one of dramatic change. It bears less resemblance to its native ancestors than any other edible plant, dramatic changes having occurred in a distant past, recognized, cherished, and distributed by Native American farmers. However momentous the changes that happened to corn in the twentieth century, they could only build on what had happened in prehistory.

In the words of corn expert, Paul Mangelsdorf (1900–1989), "the ability of the wild maize plant to respond in a spectacular fashion to freedom from competition without weeds and to high levels of fertility is undoubtedly a factor which led to its domestication . . . . [T]he corn plant is genetically plastic as well as responsive to an improved environment. There are many wild species that do not have this trait; they cannot stand prosperity."[12] The ability of corn to respond to cultivation has made it central to the cultures that grow it more than any other crop, as much of the mythology of the Native American peoples who grow it can testify; the Maya creation myth, for example, describes how the human race was made from corn.

In just one example of what appears to be a particularly well-documented quantum leap, around 1300 CE–1600 CE, there was an agricultural intensification in the upper Midwest of what is now the United States. A new corn strain, dubbed Northern Flint corn, had appeared. It was better suited to local climates and had a more appropriate photoperiodic response[13] to the latitude than previous strains; it was hardier, with greater insect resistance, and it ripened earlier.[14]

Far less well documented, indeed, positively murky, is the issue of corn itself. There is now at least consensus about the broad outlines, but for many years there were rival theories: that maize had descended from a wild grain called teosinte (*Zea mexicana*), that teosinte had descended from corn, which had evolved from the hybridization of several wild grains, and that corn and teosinte are descended from a now extinct ancestor.[15]

Teosinte and maize are so different in appearance that it did not occur to most botanists in the nineteenth century that they were related until a Mexican agronomist, José Segura, crossed them, proving a close relation-

12. Mangelsdorf 1958.
13. See technical note 9: Photoperiodism.
14. Gallagher 1989.
15. Goodman 1976.

ship. With the discovery of Mendel's principles, interest in teosinte boomed. During the 1930s George Beadle (1903–1989), a geneticist at the University of Chicago, concluded that Mexican teosinte was the ancestor of corn, which during an early phase of domestication had undergone a series of major mutations that had transformed it far more than any mutation had affected Old-World wheat or barley. He postulated that since teosinte can be "popped" like popcorn, this would have been a way in which ancient hunter-gatherers could have exploited the plant, so leading it onto the path of domestication.

That teosinte and corn were so different created a credibility gap; it seems clear looking back that some researchers simply could not have believed that such a huge number of genetic changes could have been made by either nature or "primitive" peoples, even over a few thousand years. In addition there were two dogmas of evolutionary science which got in the way of believing the unbelievable: one, that evolution cannot "go backward"—evolve from what were seen by botanists as advanced traits (such as teosinte's single female spikelets) to primitive (like maize having paired spikelets); and two, that evolution happens through small incremental changes. Current thinking has changed—now it is accepted that evolution is multidirectional and that there can be dramatic changes over short periods of time.

In 1939, Paul Mangelsdorf, with colleague Robert Reeves, proposed an alternative hypothesis: the ancestor of maize was an extinct species, and teosinte was the result of a cross between maize and another grass species from a related genus, *tripsacum*—the so-called tripartite theory. They supported this with a successful cross between maize and a *tripsacum* species—with the big proviso that the embryos from the few female spikes that were successfully pollinated had to be surgically rescued and grown on plates of agar jelly. Beadle retorted that any evidence that involved hybrids so delicate that they needed laboratory techniques to come to fruition was no evidence at all. He advanced the hypothesis that one gene controls one trait—one gene would change shattering to nonshattering cobs and another from covered to naked kernels, and so forth. Only four or five gene mutations, he suggested, would turn teosinte into a primitive corn, incapable of surviving in nature but ideal for cultivation.

Beadle could have settled in for one of those infamous academic arguments, which grumble on for decades, getting increasingly bitter. He chose not to. For him it was obvious where corn came from, there was no point in arguing about it, there were other challenges to take on, so for three decades he worked on other problems and spoke no more about corn's mysterious origins. Mangelsdorf, however, had other fish to fry; becoming the most prominent corn researcher of the century, not just an academic (a professorship at Harvard) but a distinguished plant breeder and a key member of the American team who helped modernize Mexican agricultural research in the 1940s and 1950s. If there was a "Mr. Corn," it was he.

Mangelsdorf's tripartite theory appeared to be accepted and enjoy undisputed hegemony. Around laboratory coffee machines, however, were dissenting voices; the theory seemed unnecessarily complicated—and there was plenty that did not add up. But so undisputed was Mangelsdorf's reign as the international emperor of corn that no dissent could be taken seriously. After a lifetime in genetics, Beadle retired in 1968 and returned to teosinte—not exactly a quiet retirement hobby. Growing fifty thousand maize-teosinte hybrids two generations he found that one in five hundred plants was very much like one of the parents, the kind or proportion to be expected if only four or five genes were involved. He also brought together evidence from a variety of archaeological, anthropological, and linguistic sources to support his case.

Growing vast numbers of plants to prove a point would have been second nature to Beadle, who grew up on a farm in Nebraska and was a lifelong gardener of both vegetables and flowers. Corn he loved and was proud of the fact that his garden always had the first sweet corn of the summer—a Mexican variety. From 1961 to 1968, he was president of the University of Chicago, where he established a particularly spectacular garden around his official residence, doing most of the work himself; as a consequence he was mistaken for a hired hand on more than one occasion.[16]

Despite a largely fruitless plant hunting trip to Mexico in 1971, looking for teosinte mutations in the Balsas River Valley, he began to lecture on the "teosinte hypothesis" and make converts, again drawing on evidence from a huge range of sources, including animal behavior (he fed corn and teosinte to squirrels to see if they ate it) and gastronomy (he cooked and popped many a teosinte and teosinte hybrid to see if they were edible). Mangelsdorf, also now retired, lectured frequently too. The origin of corn had become one of the great debates in biology between two of the most eminent figures in the field of genetics. But in the 1990s, with molecular genetics, a new era in the study of plant evolution and relationships began; it has provided plenty of support for Beadle's ideas and almost none for Mangelsdorf's. By the start of the twenty-first century the majority of geneticists and archaeologists had come around to supporting the view that corn is a domesticated

16. See Horowitz 1990 for an obituary of Beadle.

teosinte. Several times, Beadle and Mangelsdorf engaged in public debates but never sank to personal attacks, always staying civil with each other and polite with their opponent's supporters.

AS THE FARMER DRIVES THE PLOW—EVOLUTION IN EARLY AGRICULTURE

Of over two hundred thousand species of flowering plant, only around three hundred have been used in agriculture.[17] A larger number, one that does not seem to have been quantified, have been cultivated for some food or medicinal use or other economic purpose. In some areas of the tropics these cultivated plants, practically genetically identical to their wild ancestors, may total in the thousands. Those that have been cultivated most extensively or for the longest period are among those that show the greatest change between wild and cultivated forms. But by no means all—some long-lived tree crops are propagated clonally (usually by cuttings or grafting) so genes are not reshuffled, which means that present-day varieties may be incredibly old. The pomegranate (*Punica granatum*), olive (*Olea europaea*), fig (*Ficus carica*), and grape (*Vitis vinifera*), all domesticated in the Middle East, are examples, their ecological requirements still very similar to that of their wild relatives.[18] The 'Sari Lop' fig is possibly two thousand years old, while 'Dottato' was mentioned by Pliny (23 CE–79 CE)[19], and the 'Black Corinth' grape, a seedless mutation[20] still grown for making baking currants, may date back to the age of classical Greece.[21] The genes in those packets of dried fruit in the supermarket or whole food store may be very old indeed.

Trees or other woody plants that can be propagated clonally underwent a kind of "instant domestication," which brought particular selections into cultivation. Such a process, however, was dependent upon certain technological advances: transplanting, propagation through cuttings, grafting—each of which needs a progressively greater leap of faith, unlike sowing from seed, which can be observed in nature and copied. Grapes are particularly prone to bud mutation (see chapter five), so a branch with superior grapes

17. Holden, Peacock, and Williams 1993.
18. Zohary and Spiegel-Roy 1975.
19. W. B. Storey 1976.
20. See technical note 3: Mutations.
21. Olmo 1976.

would have stimulated the use of cuttings, possibly in preclassical times.[22] With the Chinese invention of grafting in the second millennium BCE, it was possible for plants with good characteristics that were difficult to propagate from cuttings to be "fixed." Grafting was also used widely by the Greeks and Romans. Many pecan (*Carya illinoinensis*) clones were brought into cultivation by grafting; early settlers would find a tree they liked in the woods, collect budstock, and graft onto a tree back at the farm; for example, 'Centennial' was first grafted in 1846 by a freed slave.[23]

It is clear that these tree fruits will not differ hugely from their wild ancestors, the cultivated forms being often simply better-than-average selections. Cultivated plants vary enormously in the degree to which they have been modified over time; many tropical fruits and medicinal plants used in traditional herbal medicine have hardly been changed at all, whereas others are almost "artificial" in the degree to which they differ from their wild ancestors. Corn is the most obvious example; its changes in appearance and genes from its wild ancestors are the result of the work of Native Americans more than modern science. Another example is sugar beet (*Beta vulgaris*), a man-made crop of more recent times.

Changes in crop plants are essentially about the molding of species to suit human needs. Darwin recognized this and made a fundamental distinction between three aspects of the selection process that people apply to plants. "*Methodical selection* is that which guides a man who systematically endeavors to modify a breed according to some predetermined standard. *Unconscious selection* is that which follows from men naturally preserving the most valued and destroying the less valued individuals without any thought of altering the breed. *Natural Selection* . . . implies that the individuals which are best fitted for the complex, and in the course of ages changing conditions to which they are exposed, generally survive and procreate."[24] The early stages of plant breeding had much more to do with the latter two processes; species will change by natural selection under the conditions of agriculture, and there will be unconscious selection—Darwin's "methodical selection" will only appear when there is an awareness among cultivators that crops can and do change. There is little agreement about the extent to which early and traditional cultivators were aware of this and at what stage *unconscious* selection turned to *conscious*, if not truly methodical, selection.

22. Hyams 1971.
23. P. M. Smith 1976. See technical note 8: Grafting.
24. Darwin 1875, 177–178 (original emphasis).

Darwin also recognized that the process of domestication through cultivation was analogous to the idea of evolution he supported and that at its time was so revolutionary. Indeed, supporters of evolution have always been able to look to parallels between nature and culture in this respect and use the historically well-attested process of domestication through culture as being evidence for the theory of evolution.

Among the differences between wild plants and those in the earliest stages of domestication are some that are more or less to be expected: an increase in productivity—more grains, more leaves, bigger tubers, and vigor, especially at the seedling stage. There is nothing like growing wild and cultivated relatives side by side to appreciate the difference; loving the crisp flavor and juicy texture of wild sea beet (*Beta vulgaris*), I once grew some from seed alongside one of its descendents—Swiss chard. By the time the chard was ready to harvest, the sea beets were still tiny seedlings.

In the earliest stages of domestication a number of fundamental changes often occur in plant behavior that are deleterious to survival, reducing a plant's fitness in evolutionary terms. However they enormously increase its usefulness to humans and so increase the plant's fitness in a new sense. Indeed, if plants did not undergo these changes, they would be unattractive as cultivated plants. It has even been suggested that there is a "domestication syndrome" with some evidence that traits favorable to domestication are genetically linked so that an increased dependence on humans for reproduction is linked compact size and a tendency towards self-pollination,[25] for example, in tomatoes and peppers.

Evolution driven by nature and the new form of evolution imposed by agriculture force plants along different paths. Human requirements are totally instrumental—we grow plants for a purpose; if a plant species is to survive and prosper in cultivation it needs to be able to adapt to provide what we expect of it. As a consequence the story of plant breeding is very much about enabling genes which serve our purposes to be expressed.

Human demands are complex; what is desirable for one place and crop may not be for another, as can be appreciated by looking at natural plant toxins (which usually act as pesticides). Wild celery's (*Apium graveolens*) toxins make it acrid and bitter, bringing a numb and sore mouth to those who dine on it; although used by the Romans as an herb, it was not cultivated as a vegetable until a milder strain was developed in medieval Italy.[26] However

25. See technical note 6: Inbreeding and Outbreeding

26. Hyams 1971.

Some of the most fundamental changes between wild and cultivated plants are listed here, with an outline of their disadvantages in nature and advantages in culture.

1. Loss of spontaneous shattering of the seed head when seed ripens (in grains). Wild grasses, including the ancestors of cultivated grains, have seed heads that shatter as soon as the seed is ripe, allowing seed to fall away and spread—clearly essential for the propagation of the plant. It does not take much imagination to realize that it was difficult for early farmers to harvest grain which fell onto the ground as soon as it was ripe.
   Nature—Shattering vital for effective distribution of seed.
   Culture—Nonshattering enables crop to be harvested.

2. Greater uniformity of seed ripening and germination.
   Nature—Uneven ripening reduces losses to predation from seed-eating birds and mammals; uneven germination reduces losses to predation from grazers like slugs, and poor weather conditions.
   Culture—Even ripening aids harvesting; even germination aids cultivation.

3. Plant lifespan.
   Nature—Plant lifespans related to ecological conditions.
   Culture—Tendency for lifespans to be reduced as farmers select for rapid production of useful crop.

4. Loss of spines or hairs on the seed coat.
   Nature—Spines aid distribution through clinging to animal fur.
   Culture—Impossible to sow seeds that cling together when collected.

5. Loss of toxic compounds in parts of the plant which are eaten.
   Nature—Toxins protect plant from predation.
   Culture—Loss of toxins vital if plant is to be used for food.

6. The "sunflower effect": the reduction of many small heads to fewer large heads.
   Nature—"All eggs in one basket" increases vulnerability to damage.
   Culture—Easier to harvest large heads, higher proportion of grain to stem.

the tuber manioc (*Manihot esculenta*, also known as cassava), grown throughout the tropics, is one of several warm-climate crops where the presence of highly toxic hydrocyanic acid might actually have been selected for as protection against pests and wild animals, at least once the technology for removing it from the tubers had been perfected.[27]

Wheat illustrates many of the important points of the changes undergone in domestication, and it has also been one of the most intensely researched crops. Modern-day bread wheat evolved from four species of wild annual grasses found in the Middle East. Present-day Iraq is the site of some of the oldest archaeological finds of wheat grains: *Triticum monococcum* var. *boeticum* ('einkorn,' with a gene type designation of AA)[28] is dated to be approximately ten thousand years old. Wild einkorn is a large-grained wild grass with brittle ears, which over the first millennium of its cultivation appears to have become less brittle. Possibly around the same time a tetraploid hybrid developed, *T. turgidum* var. *dicoccoides* (AABB), the result of a natural cross between einkorn and spelt (*T. speltoides*), and possibly some other *Triticum* species. This hybrid gave rise to 'emmer,' *T. turgidum* var. *dicoccum*, which through further domestication led to the durum wheats, the basis of pasta and couscous. Bread wheat originated when emmer crossed with another wild grass *T. tauschii*, probably in northwest Iran or northeast Turkey; the result was the hexaploid *T. aestivum* (AABBDD).[29]

The most critical event in the early days of the domestication of wheat, and indeed all cereals, was the selection by early farmers of forms whose heads of grain did not shatter. It is likely that it took considerable time, maybe even as long as a thousand years, before there was a high enough proportion of nonshattering heads in the population of a grain crop for early farmers to take notice and to consciously select them.[30] This process would have involved several thought processes. Imagine the moment when someone working in the field realized that the heads they were harvesting were not only perfectly ripe, but didn't immediately break up when they cut the

27. Harlan 1992.

28. See technical note 4: Chromosome Sets and Polyploidy.

29. The scientific names given here are the traditional ones, i.e. not the names given to wheats by researchers working at the level of genetic analysis, who have suggested a "genetic" system of classification. Both systems are regarded as equally valid. See "Wheat," Wikipedia, http://en.wikipedia.org/wiki/Wheat_taxonomy (accessed August 16, 2006); Wheat Genetics and Genomics Resources Centre 2001.

30. Feldman 2001.

stem with their sickle. The next step would be to realize that there was a distinction between shattering and nonshattering heads, the step after to harvest them separately so that they could be sown as a different batch the following spring. As can be appreciated in the next chapter, traditional farmers are actually very good at making subtle distinctions like this and developing distinct varieties.

Once a grain was dependent on humans to harvest and sow its seed, it became dependent upon humanity for its survival, and the human race and the new races of domesticated plant entered a new era—with a mutually-dependent symbiotic relationship.

Archaeological evidence suggests that there was a striking analogy to the development of nonshattering wheat with South American quinoa[31] (*Chenopodium quinoa*) but not with corn, whose development from wild plant to crop must have been much slower. In North America, however, nonshattering grains were not developed. Native people did not use sickles to cut bunches of grain (no pre-Columbian peoples ever developed metal refining), but instead shook loose grain into baskets. As an example, the dark, elongated grains of the marsh grass known as 'wild rice' (*Zizania palustris*)—now very popular as a tasty additive to ordinary rice—was harvested using a flail and basket from a canoe. Nonshattering grain would not have made harvesting with this method any easier; in fact, it would have made it more difficult. One possible implication is that this is the reason North American grasses were never domesticated. With the discovery of strains of wild rice with nonshattering heads in the 1960s, it at last became possible to develop the crop commercially.[32]

It was not only crops that developed nonshattering heads. From the very beginning of agriculture, crops have had rivals—weeds. In the case of some grassy weeds, their development of nonshattering heads meant that they could join the seed crop and be cared for through the winter in a nice cozy tribal hut, to be sown in optimum conditions the following spring. A kind of "domestication by indifference" was the result. Ethiopian oats (*Avena abyssinica*) are an example of a weed that has partly become a crop—never sown for itself, it has historically been tolerated as a weed in barley or emmer, so the resulting grains are inevitably a mixture. It has also been suggested that European oats (*A. sativa-byzantina* hexaploids) are the result of weeds infiltrating grain crops and eventually being recognized as having value of their own.

31. Simmonds 1976d.
32. Oelke et al. 1997.

Once domesticated, farmers would have selected (usually unconsciously) for high ratios of grain collected to grain sown, that is, plants that tiller, or produce offsets, a large number of grains per spike and large grains. Increased grain size tends to reflect a higher proportion of carbohydrate at the expense of protein—nutritional quality therefore tends to decline; however, this is one of the trade-offs of domestication not likely to have been recognized. Another trade-off, which early farmers could have been aware of, is that between total yield and stability; they would not have had the technology to minimize effects of annual fluctuations in weather and other conditions from one year to another. Consequently, plants with resilience and an ability to produce under diverse conditions would have been unconsciously selected for, rather than high yield, which one might expect. Once farmers recognized that higher yields were possible they would not necessarily select for them—year-to-year stability was more important. In cultures where grain storage or trading with neighbors in bad years was difficult or impossible, a reliable strain that produced food even in a bad year was far, far more important than one that produced bumper crops but failed in inclement conditions. As we shall see, this is a vitally important part of the highly contested transition from traditional to commercial farming today.

Researchers have attempted to reproduce some of the processes of domestication to see how changes occur and at what rate. In the nineteenth century a French seedsman, Henri de Vilmorin (1843–1899), attempted to domesticate the edible but rarely eaten European wildflower *Anthriscus sylvestris*. He managed to achieve straight and smooth roots, as opposed to the normal forked ones, after more than ten years' selection, the rate of change in root quality escalating over time.[33] Work done during the twentieth century suggested that some steps in the domestication process can occur quite quickly—work with a wild millet has shown that seed dormancy can be substantially reduced in only four generations.[34] Another trial, working with wild-type wheats and barleys, followed by computer simulations to fast-forward the next few centuries showed that domestication could occur within two hundred years, and possibly as little as twenty to thirty.[35]

One trial to explore domestication was set up by Jack Harlan's father, the great breeder of barley H. V. Harlan (1882–1944), at the University of Cali-

33. H. J. Webber 1898.

34. Harlan 1992.

35. Hillman and Davies 1983.

fornia at Davis in 1928 (when it was the Farm School) and still continues today. Twenty-eight varieties of barley were crossed in all possible combinations, their progeny sown, theirs were sown next year, and so on—without any deliberate selection. A variety of changes were observed, and interestingly yield increased, rapidly at first, then more slowly, keeping at around 95 percent of the best efforts of deliberate breeding. This demonstrates that simply growing a seed-grown crop from year to year changes it over time—farming *is* breeding. And so is gardening.[36]

SERENDIPITY—INTERSPECIFIC HYBRIDIZATION

Southern India has hundreds of banana varieties, bewildering and beguiling to European or North American visitors who are only familiar with one kind of supermarket-purchased banana. They provide a never-ending and delicious journey through tiny and delicious fingers, each variety with its own distinct flavor, all too thin-skinned to travel far: the esteemed red banana (actually the fruit is orange), 'Virupakshi,' which only develops its true flavor in the hills, and 'Kathali,' highly prized and so used for temple offerings. The region can be regarded as a secondary center of origin for the crop, the result of an ancient cross between two species.

In the case of clonal, vegetatively propagated crops such as bananas, hybridization can lead to a variety which can be propagated by division of tubers or suckers and is therefore capable of widespread distribution. In nature, the banana is a mass of pulp around hard seeds, which need to be carefully picked out if the fruit is to be eaten; images of monkeys in jungles swinging from trees skinning and tucking into juicy bananas is one that belongs firmly in children's cartoons. However, occasional mutations in the Southeast Asian *Musa acuminata* have resulted in the production of fruit without seed formation and consequently sterile. The result is juicy, ready-to-eat fruit, and since the plants are easy to propagate from suckers pulled off parent plants, such occasional mutants could have had relatively rapid distribution after discovery. When taken to India millennia ago, edible clones crossed with *M. balbisiana* and gave rise to a host of new varieties. *M. balbisiana* is adapted to a climate with a monsoon and a dry season, whereas *M. acuminata* is a plant of the humid tropics; hybrid clones are often better adapted for seasonal climates than the species, so hybridization in-

36. Harlan 1992.

creased the possibility of using the crops in more regions. In addition, many of the hybrids were more vigorous.

The stories of bananas and of wheat's complex genetic history (one that still defies clarification) illustrate what has frequently happened in the history of agriculture—hybridization between species. It is possible that early farmers may have spotted the potential of rare naturally occurring hybrids; far more likely, however, is that human migration brought together plant species across their natural boundaries. Wild plant species are separated by boundaries of geography and habitat: oceans and mountain ranges, forests and deserts. As people moved from one region to another and traded with one another, they took their crops with them, which where then able to cross with the species of their new home. Hybrids often had advantages; they combined the desirable characteristics of more than one species and often showed increased vigor—successful variants would have been selected out by observant people and maintained as new varieties. This crossing between species, or between regional varieties within species, has been a major element in history's unconscious plant breeding; many crops are hybrid strains, often of extremely complex ancestry. That the progeny of these hybrids is identical ensured that their mixed origins were never understood; it was only in the nineteenth century that naturalists and scientists began to realize that many of the most common and valuable farm and garden plants are in fact hybrids of one or more naturally occurring species.[37]

Some interspecific hybridization happened when crops were transported over wide distances or were relocated to very different habitats. Two species or races which would never normally have met have the chance to cross, and a whole new level of productivity or adaptation to a different environment resulted. This is the dramatic case, and can perhaps be illustrated by the banana story.

Less dramatic is the gradual and long-term process of the *introgression* of the genes of wild relatives into cultivated crops. Where a crop is growing in the vicinity of its wild ancestors or related species, there is the chance that they will crossbreed, resulting in wild genes joining the genes of the cultivated crop. Although it has been suggested that this process may have slowed down the rate of change under domestication,[38] once the crop had been moved out of its immediate region the introgression would have been

37. See technical note 5: Hybrids.
38. Feldman 2001.

of genes of less–closely related races and of different but closely related species. In the case of wheat, which is largely self-pollinated, the result of such crosses would be numerous relatively stable new combinations of characters, providing farmers with new genotypes for selection. A process of genetic isolation followed by an introduction of "new blood" could have been a potent force in crop evolution.[39]

Traditional farmers are sometimes aware of the relationship between wild and cultivated forms of crop plants. This understanding is often expressed through religious belief. The Ari people of Ethiopia, whose staple is the banana relative *Ensete ventricosum,* have sacred areas, called *kaiduma*, where wild ensete plants grow; if these are disturbed divine punishment will be visited on whomever has damaged the plants. The sanctuaries act as a source of diversity for the various forms of ensete cultivated locally, allowing occasional cross-pollination.[40]

The genetic changes that occurred in early agricultural systems have to been seen in the context of the very common practice of shifting cultivation, or swidden agriculture. This can still be seen today in much of Africa and hill country in Asia. Often known by the rather derogatory term "slash and burn," it is part of a seminomadic lifestyle, involving the clearing of natural habitat, generally woodland, burning it, and planting crops on the cleared land, which for a couple of years benefit from the nutrients left behind by the humus and ash of the natural vegetation and from the absence of weed populations. After a number of years soil fertility declines, and the farming community moves on to burn another area of woodland. Swidden can be a very rational and sustainable means of land-management,[41] but only when there is a low population in a large area; nowadays this is rare, and the practice has become inevitably destructive. Given the importance of shifting cultivation to emerging agricultural societies, it can be assumed that many of our most oldest crops evolved under such conditions.

If crops are grown in the same place year after year, the germination of the previous year's spilt crop would have maintained the old gene pool, whereas shifting cultivation would concentrate genes from those selected for harvesting—from which the seed to sow the next crop would have been derived; the overall effect could therefore be to push the gene pool in the

39. See technical note 6: Inbreeding and Outbreeding.

40. Brush 2004.

41. See Eisenberg 1998 for a discussion of its benefits.

direction of characteristics that favored harvesting. Shifting cultivation might also have the effect of continually exposing a crop to wild relatives, at least if they were found in nearby habitat; it also leaves large areas of abandoned land, which is often good habitat for these wild relatives. In addition, there are also often survivors of crop plants—the results are that extensive crossing may then take place in ex-farmland and continue to effect the crops to be grown the next time the cycle is repeated. This continuous process of genetic exchange has been observed in sorghums (*Sorghum* species) in Africa and can have assumed to have happened in many other places.[42] In areas where shifting cultivation still occurs, this process is still going on; even if there are now wild relatives, volunteers from previous cropping cycles often survive to provide a source of genetic diversity for a current cycle.[43] Where permanent agriculture and natural habitat create similar growing conditions and where there is a wide interface between them, such introgressive hybridization has been almost continuous; this has been shown to have happened with rice in the Orissa region of India, where wild rices are suspected of crossing continually with cultivated strains over millennia.[44] However, the increasing destruction of natural habitat by a successful, and progressively more permanent, agriculture gradually reduces the opportunities for cross-fertilization.

In American settler history a similar process had happened whereby introgression resulted in chance hybrids of cultivated and wild vines, some of which were recognized as having value, better suited to local conditions, often very different to that experienced by Mediterranean grape varieties. Many early American grape varieties were selected from spontaneous hybrids with wild vines in this way, for example, the well-known 'Concord' grape was propagated from a vine found at the edge of his pasture by a Mr. E. W. Bull of Concord, Massachusetts, in 1843.[45]

Hybrids, in so many cases, made it possible for agriculture to feed more people more efficiently, so enabling human societies to progress. Crucial though it was, hybridization was not understood for millennia. A complete picture emerged only once the concept of species was introduced in the early modern era. Many cultures, however, did display some understanding

42. Hutchinson 1974b.
43. Kizito et al. n.d.. See technical note 7: Genetic Diversity.
44. Shastry and Sharma 1974.
45. Harlan 1992; C.A. Cahoon 1986.

that mixing different strains of crops could have desirable outcomes. Traditional Mexican farmers for example have been observed to interplant corn of different varieties, effectively building hybrid races.

PLANTING CITIES AND STREETS—DOMESTICATION AND CIVILIZATION

One of the great questions of human history is why agriculture, human settlement, and civilization happened in some places and not others. To what extent did the genetic raw material of the peoples who practiced early agriculture and protoculture affect the course of history? Jared Diamond takes the view that it does. In particular he draws attention to the fact that the Fertile Crescent of the Middle East has an extraordinary concentration of wild grasses with large seeds, quoting a study showing that of the fifty-six grass species with the largest seeds, thirty-two are from this region. Furthermore, he points out that although wheat may be the result of hybridization between species and between a variety of interspecific crosses, it generally self-pollinates; so that when a genetic change occurred which protofarmers or farmers found useful, the characteristic would be more or less stable, which would encourage active selection and therefore early active plant breeding. Cross-pollinating species would not have this stability, and therefore the development of proto-crops would have been slowed down, as good characteristics would have been lost far more frequently, and early farmers would not have the same level of incentive to select and save seed from promising individuals. Corn is a cross-pollinator, and the implication is that the development of agriculture and therefore civilization in the Americas was held back.

There are a great many other reasons (which Diamond recognizes) for the early development of settlement and civilization in the Middle East. Several animal species proved relatively easy to domesticate, but the presence of grass species with a high potential for domestication, a high nutritional value (protein as well as carbohydrate) may have been a major contributing factor. Societies that depended on vegiculture may well have had an easier path to early agriculture, but using clones for crop propagation would have greatly restricted genetic development, as only propagation by seed opens the box of chance events that is offered by sexual reproduction. And only once this magic box was opened could early farmers have had the opportunity to interact with their crops to develop plant breeding as a skill. With the possible exception of west African cultures such as Benin and Ashanti, all the world's civilizations were founded on the cultivation of a grain. Grain is easily transported and can be stored for longer than roots and tubers—and it lends itself to trade and so to the development of civilization.[46]

The exact relationships may be disputed, but the implications are clear: civilization is founded upon crops. While hunter-gatherers need around twenty kilometers^2 per person in order to be able to hunt and gather enough to survive, the same area can support six thousand farmers.[47] Once enough food can be produced from a given area of land, the surplus can feed craftsmen, soldiers, priests, and others who never need to till the land. From now on, the process of civilization can be seen as a positive feedback loop; the skills of the non-farmers make it possible to produce more (soldiers to provide security, irrigation engineers to dig canals, blacksmiths to make tools, etc.). Domestication of plants enabled civilization to start, but once domesticated, plants were improved only very slowly, as with no clear knowledge of plant reproduction, and no notion that improvement and development were possible, premodern civilizations were in no position to develop anything resembling conscious plant breeding. That creation myths often featured the gift of particular crops to a people took the idea of crop improvement out of the realm of the possible and the practical and into the realm of the gods. Only they could be the breeders.

SUMMARY

As the first farmers tilled the first fields or planted the first gardens, they started breeding. In simply selecting seed or tubers from particular individuals for cultivation, they were making a choice, the beginning of a process of human driven evolution. A key distinction needs to be made between vegiculture, which involved clonally propagated crops, and true agriculture, in which crops were cultivated that had to be repropagated every year from seed. In the case of the latter, the process of collecting and sowing seed every year initiated a process of constant genetic change. The corollary of this genetic change is adaptation to human needs. As early farmers moved, traded, migrated, or fled, they transported the seed of their crops. The genetic complexity of these crops tended to grow as they are moved, with some of our most important crops the result of several species combining to form hy-

46. Diamond 1997.
47. Schwanitz 1966.

brids. The skill of tribal people was not in consciously driving breeding but in recognizing the distinctiveness of new varieties as they arose, in particular, of sudden mutations or hybrids (corn being the supreme example).

The clonal propagation of staple crops in vegiculture may not have led to much genetic progress—although the banana is perhaps an exception. Clonal propagation of fruit and nuts, however, was a way of maintaining and distributing chance finds. Here, the unrecorded history of the discovery of the art of propagation was allied to the ability of our ancestors to recognize better than average individual plants.

TWO ❦ LANDRACES ❦ BEDROCK OF TRADITIONAL AGRICULTURE

The Lahu people of northern Thailand are one of several tribal groups who live in the hills of a region stretching from northern India across to Vietnam and China. They live by shifting agriculture and foraging, the former activity having resulted in major deforestation and landscape degradation, issues now being addressed by the Thai authorities and nongovernmental organizations (NGOs). Much of the deforested landscape is now made up of bamboo scrub, a material the Lahu make excellent and extraordinarily skilled use of. Between the bamboo and associated vegetation are fields, often steeply sloping, planted with dryland rice—the Lahu staple. The plants are widely spaced and the heads of grain even more so. Mixed in with the rice are a number of other plant species, clearly not weeds, as heaps of discarded vegetation around the outer edges of the crop margins indicate recent weeding.

What may look inefficient to the modern eye has suited the Lahu very well, at least until increasing population pressures resulted in destructive forest clearance. After all, modern agriculture, although much more "efficient" in its use of space, is often notably less sustainable on other accounts. In many ways the Lahu seem to have gotten things right: their villages are simple, but everyone looks well fed and healthy. Tending the fields does not appear to take too much work; like much arable farming, work happens in bursts. Life on the whole looks good; many Westerners might well fantasize about coming to live here for a while.

The sophistication of tribal farming communities can all too often surprise us—perhaps a reflection of our own assumed superiority. Ethnographic literature has plenty of examples of "primitive" peoples with empirically based and experimentally orientated crop selection processes and sophisticated systems of classifying and naming crop varieties. There is no doubt our ancestors had a system of plant breeding, as advanced as was possible in a prescientific age; this was often expressed in the language of myth, superstition, or spirituality. "Irrational" language may often hide descriptions of evidence-based techniques. Many might argue that there is plenty of irrationality in today's corporate-dominated agriculture—it is just that the language is different.

UNIQUE TO TIME AND PLACE—LANDRACES

During the 1940s, F. A. Squire, an officer in the Agriculture Department of Britain's West African colony of Sierra Leone, made a collection of rice varieties grown by the Mende people; there were at least fourteen distinct varieties that "farmers can recognize at once and unerringly when shown samples."[1] Great efforts were made to keep seed pure, such as collecting only seed from the center of a plot for resowing instead of at the borders where it was likely to be mixed. Roguing (weeding out inferior individuals) was done by women and children, there being widespread community awareness of the different varieties. Some were regarded as better for certain types of ground, and there was variation in the time of cropping. More recent research (in the 1980s) shows how varieties released by the government are incorporated into the local planting stock and so after a few generations made more suitable for local conditions. Studies showed how villagers would try out new varieties, and the verb "to experiment" in the local language was derived from "trying out the rice." Input-output trials were customary, as the same bowl would be used for measuring seed for planting and at harvest time.

When farmers choose plants from which to collect seed, they are undertaking the all-important process of selection, both subconsciously and consciously. They may have particular traits they want from the parent plants—high yield or early maturation, for example. But the plants will also have undergone a process of selection in the environment of the field or the garden. Those whose genes, operating in their given environment, have produced the best plants and the most seed will often be the ones whose seed is selected for the next generation—Darwin's "natural selection." The environment, with its pests and diseases (*biotic* stresses) and droughts, frosts, and gales (*abiotic* stresses), will have done much winnowing: those individuals that succumbed in a spring drought, or got eaten by caterpillars, or that fell over in a storm just before harvest. Simply by collecting the seed of the survivors, the collector will be acting as an agent of the environment as an evolutionary pressure. By farther narrowing down the criteria for which they collect seeds, the seed collector is exerting further evolutionary pressure, but a distinctly anthropogenic one.

Over time, the gene pool of the crop will be narrowed, and a distinct "variety" will result. Traditional farmers who select crop varieties over genera-

1. Quoted in Richards 1985, 144.

tions end up with seed strains that are known as *landraces*. Landraces vary from area to area, from valley to valley, and from village to village—the name stresses the link between land and genetic relationship. Landraces also vary over time, as seed is generally saved from one year to sow the next, so every year's evolutionary pressures will result in a subtly different gene pool available for the next. If a farming people moves, and takes seed with them, the resulting crops would have to adapt to new environment, and rapid evolution will take place, often resulting in a distinctly different landrace.

A landrace is characterized by the following:

- It is morphologically distinct and recognized by local farmers as such with a distinct name.
- It has a certain genetic integrity, but there is a distinct level of variation.
- Is genetically dynamic, so the genetic makeup of the whole population is likely to be different from year to year.
- Is understood to be different to other local landraces in terms of its form, date of maturity, performance under different conditions, resistance to pests, quality, and so forth.

"Modern" plant breeding, based on Mendelian genetics, is a little more than a century old. In terms of time, the history of plant breeding has been the history of landraces, but it is a history of which very little has been recorded, and during which the processes of breeding have remained largely unchanged. Now that we are in an era where the majority of the world's population lives in cities there is a danger that we will lose touch with the traditional landrace-based agriculture that used to feed us. There has been very little research into landraces, how they develop, or how they are maintained; most of what has been carried out has been in the context of preserving genetic diversity. At first sight it seems as if the world of landraces is in danger of becoming obsolete. However, there are signs that there is more life and a brighter future in landrace-based farming than might be thought.

A landrace that is regularly grown from seed is inevitably variable, which can be a good thing and a bad thing. In a mixed population there are genes for coping with a wide variety of different conditions. If there is a drought one year, then there will be those individuals whose genes give them greater drought tolerance; a plague of caterpillars, and there will be those whose genes give them a greater ability to resist predation, such as hairier leaves which tend to deter insects. If the bad conditions continue from year to year, the gene pool of the entire landrace will gradually change, but at the same time it will maintain the breadth of genetic resources which will give it the resources to deal with new problems. Traditionally, variability offered a buffer against famine. Reliability was more important than yield. But increasingly, famine (as opposed to chronic malnutrition) is a thing of the past; transport, city-based relatives, and international agencies can help overcome occasional crises. The demands of a cash economy are now more likely to make farmers think about selling their produce in the marketplace, so yield becomes more important, as does the predictability of a particular quality that consumers expect. The variability of landraces, which was once a strength, is now more likely to be a weakness. For many, one of the great liberating aspects of modern varieties is their predictability, as every seed can potentially be the bearer of a good crop.

WEAVING A COAT OF MANY COLORS— THE ORIGINS OF DIVERSITY

The level of landrace diversity developed by traditional farmers astounds at least those from the developed world, where commercial agriculture tends to work with relatively few varieties. It may be a good deal less remarkable to keen gardeners, amateur or professional, who are aware of the incredible range of variety available amongst ornamental plants.

Since most crops were domesticated from only a very restricted part of the species gene pool it is perhaps surprising that traditional farmers have brought forward such variety.[2] Certainly the level of interspecies crossing that is part of the early history of many crops may account for much of this. Other factors have contributed; farming communities have created a great array of the genetic barriers which are necessary for the formation of distinct genetic entities, so that variation can accumulate faster than natural selection can reduce it. Geographical separation is an obvious one, particularly when crops are grown in places beyond the natural habitat of the parent species. It is also worth noting, as Nikolai Vavilov did, that the greatest levels of diversity are to be found in mountain areas, where a host of different environments confront farmers and their crops, and necessitate varieties adapted to variable conditions. Human demands for races that crop at

2. See technical note 10: Gene Pool.

various times, have different cooking qualities, serve unique functions, and so forth, all provide opportunities for genetic variations to be expressed. The process of selection, in other words, constantly identifies differences between plants that may be useful; these attributes are then used as a basis for the separation of part of the crop from the other, and the process of genetic isolation begins. Differences between plants that might have no consequence in nature may have great importance for the human user.

Isolation leads to variation, both in nature and among cultivated plants, because of the phenomenon of *genetic drift*.[3] As genes that rarely get a chance to be expressed are able to do so, interesting things can happen. "Waxy" corn is one such; a defect in the endosperm, very rare in the Americas, it appeared in several places in Asia (in China and Assam, India) during the early twentieth century and was then found to be a useful source of a chemical used in the food, textile, and paper industries.[4] Vavilov described a relatively low gene pool allowing the waxy gene to be "emancipated" in Asia. Flour and sweet corns are also thought to have originated in this way. Flour corn is useful, as it is easier to grind than the older, harder type, known as flint corn; while sweet corn, with a higher than normal proportion of sugar to starch, was originally used to make Mexican Indian beer, *chica*, and a confectionary, *pinole*.

What would be seen as a "high ratio of input to output" in a modern agricultural system was probably an additional reason for the high level of genetic variation: the high proportion of seed that is kept by traditional farmers for sowing the next year's crop. This would have the effect of passing on a relatively wide level of variation from one generation to another. High levels of variation provide opportunities for fine-tuning the range of varieties, so that particular strains could be discriminated for very specific purposes.

The selection criteria used by traditional farmers in developing distinct varieties can be seen as either production related or use related. Production-related varieties are chosen for their ability to thrive under particular conditions: pest and disease resistance, resistance to abiotic stresses, or timing of the crop. As a general rule, landraces tend to be better at producing a crop under difficult circumstances than modern varieties, and there is inevitably a good "fit" between the landrace's environmental tolerances and the ecological conditions prevailing in its territory. Landraces have also developed

3. See technical note 11: Genetic Drift.

4. "Waxy" corn has a different chemical makeup; it was first noticed because it produces a flour which makes a glutinous porridge when mixed with water.

for very precise uses; for example, in corn there are varieties for cooking by "popping," coarse milling (for making polenta), eating straight off the cob, making flour, making animal feed, and so on. Traditional cultures often recognize very precise characteristics, for example, one variant for pregnant women and another for children. Functionality can also cover what may appear to the cold eye of the modern observer as being nonessential uses: ornament and decoration or for use in particular religious rituals. What may appear to be ornamental to the modern Western eye may well have a spiritual function to a traditional community, to whom an absence may be profoundly unsettling.

Traditional communities label landraces according to criteria which are subjective and often strongly determined by cultural preferences and belief systems. These labels may bear little relationship to genetically determined differences, which makes the classification of landraces, vital for their conservation, something of a maze for those working with them.[5]

Mechanisms of selection among traditional communities appear to vary greatly. Selection at harvest, the picking out of grains to be next year's seed corn, is universal; preharvest selection, which involves marking out plants while still growing is more time-consuming and systematic and does not appear to be so frequent. Indeed, there is disagreement among specialists about how much preharvest awareness of a crop there is. The ability of a community to actively shape its plant breeding programs probably varies greatly through time; one suspects that many communities have a burst of selection activity every few human generations, the work of one or two individual farmers, or the pressure of changed circumstance. Other generations may do little more than roguing out obvious failures or deviations—which is plant breeding at the lowest level. Who carries out selection and breeding is also crucial to the plant breeding process, and varies enormously from community to community; the most fundamental question here is whether it is done by men, who may or may not be involved in growing but are often not involved in the end use, or by women, who are virtually always far more involved in the end use of a crop.

It was the genius of traditional cultures to have made identifications of utility when faced with a novel variant of a familiar crop. Maintaining it would have presented many problems and likely would have inspired farmers to keep different varieties separate. The Hopi (a people of the southwest United States) are known to separate different strains of corn, each strain

5. An example regarding cassava can be found in Kizito et al. n.d.

having a distinct color. Traditionally, ears that showed mixtures were not saved for seed, while families preserved varieties jealously, passing them down from generation to generation.[6] When each family resows from its own stock in this way, landraces can split into countless "family races," as was the case with Japanese rice until only a few decades ago.[7]

Sometimes great sophistication is needed to maintain particular varieties. In China, millet (*Setaria italica*)[8] occurs in a large number of varieties, including one with a glutinous endosperm which can be used to make wine—this characteristic is due to a recessive mutation of a single gene.[9] It is known that millet wine was used in Shang era ceremonial (ca. 1766 BCE–ca.1122 BCE), which indicates that there must have been a remarkably high level of genetic knowledge present at the time, in order to maintain the expression of this recessive gene.[10]

Preliterate cultures could record knowledge about crop varieties only orally. But the beginning of writing gives us a window into the past, and in the case of descriptions of crops, into the factors which farmers and consumers regarded as important in distinguishing them. The Sanskrit medical scholar Charaka, writing ca.700 BCE, lists various varieties of rice, including:

Raktashali—red
Mahashali—large grain, fragrant
Kalam—thick stem
Shakunarhita—curved grain
Turnaka—quick maturing
Tapiniya—golden, maturing in hot weather
Kardamaka—for slimy soil

Various varieties were also described as being good for particular medical conditions.[11]

Prescriptions to not mix varieties have been laid down in the custom and

6. Wallace and Brown 1956.

7. Ohnuki-Tierney 1993.

8. "Millet" can refer to one of a number of different small-seeded grasses. The species discussed is often known as 'foxtail' millet. The most widely grown is 'pearl' millet (*Pennisetum glaucum*), found in Africa and India. 'Proso' millet (*Panicum miliaceum*) is the third most important species, domesticated originally in a number of locations across Asia.

9. See technical note 12: Recessive Gene.

10. Needham and Bray 1984.

11. Nene 2005.

laws of many communities, including the ancient Israelites; the Mosaic Law states "Thou shalt not let your cattle gender with diverse kind, nor sow thy field with mingled seed" (Lev. 19:19). Such a warning may explain the unease felt in many quarters when deliberate hybridization began in the eighteenth century.

There is a tendency to think of traditional communities as being very dependent upon nature, forced into eating whatever nature can provide. Cultural factors, however, can make a considerable difference to what is grown and developed and what is not; for example, it has been suggested that in Africa tribal custom is more important than environmental factors in deciding which crops are grown in a particular area. Particular cultural groups are linked to particular crops or varieties of crops; indeed, they may define themselves in terms of either ethnicity or village through what they eat, which may even extend to particular varieties. In the case of chilies (*Capsicum* species), the fact that cultivated varieties are far more likely to be self-pollinating than wild species enables a mixture of different varieties to be grown in a small area with minimal cross-pollination; in the highlands of Guatemala, varieties of chili, along with those of beans and corn, are very geographically restricted; local people report that they would fail if grown anywhere else.[12] It is often believed, at least by those sympathetic to the maintenance of high-diversity, landrace-based agricultural systems, that landraces have evolved to closely suit the prevailing conditions of their environment—which reflects a similar belief among some ecologists into what they believe is a similar adaptation to geographically limited regions among flowering plant species. Little research has been carried out in this area, but given that weather conditions in particular often change drastically from year to year, and landraces have been selected to give a crop under a variety of conditions, this may be an unwarranted claim.

Variation does seem to have been valued for its own sake by many communities; indeed, a love of diversity does seem to be part of the human condition. For some communities, the maintenance of a diverse gene bank appears to be part of the role of particular individuals; there is the example of a shaman in an Amazonian community, the Amuesha, with 53 manioc varieties in his plot, compared with only 5.9 in the average household's.[13]

Try as they might to keep varieties separate, combinations occurred, and new varieties would result from hybridization. Whenever a new landrace

12. Pickersgill 1969.
13. Brush 2004.

was introduced to an area, there was the possibility that this would happen. Occasional breakthroughs resulting from the shuffling and recombination of genes might enable a crop to yield substantially more, grow in conditions it had not been able to before, or resist a disease.

Such breakthroughs happened slowly, however, and many factors controlled how quickly or extensively they spread. Trade routes played a major part, and it is no coincidence that the areas in which many crops were first domesticated are also areas that could be described as "hubs," where tradesmen, wandering shamans or holy men, tribes on the move, and invaders would all pass. Mesoamerica, the Fertile Crescent, and central Asia were all such regions. The Guatemalan highlands are a more specific example in Mesoamerica, where there is as much landrace diversity among corn as there is in all of the present-day United States and Canada, as well as a good part of Mexico; broad valleys leading through the area from the coast provided a natural invasion route for the conquistadors and pre-Columbian tribes.[14] Much of the Flint corn grown by Native Americans at the time of European settlement in North America was clearly related to landraces from this region, not from geographically closer Mexico.

That the Chinese had wheat by 1500 BCE, and that oranges (*Citrus sinensis*) and peaches (*Prunus persica*) had arrived in Europe from China by Roman times attests to the power of the trade route. In many cases, we must assume that introductions were made through the initiative of traders, taking a risk that someone somewhere might be willing to pay a high price for seed or root of a new and exotic plant. In some cases, however, introduction appears to have been deliberately carried out, as happened in the second century BCE, when Queen Hatshepsut of Egypt sent an expedition to Punt (probably present-day Eritrea or Ethiopia); temple carvings illustrate five ships returning with trees in pots. Royal, and later, monastery, gardens also played a part in trialing and evaluating new crops. A faint historic echo of what might have been a royal initiative in crop introductions and trials can be heard in the story of the Assyrian king Merodachbaladam (ca.700 BCE); monumental inscriptions claim that he introduced alfalfa (*Medicago sativum*), garlic (*Allium sativum*), leeks (*Allium ampeloprasum* var. *porrum*), cress (*Lepidum sativum*), lettuce (*Lactuca sativa*), and many spices into the region.[15]

Although introduction took crop species on long journeys, crop varieties or landraces would not necessarily travel so far; a variety would have to be

14. Anderson 1967.
15. Ambrosoli 1997.

very distinctively different for anyone to want to transport it a long distance in an age when the difficulty of long-distance trade restricted commerce to only cargos valuable enough to be clearly worth the trouble.

For all of the premodern era, and indeed well into the nineteenth century, introduction was seen as the only way to achieve distinctly different new varieties. The question has to be asked—since many premodern civilizations achieved considerable technological progress—why was so little progress made with plant breeding?

The most obvious answer to this question is one we have already touched on—that progress in breeding benefits from a knowledge of basic genetics that simply was not present in the premodern world, and that there has to be an awareness that it was possible for humans to make changes to living organisms. But there are also other reasons; plant breeding was nested at the center of the whole problem of why premodern societies that had moved beyond tribal systems to urban-based civilizations had a slow rate of innovation.

Premodern, precapitalist civilizations were almost invariably divided between a tiny wealthy elite and a mass of very poor craft workers and farmers. The desire of the elite for luxury goods used up a very large part of what we would now call the gross domestic product, as well as an overwhelming part of the growth rate. Far and away the easiest way to increase wealth was by conquest, appropriating the crops, products, and populations of subject peoples. Technical advances in manufacturing were so slow as to produce only the smallest economic growth. Elites, quite rationally, invested in conquest and the military in order to gain wealth. Trade and manufacture, which earned far less, came to be despised by the elite, which made technical progress even less likely. The position of agriculture was even lower. Increasing agricultural production was achieved overwhelmingly through taking in more land from natural habitat, which in dry climates involved irrigation (which explains why irrigation technology was quite sophisticated). Some cultures also appeared to have a good understanding of how to maintain soil fertility, notably ancient Rome and China, but on the whole, the gains to be made in innovation in agriculture were so low as to be not worth bothering with—which must account for a lot of the low social status of agriculture in the premodern world. Since conscious plant breeding was limited to simply selecting and keeping pure the best strains of crops, progress would have been immensely slow—far too slow for anyone in the elite to recognize it as a source of tax revenue. The only people who knew about plant selection were the peasants or serfs, very often subject to such heavy taxation or other exactions that they had little incentive to improve their crops. So, of all fields of possible innovation, plant breeding was low on the list.

SEED AND SPIRIT—DIVERSITY AND LANDRACE POLITICS

The last few decades have seen the widespread erosion of landrace diversity as farmers abandon traditional varieties in favor of modern, higher-yielding ones. This process is part and parcel of the transition from traditional economies, often based on subsistence agriculture, to cash-based commercial ones, and from largely locally orientated economies to the global scale. The loss of diversity, which has seen massive levels of landrace extinction in some regions, and its associated effects must be recognized as one of the key political issues facing plant breeders. It might appear that there is no future for landraces except as museum crops, tended by peasants in traditional costumes for the benefit of wealthy tourists from the developed world, or as packets of seed chilled in seed banks, awaiting a time when their genes might be found useful for breeding into a modern variety. But is the future necessarily such a bleak one?

Many plant breeders throughout the twentieth century have been well aware that their work is inevitably destructive—that new and more productive varieties would drive old ones into extinction—along with their genes, which could be vital in the future for the creation of new varieties. This awareness has led to the establishment of international seed banks, and a movement towards *in situ* conservation in regions of high diversity (to be discussed in chapter 14). Many now in the plant breeding community do seem to see landraces as simply germplasm, whose conservation is a vital task though they have no economic role; so it would seem that the role of the people who sow, till, and harvest them is no more than that of genetic guardians.[16]

Opposite, and often opposing, commercial and mainstream plant breeders are those who see landraces as vital for the communities who grow them, and who either oppose the commercialization of traditional farming practices or who are deeply skeptical about it. This movement is all too easy to dismiss as backward looking and negative; however, it may help to give a new lease of life to landrace agriculture, especially valuable for marginal communities for whom joining the world of commercial agriculture is not a viable option or is too risky. Navdanya in India is one of the most

16. See technical note 13: Germplasm.

highly organized such groups: an alliance of middle-class activists and low-income farmer groups cast very much in the Gandhian tradition of developing village-based economies around self-help and community control of resources. There is also a streak of nationalism and a stress on the preservation of indigenous culture through promoting traditional Indian agricultural practices. With branches or associated organizations all round India, it has since the early 1980s been campaigning on a number of issues (anti-GM, organic agriculture, against "bio-piracy," etc.) but also on organizing practically around what it calls "seed sovereignty." Its members encourage farmers to save landraces and help educate them in ways of preserving seed and how to select the best seed. This is done partly through "community seed producers" (*beej utpadak*) who multiply landrace varieties and make them available to farmers who want to go back to traditional varieties after bad experiences with modern ones. Events are organized where peasant groups can exchange landrace seeds.

The importance of organizations like Navdanya is that they can potentially provide an alternative to the trend to commercialization—a safety net for marginal farmers in areas where yields may always be uncertain, as well as for a devolved network of landrace conservation protected from commercial pressures. The networking involved in seed exchanges could potentially set off a new train of landrace evolution, beneficial perhaps to those involved but paradoxically eroding landrace diversity farther through mixing genes formerly kept apart.

Landrace conservation may appear to make sense to many marginal farmers who do not feel confident about using modern varieties; indeed, this may be an issue almost of life and death for them. But for those who are not farmers, who are engaged with campaigns for landrace conservation (and, by extension, traditional agricultural systems) for political reasons, there is a clear element of romantic politics: the appeal of unity of land, crop, and farmer, of the supposed integrity of indigenous culture, the notion that until outside commercially driven forces arrived everything was somehow in balance. However, as is so often the case with anything described as "traditional," there is a strong element of ideologically driven wishful thinking, and the "traditional" may turn out to be surprisingly modern. One of the key aspects of landraces is that they are dynamic, always changing; farmers have throughout history sought out new crop varieties and grew them with the result that their genes have joined the original landrace. Green revolution high-yielding variety genes began to move into landraces as soon as they produced their first pollen, with the result that, at least with rice and corn, many landraces are now heavily influenced by modernity.

Peasant communities that adopt HYVs may make attempts to save landraces for a variety of reasons, ranging from the ritual and spiritual to the gastronomic or simply as an insurance policy. They may also welcome landrace and HYV crossbreeding. An example of this is provided by farmers studied in Chiapas, Mexico, where traditional communities have shown a strong desire to retain corn landrace diversity, as they recognize that each has advantages and disadvantages in different growing conditions and different levels of risk. Higher-income farmers apparently were more likely to maintain older varieties. That their traditional corns were crossbreeding with HYVs was recognized; there was even a word for it—*acriollado*—becoming a Creole crop—implying that it was better adapted for local conditions but had lost some of the distinctive traits of HYVs, notably, short stature.[17] Not everyone in the region agrees, however; the Zapatista movement, whose politics are a blend of indigenous nationalism and Marxism, appears to see all nonlocal corn as a threat and has organized cold-temperature seed saving so that seed can be "safe from contamination and annihilation by the invasion of foreign corn strains."[18]

That supposedly ancient landraces may not always be so ancient in their genetic heritage is a useful perspective from which to look at claims of "genetic pollution" of landraces by GM genes, and indeed, the whole relationship between landraces and modern varieties. Such claims, including those that this "genetic pollution" was deliberately engineered, are regularly made by activists in Mexico. That corn should have become such a politically contested area should not surprise us; its centrality to the worldview of many Native American cultures means that it not just a crop and a staple but a cultural icon. Responses to corn are likely to be highly emotive, and given the economic and political marginalization of Native American peoples throughout the Americas, corn is almost guaranteed to become a hot political topic. One essay on the topic of genetic pollution alleges:

> Corn is in the fix it is today because of a long process of aggression not only against corn itself, but against all the social ways and means that made it possible, and in particular against the peoples who have created and nurtured it for so many centuries. Examples of these attacks include

17. Bellon and Brush 1994.
18. AScribe Newswire 2002.

ignoring the rich and sophisticated knowledge that sustains local corn varieties, imposing ultra-simplified cropping and consumption habits, destroying local systems that maintain, breed, and distribute seeds and, above all, destroying its sacred and life-giving nature.[19]

The issue here is clearly one of lack of ownership over new varieties and new technologies, inevitable among impoverished and marginalized communities. It is indeed difficult for any evidence-based discourse to engage with one conducted which uses terms like "sacred," but it should be borne in mind that communities, or their spokespeople, whether elected or self-appointed, generally seek refuge in such emotive language because of their economic and political marginalization.

The quotation above highlights some central problems in the contemporary politics of plant breeding, an understanding of which requires awareness of social and historical background. The cultural importance of certain crops and their role in ethnic identity also needs to be appreciated. Rice, for example, is intrinsic to Japanese self-identity, although it is not a native plant and took many centuries to become adapted to northern areas; the fact that many communities may have been millet or buckwheat (*Fagopyrum esculentum*) rather than rice eaters until relatively recently also appears to have been deliberately ignored by nationalist historians. Even the mutation that led to the characteristically glutinous nature of Japanese rice occurred in Southeast Asia.[20] In Latin America, corn expert Paul Mangelsdorf alleged that he was nearly lynched in Mexico city when he suggested that corn might have originated in Peru, and again in Peru when he speculated that the famous 'Cuzco Gigante,' the race with the largest kernels, might have been originally Mexican.[21]

Food is central to much ethnic identity, and so it should not be surprising that landrace crops have become part of this identity and that their antiquity and purity has been stressed by activists. When communities are afflicted by poverty and their cultures and languages relegated to lower status, it is almost inevitable that items of ethnic identity such as landraces should be given a highly symbolic value. To the outsider, however, such claims may seem overinflated and irrational. Worldwide there are communities that have been left out by advances in agriculture and economic development.

19. GRAIN 2003.
20. Ohnuki-Tierney 1993; North Carolina State University 2002; Chang 1976.
21. Fussel 1992.

The political responses to this may take forms which are in the end highly counterproductive and deepen their marginalization; the turn to guerrilla warfare and violent rural insurrection by small groups being the most extreme response. Much of this marginalization has an ethnic flavor; in Latin America the repression of Native American peoples by Spanish-speaking elites of European descent is well documented. Assertions of ethnic identity the world over involve the creation of belief structures that call upon tradition to lend authority to ambitious political leaderships, so there is a strong tendency to overemphasize the antiquity of such traditions; this could be called the cultural politics of "roots."

Fundamentally it is perhaps not GM or the integrity of landrace or any other crops which are the real issue, but that they have become a trigger for peasant anger because of their symbolic status. The fact is that many communities that continue to use landraces have been left out of the development process because of the operation of unjust economic, social, and political systems. Plant breeding as a discipline cannot pretend that these systems do not exist, that they operate "above politics"; breeders have a responsibility to consider how their work impacts on issues of social and economic justice. It is only with such justice that there is any hope of integrating such communities into a broader national and global community and marketplace. Once integrated, treated with respect, and above all, making economic progress, such communities are almost certain to relate positively to breeding advances.

Organizations like Navdanya will in all probability have an impact only in the economic margins, which is not to dismiss the potential value of their work in safeguarding landraces. A more mainstream future for landraces lies in some other recent developments—the power of the consumer in demanding food diversity. There is some evidence that levels of diversity in modern agricultural systems are increasing for a variety of reasons (see chapter 15), and more discriminating consumers are demanding more variety. The Slow Food Movement is just one example of a growing trend, partly a reaction against what is seen as the bland offerings of global agribusiness, partly an interest in food for its own sake—in particular, food from local producers that reflects a sense of place.[22] The 1990s and early 2000s have seen a spread of "farmers' markets" (just old-fashioned markets, really, but rebranded for higher-income and aware, mostly urban consumers) and artisan food pro-

22. See http://www.slowfood.com.

ducers in not only the developed world but in higher income parts of the nonindustrialized world.

Where once supermarkets just offered potatoes and baking potatoes (and new potatoes in season), now they offer bags of named variety potatoes: potatoes for baking, for salads, and for mash, as well as niche varieties such as the oddly shaped 'Pink Fir Apples.' Led perhaps by the New World wine industry, which has put varietal names in the foreground (who these days doesn't know the difference between a merlot and a cabernet sauvignon?), the movement toward single-variety products is growing.

Farmers in developing countries and regions still have access to landraces that have been fine-tuned for very specific purposes. In Anatolia, Turkey, modern varieties earn money, but landraces are still grown in order to make a holiday food, *aşure*, a kind of sweet compote of wheat and fruit. Many people in Turkey prefer to make bread or other baked goods from landrace varieties; they are often superior in flavor, so they continue to grow them for their own use or for trading in the village. Once sophisticated urban consumers (or just former rural dwellers with a taste for the old days and the money to indulge it) start to demand "old-fashioned flavor," a role for landraces in the market economy is assured. Old corn varieties are apparently still very popular in the countryside around Mexico City, while in Peru the interest of urban consumers in traditional, distinctively flavored varieties, ensures that there is still an economic reason to grow them.

As food cultures become more sophisticated, there is increased scope for diversity, and so more room for distinctive landrace varieties—or indeed more advanced versions of them. Globalization is as much about the creation of space for multiple identities as it is about the development of a common culture, so possibly making new places for landrace crops.

SUMMARY

Once established as a cultivated plant, seed-raised crops would have developed into distinct landraces—discrete varieties, usually strongly linked to a particular geographic region. Landraces can be extraordinarily diverse—some crops develop distinct varieties for every microhabitat. Landrace diversity was also a means by which traditional farmers could recognize and preserve varieties for particular purposes. Landraces are also diverse internally; each one comprises a range of genetic variation, which allows the crop as a whole to be adaptable in the face of whatever climatic variation or the annual ebb and flow of pests and disease populations might throw at it. This variation is the reason landraces are still valued by many traditional farmers, who place importance on security above all else, and are given a particularly high value amongst those who see themselves as defending traditional communities. Among more "modern" consumers, landraces are still valued for particular purposes or qualities; it is perhaps wrong to see them as intrinsically old-fashioned.

THREE ❦ "IMPROVEMENT" ❦ THE AGRICULTURAL REVOLUTION

In the quiet southern German town of Eichstätt is a rather remarkable garden. It lies high up above the town, enclosed by the massive ramparts of the Willibaldsburg fortress. A large collection of plants is laid out in long rectangular beds, each one defined by low box hedging (*Buxus sempervirens*). Nothing is placed artistically; instead, plants (including shrubs, heavily pruned trees, herbaceous perennials, bulbs, and annuals) all stand in serried ranks like the exhibits in the glass case of a rather old-fashioned museum. Although an entirely modern creation, it is designed to evoke a garden which existed here during the late sixteenth and early seventeenth centuries. All the plants here were illustrated in a remarkable book published in 1613—*Hortus Eystettensis*. Commissioned by Prince-Bishop Johann Conrad von Gemmingen (1561–1612), the book illustrates a collection of plants which had been gathered together in the castle garden. Both are a fascinating insight into the worldview of the time; here are gathered together all the plants considered interesting or beautiful at the time—mostly European or from around the Mediterranean, but with a scattering of species from the newly discovered Americas.

What makes Gemmingen's garden and book special is that for the first time in Western history, plants were being collected and celebrated not for their use but for their own inherent interest. This was not a collection of medicinal herbs or vegetables (although many were included) but something altogether more modern. Many ornamentals are included, while other species must have been included for their "curiosity" value—they were considered interesting because they came from far away lands, or were sufficiently different from the normal run of south German vegetation as to cause comment among those who might have the leisure to notice such things.[1]

A TIME OF FERMENT—THE AGE OF MULTIPLE REVOLUTIONS

The garden at Eichstätt is a peaceful place in which to reflect on the intellectual ferment which lay behind it, and the revolution that was to fol-

1. Dressendorfer 1998. The book itself is available as a facsimile: Besler 2000.

low. The sixteenth and seventeenth centuries saw a change in intellectual attitudes about the natural world, at the core of which was a curiosity that began to develop into what has become known as the Scientific Revolution. The English Francis Bacon (1561–1626) was a key figure; credited with key developments in scientific methodology, his argument was that if nature could be understood then it could be managed to the benefit of humanity. From here it is only a short conceptual leap to questioning the fatalism and determinism of traditional religious belief—such questioning was central to the Enlightenment, the term which has become something of a catchall for the philosophical rationalism which came to be most clearly articulated during the eighteenth century. The Enlightenment unleashed a strong feeling of optimism among the educated of Europe and America—the idea that political and religious liberty would advance hand in hand with technological progress. The Enlightenment, in some form or another, underlies the entire conceptual edifice of modern society and culture.

While the Scientific Revolution took a long time to mature—and even longer to make much impact on how people lived their lives, other developments in the nature and pace of technological change were unleashing a transformation in western Europe and North America. These changes have been described under the umbrella term "the Industrial Enlightenment." Techniques (in any field of practical work) that were developed by one person or one locality were communicated to others who then tested them, cataloged them, modified them, and passed them on. Attempts were made to understand why some techniques worked and others did not. Observations of the natural world, followed by their cataloging and categorization—those of Linnaeus being the prime example—were also a vital part of this movement. The results of experimentation and observation were gradually accumulated to form a body of knowledge, which others could draw on, which novices and apprentices learn from, and which could be amended and passed on—through the spoken word or through printing and translation. The decreasing cost and increasing availability of the printed word played a huge part in this process. Experimentation and the accumulation of knowledge assumed an orderly world, governed by laws, not subject to the caprice of an unpredictable and willful deity. The next step from this assumption of an orderly world, was faith in the ability of people to change that world.

Alongside revolution in the world of ideas and technology, the sixteenth and seventeenth centuries saw major developments in how business was conducted; this commercial revolution was centered on the Netherlands, a new nation that had emerged from its landscape of reeds and flat fields to become a major trading power. All these changes were interlinked—in that it is difficult to see how one could have happened without the other. And none could have happened without a revolution in agriculture, too. Farmers were increasingly able to produce more food from the same area of land, as well as bring new areas under cultivation; increases in food supply resulted in increases in population and greater food security, especially for workers in the rising number of workshops and factories. Greater populations then became a major driver for urbanization and increasing specialization in employment, helping to fuel the industrial revolution, which began in Britain in the late eighteenth century.

The Agricultural Revolution of the latter half of the eighteenth century and the first half of the nineteenth was a complex phenomenon, centered on Britain, as well as the United Provinces and Flanders (present-day Netherlands and Belgium). It was the culmination of a long process of advancement, much of which had happened elsewhere in western or southern Europe. It laid the foundations for similar advances in farming in the rest of Europe and North America. Its main physical manifestation was an improved ability to add fertility to the land; but just as important it was about a change in attitudes among landowners and farmers.

During the eighteenth century new crops, for example, turnips (*Brassica rapa*, which enabled more animals to be fed over the winter), more complex and efficient rotation systems, and enclosure (which rationalized complex and archaic systems of ownership) all contributed to great improvements in productivity. It has been calculated that the new four-course rotation produced 80 percent more food. Mechanization did not contribute much until the later years of the period, and neither did plant breeding as such, although the introduction of new crops did. The breeding of cattle and sheep was of more interest to most farmers and initially made more impact and probably contributed to the spread of the idea of the *possibility of* the improvement of plant crops. In Britain, the breeding of fine livestock became an aristocratic pastime; there was a growing interest in pedigrees and genealogy and, by extension, the possibility of perfecting and improving of all kinds of nature. It is important to stress the word *possibility*—as until then we must not assume that people really understood that improvement was possible, in animals and crops, or indeed in any other sphere of human life. Until the late eighteenth century, change had been slow, to an extent that we in our sometimes exhaustingly fast-moving times find very difficult to imagine. The Agricultural Revolution of these years, as with the Industrial Revolution or the Scientific Revolution could not have happened without some people seeing either a need for change or simply the possibility that change was possible, The changes that these revolutions are primarily about were all linked to this sea-change in attitudes among the educated; attitudes that looked to the idea of "improvement"—the word is a key one for understanding this period.

As a general rule, the elites of traditional societies are not much interested in agriculture—it is dirty work conducted by social inferiors, its importance realized only in times of famine, its everyday existence occasionally registered by art when it can be picturesquely romanticized as happy peasants standing by sheaves of wheat glowing in harvest sunshine. Few of the educated or powerful took much hands-on interest in improving agricultural productivity, and if they did they did not communicate with others. The seventeenth and eighteenth centuries saw a change however, mostly in northern Europe and the American colonies. Landowners began to take more interest in their estates, and the richer and more educated (that is, literate) farmers began to read and write more about their work. In Britain from around 1640, there was a sudden increase in the amount of agricultural literature; books began to be written by members of the gentry who were interested in the methods of the classical civilizations and what was being done on the European continent. Gervase Markham of Nottinghamshire (1568?–1637) was one such person who supervised or promoted translations from French and German authors. By 1640 it was possible for an English landowner to have quite a serviceable library on farming matters. Interest in farming as relaxation from involvement in politics or the court or law began to grow. Gardeners, herbalists, and botanists were becoming more numerous and they too wrote and corresponded. A newly wealthy class of urban dwellers began to become interested in gardening, in growing newly imported plant novelties for ornament; luxuries became more widely available, such as peaches, vines, and mulberries (*Morus nigra* and *M. rubra*). Farming and growing things began to lose its low social status. In Britain agricultural improvement had even become something of a craze by the mid-eighteenth century; it is said that the first British prime minister, Robert Walpole, read letters from his farm steward before state papers. Agriculture and country life also came to be seen as morally superior to that of urban life—the pastoral poetry of the Greek poet Virgil enjoyed a surge in popularity. The romantic movement in literature and the arts was also arguably fed by the enthusiasm for rural affairs.[2]

Probably the main reason for the enthusiasm for all things growing was

2. Johnstone 1938.

that agriculture was being drawn into the orbit of a developing capitalist economy. Landowners increasingly had to see themselves as agriculturalists and as entrepreneurs able to institute changes and improvements rather than as feudal grandees overseeing an unchanging and rigid social order. The rural poor were also increasingly being forced to see themselves in a different light, too. As rents payable in products or labor were transmuted into rents payable in cash, tenants were pulled into the market economy, either having to sell the crops they grew or work as landless laborers. Britain was the first country to experience the monumental changes which were to transform the nature of farming everywhere: the change from an agrarian society where the majority of the farmed landscape was peopled by those who consumed what they themselves grew to one where it was dominated by those who grew to sell. In the past the majority of the population would have been subsistence peasant or tenant farmers; now those who worked on the land would become a minority. Once the act of growing food became a commercial activity, the pressure increased to look at it with the eye of an accountant, to note inputs and outputs—one such input being the quality and productivity of the crop.

Driven by commercial pressures, innovative farmers and landowners began to correspond and network; one of the results was that new ideas and plants would spread much more quickly than before. Samuel Hartlib (ca.1600–1662), a London publisher and one of the great intellectual figures of the period, acted as a key networker, a veritable spider in a web of agricultural innovators, authors, and travelers. Through him, a German physician, Johann Brun, introduced new apple varieties to England, and Francis Lodwick, a merchant, philosopher, and pioneer linguist, introduced black cherries (*Prunus avium* cultivars) from Flanders. He published books on fruit and vegetable growing, bees, and husbandry in Flanders, as well as William Howe's *Phytologia Britannica* (1650), which listed grass species and made distinctions between them.[3] Starting as informal clubs or corresponding societies, fraternities of progressive farmers and landowners began to form agricultural societies from the mid-eighteenth century onward. Botanical gardens, which had originated in the need for herbal medicines, began during this time to become more orientated towards the practical needs of farmers and market gardeners—the small-scale producers of fruit and vegetables who catered for regional, and increasingly urban, markets.

A key issue among improving agriculturalists which was established by

3. Thirsk 1985.

the Hartlib circle, and which has remained an abiding principle, was that of diversification: of different practices, of uses of land, and of crops. Parallel to the ornamental gardener's search for the new and exotic there was a thirst among farmers for new vegetables and crops of which the most enthusiasm was for clover and other fodder crops.

FROM FODDER TO FRUIT—NEW CROPS AND NEW MARKETS

Fodder crops first made their appearance in Roman times. Lucerne, or alfalfa—the word derived originally from an old Iranian word meaning "horse fodder"—spread with nomads from central Asia in the second millennium BCE. A vigorous tetraploid form had almost certainly been noticed early on and brought into cultivation. The Greeks got the plant from the Persians and passed it into onto the Romans, who took it with them wherever they went.[4] Following the collapse of the Roman Empire, the knowledge of using alfalfa and other fodder crops went with them, but with the Renaissance, much classical knowledge was rediscovered, often through Arab writers or practice. Forward-looking estate owners throughout Europe drew on ancient agricultural texts to learn about how to use them.[5]

Leguminous fodder crops provided not just feed for animals but also soil fertility—all legumes have bacteria in root nodules which can "fix" atmospheric nitrogen and turn it into a soluble form which all plants can then utilize. Medieval agriculture was capable of maintaining low levels of fertility but not building on them—which only began with the use of leguminous crops, which at last enabled fertility to be increased if used in rotation with cereals. Fodder crops appeared in Flanders in the sixteenth century, and by the eighteenth, alfalfa, sainfoin (*Onobrychis vicifolia*), and clover (*Trifolium* species) were widespread in France, along with (nonleguminous) turnips. Until the seventeenth century most seed sown was that saved by growers themselves, the little seed trade there was being conducted almost entirely at country fairs. During the seventeenth and eighteenth centuries a growing trade in seed developed, particularly of fodder and "garden" crops (i.e., vegetables), generally from Italy, southern France, and Switzerland to northern Europe; there was even some importation from Turkey and Syria. Not surprisingly there was an increasing problem with poor-quality seed, either because it was old or had been stored in unsuitable conditions or be-

4. Lesins 1976.
5. Kjærgaard 2003.

cause of the nature of the variety. One consequence of this was a growing interest during the latter part of the eighteenth century in the distinctions between different varieties of clover and alfalfa and in finding good local sources of supply. The farmer in Somerset, England, who in the early 1700s had begun to sell clover seed from plants on his land, known locally as 'Marl-grass,' got a good price for it in London when it was rebranded as 'perennial red-flowering clover.'

In the American colonies, most seedsmen were Europeans who imported much of their stock. A key role in the development of an American seed trade was played by the Shakers, a religious sect that originated in England in 1772. The Shakers were very successful in the United States during the nineteenth century, largely because of their efficient farming and disciplined communal lifestyle. In particular they were pioneers in the seed trade, growing crops for seed production and marketing all over the United States, the largest seed producer being the community at New Lebanon, New York. One of their most lasting innovations of the early nineteenth century was the seed packet; previously seed had been poured out of a bag or scooped out of a barrel into whatever receptacle was handy. In 1835, they introduced another innovation that has lasted: the seed catalogue, listing not just plants and prices, but detailed cultivation instructions, too. The Shakers were almost certainly involved in early selection work, primarily among garden vegetables. Their religion forbade grafting, but the improvement of fruit and crop varieties was seen as creating perfection, and therefore as part of God's work.[6]

In Europe, the Low Countries (present-day Netherlands and Belgium) seemed to have been the leading seed producers, and it was a close-knit community of Dutch and Flemish immigrants who started the business of growing seed in England, around Sandwich in Kent, during the seventeenth century. Although active selection was probably not practiced, their seed gained a reputation for quality. In the case of carrots this almost certainly was because of varietal superiority; by 1610, the 'Sandwich carrot' was renowned throughout the country. The seed trade slowly built up its strength, diversity, and geographical spread; interestingly, unlike other areas of agricultural and horticultural progress, which were led by the gentry, the seed trade was built up by small tradesmen and the growers themselves.

New networks of progressive growers did not, at least until the nineteenth century, turn serious attention to plant breeding, in the sense of systematic attempts at selection. While there was a great deal of interest in the

6. Sommer 1966.

introduction of new species from abroad, the notion that anyone could actively search for new crop varieties among existing crop species was slow to catch on. The reason is probably that there was no need; increasing populations in Europe and the American Colonies could be met from the enclosure of what contemporary writers called "waste"—what we would today call "nature" or "good wildlife habitat." Substantial increases in productivity resulting from a greater use of animal manure, marl, lime, and leguminous forage crops would also have ensured that there was perhaps little pressure to turn to systematic attempts at increasing yield from crops themselves. The new fodder crops, being mostly annuals or at least bearing seed rapidly, would also have adapted to their new homes over time in the way that landraces always adapt when moved. However, active networks of correspondents and an increasingly lively seed market meant that if new varieties were found, they spread much more readily than before.

Variety selection appears to have been strongest among fruit and vegetables, and probably reflected the growing strength of the market gardening (i.e., truck farming) sector. Traditionally, everyone would have grown their own and traded only very locally. But with a growing commercialization of life, linked to urbanization, specialist growers of fruit and vegetables began to become important around towns and cities. Fruit varieties seemed to be the first to be given names to distinguish them from each other, a sign that at a time when consumers were becoming separated from producers, some indication of difference between varieties was needed in the market place.

A particularly important country for the growth of a distinctively horticultural scale and approach to growing food crops was France. While the Dutch were arguably the technical leaders in new ways of protecting plants for early season production, pruning, feeding, seed production and propagation, it was in France that fruit and vegetable production became almost a high art, culminating in the *jardin potager*, an elaborate garden within a garden where utility and order created pleasing scenes of what could be best described as a beauty of utility: vegetables organized into neat rows or patterns and fruit trained as cordons or espaliers looking almost as if they had been choreographed by the court ballet. It was during the reign of Henry IV (1553–1610) that the origins of this systematic approach to gardening and to estate management in general can be traced. Intriguingly, Henry and several of the key writers on gardening and estate management during this period, such as Olivier de Serres (1539–1619), were Protestant. Links between the origins of several key elements of modernity, such as capitalism and science have been linked to the rise of Protestantism, but they are singularly difficult to establish.[7]

Catholicism itself was no barrier to plant breeding, as can be appreciated from the great number of vegetable varieties which originated from Catholic Italy. Keen vegetable growers will know just how many of our vegetable varieties of today are Italian in origin. The country's rich vegetable and salad culture started with the Romans. Among other crops, they refined various wild cabbages, which were then distributed around the Mediterranean—new forms then tended to come back to Italy, as Italian traders were always active in their dealings with whoever governed the other shores of the sea. The cauliflower, for example, was bred in the Arab Levant (modern Lebanon) and introduced into Europe by Italian traders at some stage in the late medieval period. Broccoli, another flowering cabbage, originated in Italy, possibly in Roman times.[8]

New technologies of cultivation and the growing popularity of well-ordered kitchen gardens stimulated interest in variety selection, particularly of plants where an extra effort in cultivation had to be made—to be able to afford to do this gave such crops cachet and status. Pears (*Pyrus communis*) in seventeenth century and eighteenth century France are a case in point; the fruit became something of a craze, and a 1693 book by the head gardener[9] at the French royal court of Versailles mentioned sixty-seven varieties.

It is around this time that many of our most familiar vegetables were developing a form we might recognize today. The carrot is a good example. It is thought that Afghanistan is the center of diversity of the 'anthocyanin' carrot (i.e., strains containing one of several related strongly colored pigments); most wild carrots are only dirty white or muddy yellow. Carrots appeared in Asia Minor (modern-day Turkey) in the tenth century, in Arab Spain in the twelfth, in China in the fourteenth, and in Japan in the seventeenth. European carrots before the sixteenth century were purple or yellow; when cooked, the purple root exuded anthocyanin, which made sauces

7. Max Weber's *The Protestant Ethic and the Spirit of Capitalism* (1904) is the classic study of this kind. The issue, regarding science, is touched on in Henry 1997.

8. Lovelock 1972.

9. The gardener was Jean de Quintinie, and his book was translated by the great English gardener John Evelyn as *The Compleat Gard'ner or Directions for Cultivating and Right Ordering of Fruit-Gardens and Kitchen-Gardens with Divers Reflections on Several parts of Husbandry.*

and soups muddy purple, so the paler yellow variant was favored. In the Netherlands around 1600, orange types began to be selected from the yellow, the color gradually deepening, so that by the late eighteenth century a number of orange types were known. The purple and yellow carrots almost disappeared but were revived in the dying years of the twentieth century as a novelty in seedsmen's catalogs.[10]

NEW RICE, NEW METHODS—SONG CHINA'S "GREEN REVOLUTION"

Before leaving this period of major change, it is worthwhile to take a brief look at China. It is now well established that major advances in technology were made in China before they were made in Europe; indeed, travel along the Silk Road may have brought inventions such as paper and gunpowder to Europe. What Chinese civilization did not do, however, was develop a concept of experimental science or a lively commercial system independent of the state—so many technologies were not developed beyond their early stages. The field of plant breeding and genetics is no exception.

It is a period of the Song dynasty (960 CE–1279 CE), which today reads as one of the most remarkable episodes in agricultural history, an early example of a government-sponsored "green revolution." Needing to increase production to feed an increasing population, especially in vulnerable border areas, rulers instituted a wide-ranging program of agricultural reform. The neighboring Indochinese state of Champa (modern-day Vietnam) was known for its early-maturing rices as early as the first century CE. They did not reach China until centuries later, but they spread rapidly when in 1012 the emperor Zhenzong called for the distribution of seed in the Yangtze River region. The new varieties were distributed with written instructions for their cultivation, and a system of "master farmers" established—literate individuals chosen for their skill and given a minor official post; they acted as extension agents, educating other farmers in a range of new techniques, including the use of fertilizers and irrigation and in the use of new tools. The administration also made cheap loans available, and there was reduced taxation on newly cultivated land. A modern development program could not do better.

Although this green revolution was centered in the most advanced areas

10. Banga 1976.

of the lower Yangtze, its methods spread during the next few centuries over the rest of China, assisted by selection which gradually improved yields. The results were unprecedented surpluses, which stimulated trade and local industry, the production of luxuries like silk and sugar, and the establishment of a range of cottage industries. The greater wealth of the peasants strengthened their hands with the landlords, and tenancies gradually shifted in their favor.

HEMMED BETWEEN THE MOUNTAINS AND THE SEA—JAPAN

One of Japan's eternal problems is that there is very little arable land per head of population. As a consequence, high yield has been recognized as important for perhaps longer than anywhere outside Europe; Japanese farmers could not "enclose the wastes," plowing up heathland or draining marshes as in Britain, the Netherlands, or Denmark, nor could they simply "go west" as American pioneers did. Instead, production on existing land had to be intensified.

During the Tokugawa period (1603–1867), Japan experienced some of the features of a the European agricultural revolution, with widespread publication of books on farming—some actually written by those engaged in farming rather than by scholarly commentators. While a major transformation of agriculture did not occur and could not have occurred in a society so constrained by feudal codes and social stratification, the scene was set for the development of one of the most dramatic genetic breakthroughs in the battle for agricultural efficiency, an event which did not play out its full effects until the latter half of the twentieth century—the selection of semi-dwarf strains of wheat. However, it is worth noting that Japanese farmers increased yields of rice from 1.3 metric tons per hectare in 900 CE to 2.5 at the end nineteenth century.

With the Meiji restoration of 1868, farmers were freed from the restrictions of feudalism, with the result that many of the technical developments in farming and horticulture that had been brewing over the last few centuries in particular localities could now be tried in other areas. Farmers were now free to decide what to grow and where to live, had greater liberty to buy and sell land, and were less burdened by feudal rents. One result was the redistribution of landraces from one area to another. Previously, restrictions on personal freedom and communication had been almost incomparably greater than in Europe; one village, Maesawa in Iwate Prefecture, had apparently been so anxious to stop its own rice from being grown by others that a guard was placed at its border on the highway. Restrictions or not, the late Tokugawa era was noted for figures analogous to the progressive gentleman farmers of western Europe and North America, men who experimented, noted their results, corresponded, and wrote books. The most important one for variety selection was Naozo Nakamura (1819–1882) from Nara Prefecture, who collected rice from all over Japan, assessed each variety and then distributed the best, enabling yields to rise as farmers dropped less productive landraces. Peasants too were active in variety selection, particularly in varieties that matured early, that would allow double-cropping, or that would mature before the typhoons of late summer. Japanese peasants were unusual in that they appeared to be prepared to articulate the differences between strains more than traditional European farmers, naming rice varieties, usually after people, local deities and localities. Some varieties may have been jealously guarded, others were exchanged between villages.

The communication of new techniques and the distribution of high-quality or high-performing varieties were often achieved by agricultural societies or seed exchange societies, usually formed by farmers who owned their land or by landowners who worked some of their holdings themselves. Support for improvements by state institutions was assured, sometimes even to the point of enforcement "by the sabers of the police."[11] Experimental stations were also established in the Meiji era (1868–1912), one of their first tasks being the comparative assessment of landraces and the refinement of good varieties for local conditions.

Much had been done during the Tokugawa era to improve crop nutrition through the use of manure, so there was a history of selecting crop varieties, rice in particular, which showed a good response to feeding. During the Meiji era, plant food in the form of cakes made from compacted soybeans was imported from Manchuria, which encouraged the spread of selected rice landraces such as *shinriki* and *kamenoo* throughout the whole country—these were not only productive but showed good fertilizer response, were not prone to disease at high levels of nitrogen and were not prone to lodging.[12] It was from among these rices that the genes came for the development of the green revolution's high-yielding varieties. The story with wheat was similar, with an early systematic use of manure, and again selections which were short growing, and therefore less top-heavy.

11. Shand 1973, 14.

12. See technical note 14: Nitrogen and Lodging.

SUMMARY

The seventeenth and eighteenth centuries were a crucial time in world history, for this was when, in parts of Europe and the American colonies, there occurred a vast shift in all fields of life, particularly the intellectual, commercial, and technical. Agriculture and horticulture were as much a part of this period of revolution as anything else. Particularly important was the growing interest shown in farming and gardening by both traditional landowners and newly wealthy elites. Production needed to be intensified to feed expanding populations, and especially where land was in short supply, as in the Low Countries of northern Europe and parts of England, farming began to see major changes. Crop diversification was one of the earliest developments, followed by the beginning of the seed trade, accompanied by a growing realization that crops could be—and should be—improved.

Historians argue about European "exceptionalism." That European and American settler cultures were not exceptional can be appreciated by studying the remarkable developments in agriculture that also occurred in China and Japan. It was arguably other social, political, and economic developments which provided the framework for a massive expansion of European agriculture, and with it, the beginnings of systematic plant breeding.

FOUR ❦ VEGETABLE MULES ❦ THE BEGINNING OF DELIBERATE BREEDING

There is a crush of people around the tables, but all they are looking at is onions! Onions (*Allium cepa*) neatly tied, every one a perfect specimen, their usually untidy leaves neatly plaited with raffia; big single onions, much bigger than you could ever buy in the shops; bunches of spring onions, every one clean, straight, lush, perfect; there is even a special section for garlic.

Welcome to the Newent Onion Fayre, just one of thousands of events which happen throughout the summer from one end of Britain to the other, in which flowers, fruit and vegetables are exhibited in competition. Intended for amateur growers, some of the exhibitors do more or less nothing else—dedicated onion men may show the same bulb in Swansea on one weekend, in Chesterfield the next, and in Corby the week after. Newent's is the only dedicated onion fair in Britain; more common is a flower and produce show as part of a larger event, a village fete, or agricultural show, with a community hall or canvas marquee lined with tables. Placed on display are rows of carrots neatly aligned, their roots tapering to a hair, impossibly long parsnips (*Pastinaca sativa*), and polished and perfect-looking potatoes. Although by no means unique to Britain, it is here that the flower and produce show is most widespread, and most woven into the fabric of community life. Evolving out of the competitive showing of flowers that began during the seventeenth century (see chapter 13), the contemporary flower and produce show is not just a quaint custom of a notoriously eccentric nation, but the legacy of what was once a highly effective way of comparing and judging plants.

It was during the nineteenth century that the show took shape as a means of enabling growers to compete and compare. In horticulture it has always reigned supreme; in the farming world crops have always tended to be overshadowed by the big beasts of the show ring. Seed of good varieties seen in shows however would circulate informally, certain farmers and growers would acquire a reputation for a particular variety, names would be given, and increasingly, distinct varieties would emerge. Later, as horticulture and agriculture became more commercial, shows became a marketplace for seed merchants and nurseries. Under the wing of organizations, such as Britain's Royal Horticultural Society (RHS), some shows have become institutionalized; London's annual Chelsea Flower Show, for example, has been a national forum for plant introduction and marketing ever since it first pitched its marquee in 1905.

Closely related to the show system is the giving of awards to particular plant varieties; given the preeminence of Britain in matters of horticulture, the RHS Awards of Garden Merit and First Class Certificates given to fruit, vegetable, or ornamental plant varieties, instituted in 1859, are recognized worldwide. Interestingly, no comparable set of internationally recognized awards for field crops has ever evolved, although during the latter part of the nineteenth century, international shows of manufactured and agricultural products were an important feature of the commercial and industrial calendars of the world—at which awards given to crop varieties were much prized by their breeders and producers.

The show as an institution came about because growers were faced with more and more variety, and because they were under pressure to succeed commercially. It was now important to keep up, to make a profit, to choose the best for the land; what better way to make assessments of the host of new varieties than going to a social event? It would be possible to meet your peers, make new contacts, and enjoy yourself too—shows have traditionally involved plenty of eating and drinking.

That so many new varieties were there to be chosen from was the result of a newfound interest in and awareness of the opportunities offered by the plant world. Once it was realized that plant varieties had a certain malleability, it became possible to try to change them to suit human needs. After making much progress in the fields of plant nutrition, soil management, crop rotation, crop management, and propagation, the scene was set during the latter years of the eighteenth century for a much more active intervention in the heredity of the plant world.

Shows, however, played roles other than the strictly horticultural or commercial. The very act of showing seems to have become a focus for either social activity or one-upmanship; when this happened breeding became almost an end in itself. John Lindley (1799–1865), editor of the highly influential British journal *The Gardeners' Chronicle* and a leading figure in both British botany and horticulture had several times attacked the breeding of plants specifically to win prizes at shows. This was a fundamental weakness in the whole show system, and it came to be dramatically illustrated in corn shows in the American Midwest (see chapter 10). Sometimes shows became almost obsessively obscure, as with the gooseberry (*Ribes uva-crispa*) shows of nineteenth-century Britain, when workingmen in the industrial counties of northern England and the Midlands formed themselves into societies, constituted with presidents, secretaries, and stewards, for the purpose of running gooseberry shows—weight being the decisive factor. Quite why this fruit, always something of a minority taste, should become the subject of what only could be described as a cult remains a mystery. As happened with florists' shows (see chapter 13), prizes—cash or valuables such as copper kettles, teapots, cups or medals—were contributed by wealthier members.

The gooseberry growers even had an anthem, the first verse going:

Come all you jovial gardeners and listen unto me
Whilst I relate the different sorts of winning
  Gooseberries
This famous institution was founded long ago
That men might meet and drink and have a Gooseberry Show

Subsequent verses were mostly composed of the names of winning varieties. The most famous of these was the red 'London' berry, celebrated with its own verse:

This London of renown, was that famous Huntsman's son
Who was raised in a Cheshire Village, near the Maypole
  in Acton
While in bloom he was but small, yet still so fast he
  grew
That everyone admired him, for his equals are but few.[1]

The 'London' was unusual for a show gooseberry in having a short and punchy name: most tried to incorporate a mention of both their raiser and a mention of something stirring or patriotic. A list of 1800 includes: 'Boardman's Royal Oak,' 'Mason's Hercules,' 'Hill's Royal Sovereign,' 'Worthington's Glory of Eccles,' 'Parkinson's Goldfinder,' and 'Fox's Jolly Smoker.'

FLORAL FACTS OF LIFE—EARLY KNOWLEDGE ABOUT PLANTS AND SEX

The fact that life could be changed through human will was a novel one; indeed, it was potentially a deeply subversive one. Surely the Creator, who

1. Redfern 1972, song quoted from *North Cheshire Herald* and *Hyde Reporter*, Bredbury and Romiley edition. March 17, 1944, 1–3.

had made the world in six days, had created perfection! It was generally accepted that species were constant and could not be changed. Furthermore, there was no understanding of how reproduction took place, especially in plants, so there was no conception of any mechanism being responsible for the transmission of what we nowadays call genetic information. Without any such understanding it was not possible to conceive of how changes might occur, let alone that human intervention could cause such changes. Not surprisingly, the idea that changes in God's creation could happen, and especially that they could happen with human direction was strikingly novel and potentially blasphemous. It took the sea change in thinking brought about by the Enlightenment to create the intellectual climate where such change could even be considered.

During the eighteenth and nineteenth centuries, the development of plant and animal breeding progressed in very close conjunction with the slow revelation of the facts about reproduction and the transmission of genetic information from one generation to another. Agriculture and horticulture on the one hand and "pure" science on the other were two sides of the same coin. The growing body of facts that they revealed pushed forward knowledge of heredity at the same time as they had practical results in more productive wheat, hardier vines and sweeter apples—and more productive cows and sheep.

Deliberate hybridization—that is, the crossing of different varieties or species in order to produce new ones—had to wait until there was an understanding of sex in plants, which did not happen until the European early modern period. Cultivation had long been observed to alter wild plants, and before the modern period it was a widespread belief that under certain conditions one species would turn into another. To modern ears, this sounds strange, but think for a moment about how easily crops become mixed—a few oats in a field one year, their grains mixed with the wheat crop, and then gradually increasing year by year. Hybridization and mutation would have seemed to be unnecessary hypotheses.

The knowledge that date palms (*Phoenix dactylifera*) needed to reproduce sexually had been understood since ancient times; some trees clearly produce fruit, while others only pollen—they are dioecious.[2] Stone carvings dating from ancient Assyria (2400 BCE–612 BCE; now present-day Iraq) clearly show artificial fertilization going on—which would have been necessary in plantations where a maximum number of productive females rather

2. See technical note 15: Monoecious and Dioecious.

than unproductive males would have been required. Medieval Arab writers clearly understood that date palms had different sexes but did not extend the idea to other plants, even though there was a recognition that plants could "blend"; the Arab writer Abd-al-Latif, writing during the early thirteenth century, described how when oranges and lemons were grown together in Egypt, intermediate kinds would arise as seedlings.

The understanding that flowers were sex organs and that seed was the result of sexual union was slowly arrived at in Europe during the seventeenth century. Before this, it was possible even for someone as perceptive as Francis Bacon to state that "copulation extendeth not to plants."[3] Despite this, in his utopian essay of 1624, "The New Atlantis," Bacon suggests that animals and plants might change through domestication to produce totally new forms of great advantage to humanity.

It is generally accepted that the first clear indication of gender in plants came in 1694, when the Rudolph Jakob Camerer (Camerarius; 1534–1598), professor of natural philosophy at the University of Tübingen in Germany, published a memoir concerning some experiments he had made with a variety of plants. He concluded that pollen was necessary for the production of seed, and that plants, or plant parts, could be described as male or female. Incidentally, it was Camerarius who was initially responsible for the layout and planting of the garden at Eichstätt (see chapter 3). An interpretation of flowers as sex organs soon followed, and was made into the centerpiece of a whole system of plant classification by Linnaeus during the following century.

The fact that corn has male and female parts that are so clearly physically separate made it an obvious early subject for investigations into plant sex and hybridization—one pursued by several early American farmers and botanists. A major leap forward was made by Cotton Mather (1663–1728), a puritan minister in the American colony of Massachusetts, renowned for his voluminous and censorious writings on morality and his defense of the infamous Salem witchcraft trials. Another side to his personality and career was that he was also a forward-looking and incisive observer of nature. He was one of the first to accurately record the effect of the pollen of one kind of corn settling on another:

> My friend planted a row of Indian corn that was colored red and blue; the rest of the field being planted with yellow, which is the most usual color. To the windward side this red and blue so infected three or four rows

3. Leapman 2000, 27.

as to communicate the same color unto them; and part of ye fifth and some of ye sixth. But to the leeward side, no less than seven or eight rows had ye same color communicated unto them; and some small impressions were made on those that were yet farther off.[4]

Around the same time as Cotton Mather was making his observations, Paul Dudley (1675–1751), a naturalist and scholar also from Massachusetts, noticed that strains of corn tend to mix if planted near to each other, and recognized that "a wonderful copulation" takes place. Some decades later James Logan, Governor of Pennsylvania, conducted a series of experiments; Logan seems to have been the first to perform what we would regard as a formal experiment to prove that pollen was necessary to effect fertilization.

OUTRAGE! THE CAUTIOUS BEGINNINGS OF DELIBERATE HYBRIDIZATION

Linnaeus, as perhaps befits the liberated image of Scandinavia, let a fertile imagination run riot with his system of classification, including descriptions of flowers as "marriage beds" and comparisons with polygamy, concubines, and harlots; anyone who read him would have got the message about plants and sex very firmly. Linnaeus observed and recorded a number of naturally occurring hybrids and made several deliberate crosses, the first of which was made in 1757, between two species of *Tragopogon* (a dandelion relative). He recognized the potential of hybridization and made the conceptual leap that a wide variety of kitchen garden plants, such as brassicas, are the results of generations of hybridization. In the system of classification he published in 1735, Linnaeus held that the number of species which existed to be the same as it was at the Creation. His subsequent experience with the crossing of plants led him to believe that hybridization created new combinations of the traits shown by the parents. His way out of what could have been a conundrum was to suggest that the genus was the basic unit of creation—so that species could interbreed but also appear and disappear. Flexibility had begun to creep into Creation.

Linnaeus's main interest was plant classification; despite his obvious enthusiasm for hybridization, this was not an area he pursued. No one else in the academic world made a systematic attempt at doing so until Josef Gottlieb Kölreuter in the latter half of the eighteenth century. Outside the ivory

4. Zirkle 1935, 105.

tower, however, more practically minded men were addressing the issue. If a key person, plant species, and event stands out, it is Thomas Fairchild, his pink, and a meeting of London's Royal Society in February 1720.

The Royal Society is spoken of as the world's first scientific society; it was a meeting point for all the great and the good who were curious about the natural world, but its program would today be more accurately described as "natural history." Its members had an outlook on the world around them that was very characteristic of the more advanced countries in Europe and the American colonies at the time—that of "curiosity," a keen interest in natural phenomena, especially in the novel and unusual, the exceptional and atypical. On one level this comes across as an almost schoolboyish love of the weird and bizarre, but on another level it has to be recognized that through the examination of the strange the normal can often be better understood, that exceptions and oddities often point up underlying patterns or highlight explanations. The gentlemen of the Royal Society were interested in anything and everything; the world they lived in was beginning to change, new lands were being opened up, old certainties questioned. For those prepared to open their eyes and their minds, the world was full of wonderful things and happenings, all of which needed an explanation, or at the very least to be noted and discussed in convivial intelligent company after a jolly good dinner.

Thomas Fairchild (1667–1729) was a nurseryman—not quite the social caliber to be welcomed into the Royal Society and almost certainly too busy to be able to participate. His hybrid pink (a cross between a sweet william [*Dianthus barbatus*] and a carnation [*D. caryophyllus*]) was introduced to the society through a friend, Patrick Blair, a doctor and amateur botanist. Blair's paper to the Royal Society was primarily concerned with plant reproduction, establishing that plants do have male and female parts and that deliberate breeding of them was possible. The other items on the agenda the night it was read were the sighting of a meteor near Dublin, the birth of Siamese twins, and the pupae of moths. Fairchild's pink had arisen on his nursery at Hoxton, then a village on the outskirts of London. Whether it was an accidental seedling or the result of a deliberate cross, history does not relate, although he is reported to have carried out the hand-pollination of pinks.

Fairchild was one of many nurserymen catering to a growing demand for interesting and unusual plants and flowers for those inhabitants of London able to afford gardens. The growing wealth of London meant that more and more had time and money for gardening, an interest which fed on the beginning of what was to become a tidal wave of plant introductions from abroad. The late seventeenth century saw a flurry of books on gardening, and in 1688, with the publication of the first nursery catalog, consumer gardening had officially begun.

The hybrid pink, which became known as 'Fairchild's mule,' was yet another novelty with which to tempt customers. He was clearly an innovative and adventurous nurseryman, interested in pushing the boundaries of the possible, engaging in experimental grafting (such as oranges and lemons [*Citrus x limon*] on the same tree and black and green grapes on a single vine) and making a collection of variegated plants.

Until this period, farmers and gardeners had had two sources of new plants: new introductions from foreign lands, and chance discoveries of new forms, either found in the wild or from sowings of seed of cultivated plants. The idea of being proactive and deliberately trying to make new plants was a new one—Fairchild appears to have been one of those who had made this conceptual breakthrough. His "mule" and the concept of hybridization was not however widely followed up; the increasing numbers of new plants arriving from foreign lands became the main focus for gardeners for another hundred years at least.

One of the most interesting themes of the whole early modern period is the mental wrestling that was clearly taking place in many learned minds between religious beliefs and the empirical evidence being gathered about how the physical world operated and, in particular, how it might be manipulated. Those concerned with natural history were occupied with the study of the creatures of God's creation, which according to Christian doctrine had traditionally been made for man's use. But to what extent could the human race change what God had provided?

In December 1684, the English botanist John Ray read two papers to the Royal Society that summarized the existing state of knowledge regarding plant reproduction. In his mind it was clear that the numbers and categories of plants had been made and fixed by God in six days, and that while there was infinite variation within a species, the species were fixed, that is, they were God-given. Crossing two of God's species was clearly taking dominion over nature too far. The word "hybrid" is derived from the Latin *hybrida*, meaning a cross between a tame sow and a wild boar, a word possibly related to the Greek *hubris*, meaning pride but also an outrage against nature, especially when made to the Gods or connected with sex. Later on the word came to mean "mongrel," something which was neither one thing or another, and in Roman times was applied to humans, too, just as the words "half-breed" and "half-caste" have been used (often insultingly) to describe people of mixed heritage in recent times. There seems to have been a widespread belief that sexual intercourse between different species was an immoral perversion and that the making of new life forms by humanity an insult to God—almost a criticism that humanity could do better. It was believed, possibly incorrectly, that the ancient Hebrews were forbidden from breeding mules; although, as we have seen that crossbreeding of animal breeds was forbidden (chapter 2), there are some sixteen mentions of mules in the Old Testament. It should come as no surprise that early plant breeders felt they had to justify their attempts at crossing different species.[5]

The issue is even tantalizingly raised by Shakespeare in *The Winter's Tale*, with Perdita worried about introducing cultivated forms of flowers into her, otherwise "rustic," garden, but being reassured by Polixenes:

Perdita: Sir, the year growing ancient,
Not yet on summer's death nor on the birth
Of trembling winter, the fairest
flowers o' the season
Are our carnations and streaked gillyvors,
Which some call nature's bastards; of that kind
Our rustic garden's barren, and I care not
To get slips of them.
Polixenes: Wherefore, gentle maiden, do you neglect
them?
Perdita: For I have heard it said
There is an art which, in their piedness, shares
With great creating nature.
Polixenes: Say there be;
Yet nature is made better by no mean,
Which you say adds to nature, is an art
That nature makes.
You see, sweet maid, we marry
A gentler scion to the wildest stock,
And make conceive a bark of baser kind
By bud of nobler race. This is an art

5. Zirkle (1935) is one of the few writers to discuss this issue—ironically he has a reputation as a eugenicist. See also Leighton 1967.

Which does mend nature—change it rather, but
The art itself is nature.
Then make your garden rich in gillyvors
And do not call them bastards.[6]

The novel idea of flowers, fruits, vegetables, and trees having sex was a gift to the wits and bawds of the Restoration period in England, when the theater, newly freed from the puritanical control of the Commonwealth era, went in for innuendo on a huge scale. It is possible to speculate that an increasing awareness of the possibility of plants crossing natural boundaries found an echo in responses to a variety of transgressive sexual behaviors. References to plants and gardening are apparently frequent in the semi-pornographic literature of the time; Vauxhall gardens often being mentioned—a city park famous for both horticulture and prostitution. One of the best has to the 1741 *Natural History of the* Frutex Vulvaria *or Flowering Shrub, as it is Collected from the Best Botanists both Ancient and Modern*, by Phylogenes Clitorides.[7]

THE PROFESSORS—KÖLREUTER AND OTHER ACADEMIC HYBRIDISTS

In Britain and its American colonies, the gentleman amateur reigned supreme in research into the natural world, in Germany it was individuals in the universities who made major advances. Joseph Gottlieb Kölreuter (1733–1806) trained originally in medicine but ended as professor of natural history at the University of Karlsruhe and director of its botanic garden. In 1749, J. G. Gmelin (1709–1755), professor of chemistry at the University of Tübingen, suggested to Kölreuter that he perform some experiments in crossing plants with the specific aim of refuting the dogma of the constancy of species. Thus began many years of research, culminating in a series of papers outlining his work with plant crosses. He started with *Nicotiana* species (tobacco and tobacco flower), moving onto *Dianthus* (pinks and carnations), *Verbascum* (mulleins), *Hibiscus*, and many more. Until his 1764 promotion to Karlsruhe, most of his experiments were carried out in the garden of a doctor colleague in his home village of Calw in Württemburg in southern Germany.

On the whole, Kölreuter's hybrids turned out to be sterile, which he interpreted within a framework dictated by a version of the doctrine of the con-

6. Shakespeare 1623, 4.4.92–117.
7. Leapman 2000.

stancy of species; the parents of a hybrid "were not destined for each other by the wise Creator . . . is not in condition to fulfill the final object toward which otherwise all the operations demanded for development appear to be directed." It appeared to him that the consequences of unrestrained hybridization would be awful and that the apparent sterility of hybrids acted in order to safeguard the work of the creator—a kind of heavenly terminator gene. "What an astonishing confusion" he wrote, "would not the peculiar and unchanged hybrid characters, and the continually retained fertility of such plants give rise to in Nature. What evil and unavoidable consequences must these not draw after them?"[8] Despite this, Kölreuter was also a man who could see possibilities; in the third part of his essay on sex in plants, published in 1766, he wrote: "I could wish that I, or someone else, might one day be lucky enough to produce hybrids of trees, the use of whose timber might have great economic effect. Among other good properties such trees might have one would be that if the original trees needed, for example, a hundred years of full growth, the hybrid would achieve the same in half that time. At least, I do not see why they should behave differently from other hybrid plants."[9]

Kölreuter followed up his work in making crosses with an attempt to understand the physical mechanism of pollination, using a microscope to examine floral parts, meticulously measuring quantities and experimenting with mixing pollen with various oils to see if they made any difference to the ease of fertilization. He made the important observation that hybridization was only possible between closely related species and noted hybrid vigor and the fact that some characteristics were more likely to appear in offspring than others, foreshadowing Mendel's work on dominant and recessive genes.[10]

Despite being the work of a notable professor at a great university, Kölreuter's researches was largely ignored for several decades. As was the work of another German, Christian Konrad Sprengel (1750–1816), whose main official post was as an instructor at the Royal Military School in Berlin. Records show Sprengel to have been something of a larger-than-life character, his work at the Royal Military School peppered with accounts of insubordination, conflict with parents of pupils, and accusations of inhumane

8. Quoted in H. F. Roberts 1929, 45.

9. Quoted in Orel 1996, http://www.mendelweb.org/MWorel.html.

10. See technical note 16: Hybrid Vigor or Heterosis.

discipline—just the sort of character to overturn conventional wisdom in other words. His main contribution to an understanding of plant heredity was the idea that normal reproductive processes in plants bring together the different characters of both male and female parents—in other words, combinations of subtle differences is the norm in nature. Darwin took this idea further with his theory of natural selection, which operates on the basis that all individuals are slightly different. Conceptually, Sprengel had made a breakthrough: that species are not fixed and immutable but are in a state of constant flux, resulting from an endless reshuffling of characters during reproduction—what the Almighty created was not therefore unchanging. Unfortunately, without any official position in the academic world, Sprengel's work had even less chance than Kölreuter's of being noticed.

Almost certainly inspired by Kölreuter, and living in the same village of Calw in which the professor had worked, was the most prolific early hybridizer, Karl Friedrich von Gärtner (1772–1850). A doctor and the son of distinguished botanist Joseph Gärtner, who was professor of botany at Tübingen and St. Petersburg (where pupils of Linnaeus had conducted some very early work on hybrids), Karl Friedrich von Gärtner carried out nearly ten thousand separate experiments in crossing over a period of twenty-five years. He worked with seven hundred species belonging to eighty different genera of plants; in all, he obtained some three hundred and fifty different hybrid plants—all of which he recorded. Among the genera worked with were *Alcea* (hollyhock), *Antirrhinum* (snapdragon), *Aquilegia* (columbine), *Avena* (oats), *Datura*, *Delphinium*, *Dianthus*, *Digitalis* (foxgloves), *Fuchsia*, *Gladiolus*, *Hypericum*, *Lobelia*, *Lychnis*, *Malva* (mallow), *Matthiola* (stock), *Nicotiana*, *Oenothera* (evening primroses), *Papaver* (poppies), *Primula* (primroses, cowslips, polyanthus, etc.), *Ribes* (ornamental and edible currants), *Verbascum*, and *Zea*. As can be seen from this much-abbreviated list, he did not have a particular focus on agricultural or edible crops, but he did recognize that the ability to cross plants could potentially be most useful for agriculture. Gärtner noticed the fact of unusual vigor and a marked increase in the size of the flowers, and indeed an increase in general luxuriance of growth among his hybrids; he did not, however, seem to recognize that the vigor of hybrids belonged only to the hybrid generation and not to generations derived from it. Although many of Gärtner's plants were sterile, he found some that appeared to be both fertile and constant, producing offspring so similar to the hybrid that they appeared to be new species—a fact which might have alarmed the more cautious Kölreuter.

GENTLEMAN PRACTITIONERS—THOMAS ANDREW KNIGHT AND OTHERS

The first person in Britain to undertake systematic work with hybridization was Thomas Andrew Knight (1759–1838), a country gentleman by occupation whose brother, Richard Payne Knight (1750–1824), was a major figure in the English landscape movement. Both used their estates as testing grounds: for early genetic research in Thomas's case and landscape art in Richard's. Together they presented an image of quintessential English Enlightenment figures, firmly rooted in their estates but linked to a national but inevitably London-orientated intellectual world. They also lived at a time when, given a few years, their work would reach an audience on mainland Europe and North America. Thomas was educated at Oxford and early on in life began to interest himself in experiments in the raising of new varieties of fruits and vegetables at his estate at Elton in Herefordshire. A brief portrait of him in his old age was penned by a colleague: "active, and energetic in his seventy-eighth year as a man of forty, is one of those rarities among men, who know everything. . . . [H]e is one who can discern rottenness in church and state, as well as canker in a fruit-tree, and can fathom both."[11]

Thomas Knight's interest stemmed initially from working with cattle and sheep; along with many other members of the English gentry he had been inspired by the pioneer animal breeder Robert Bakewell (1725–1795), who had been the first to deliberately control the mating of his animals and so breed together individuals with particular characteristics. He also broke the biblically driven taboo against inbreeding, which enabled him to "fix" and strengthen certain desired traits.

It was in 1795 that Knight's work as a horticulturist first became known through papers read at the sessions of the Royal Society. He went on to become a key founder member of the Horticultural Society of London in 1804 (later to become the RHS), of which he was president from 1811 until his death in 1838. A prolific experimental gardener and writer, he was the author of nearly one hundred papers; he was interested in theory only in as much as it pointed to practical outcomes—a classically English stance.

Initially focusing on fruit trees, Knight turned to annuals, peas in particular, as these gave much quicker results. From his work he derived a series of principles of heredity, and it might be argued, came within a hair's breadth

11. Quoted in R. Webber 1968, 124.

of reaching the same conclusions that Mendel did around a hundred years later. Perhaps his practical bent drew him away from the pure research or abstract thinking which such a breakthrough demanded. As a good farmer in an age when the profit motive was beginning to drive agriculture, the fact that the progeny of his crossed races of peas gained extraordinarily in vigor and yield was almost certainly more important. He not only saw the potential of plant breeding, but also wrote and actively promoted it among his circle of improving landowners.

> I cannot dismiss the subject, without expressing my regret, that those who have made the science of botany their study, should have considered the improvement of those vegetables which, in their cultivated state, afford the largest portion of subsistence to mankind, and other animals, as little connected with the subject of their pursuit. Hence it has happened that whilst much attention has been paid to the improvement of every species of useful animals, the most valuable esculent plants have been almost wholly neglected. But when the extent of the benefit which would arise to the plants, which, with the same extent of soil and labor, would afford even a small increase of produce, is considered, this subject appears of no inconsiderable importance. . . . The improvement of animals is attained with much expense, and the improved kinds necessarily extend themselves slowly; but a single bushel of improved wheat or peas may in ten years be made to afford seed enough to supply the whole island.[12]

Breeding improved varieties of fruit appeared to be Knight's main concern, as it was of many gentleman amateurs of this period. Cider[13] production recently having become somewhat depressed in his home county, the breeding of new apples was one of his main practical interests; in an article of 1806 he described how he crossed domestic apples with the hardy Siberian crab apple (*Malus baccata*) in order to improve local fruit production. The same article also reveals confusion over whether acclimatizing plants to different climates results in hereditary changes or not—it was to be nearly a century before the distinction between the hereditary and the nonhereditary would be finally clarified.[14] Ironically, it was faulty knowledge which

12. Quoted in H. F. Roberts 1929, 93.

13. "Cider" here is understood as alcoholic "hard cider," a popular alternative to beer among country people in parts of northern Europe. Weak cider or beer was drunk as a safe alternative to water, which would often have been unfit for drinking.

14. Knight 1806.

led to Knight's drive to work with fruit. Given that many of the fruit trees in Herefordshire were in decline, Knight came to the conclusion that varieties aged, even when newly propagated, so that for example, budstock from a hundred-year old 'Bridgewater Pippin' apple tree would be a hundred years old too. In other words, new varieties had to be constantly created to replace older ones as they aged—which set him off on the path of creating as many new ones as possible. We now know that this is not the case, that "decline" is often associated with an accumulation of virus diseases and some apple varieties still available from suppliers of heritage fruit may date back hundreds of years.

William Herbert (1778–1847) was a contemporary of Knight's. Another Oxford-educated member of the gentry, he entered Parliament and then holy orders, finally becoming dean of Manchester in the Church of England. Clearly one of that infamous eighteenth-century English breed, "the hunting parson," he was reportedly fond of outdoor life and sport, as well as being a poet and a keen amateur naturalist—as were so many of his colleagues in the Church. He experimented mostly with florists' flowers (see chapter 13) but also with some agricultural plants, notably swedes/rutabaga, as fodder. Discovering that some hybrids were fertile, he challenged the idea of species as being fixed, concluding that as far as capacity for hybridization was concerned, species and varieties were arbitrary and artificial distinctions in the plant kingdom. Most radically, he decided to make artificial crosses between daffodil (*Narcissus*) species to see if it were possible that certain wild plants he had found in France were natural hybrids. Shortly before his death, he published a paper on hybridization that, in many ways, was very advanced for its time, suggesting on the evidence of nearly forty years of experimentation, that hybridization and speciation had considerably diversified the Almighty's original creation of the Plant Kingdom.[15]

Augustin Sageret (1763–1851), one of the founders of the Society of Horticulture of Paris, followed Knight in working with annual plants—melons (*Cucumis melo*) in particular. He was interested in following the descent of traits through several generations, discussing what came to be called segregation and was the first to use the term "dominant" for genetic traits.[16]

Sagaret was one of several researchers around this time who put experimental flesh on the bones of what Linnaeus had theorized: that the forms of cultivated plants are derived from earlier forms, often in combination, and

15. Herbert 1846.
16. See technical note 17: Segregation.

that in breeding, ancestral traits may sometimes appear. Another who realized this was Henri-Louis Duhamel du Monceau (1700–1782), one of the most systematic and practically minded of eighteenth century writers on plants. Having an encyclopedic mind of the kind for which the eighteenth century offered opportunities as no other time yet had, he was an authority on agriculture, pomology, sylviculture, metallurgy, meteorology, and shipbuilding. In his book *La physique des arbres* (1758), he addressed the variability of cultivated plants, concluding that forms appeared and disappeared over time, and that most cultivated varieties were composites of older varieties brought about by mixing of pollen.

SEX AND THE SINGLE STRAWBERRY

Strawberries have a special place in our consciousness: they are the most jewellike of fruit, and they encapsulate the highest qualities for which fruit is esteemed—a perfect combination of acidity, sweetness, and aroma. Their associations are vivid: finding wild ones on family walks in the mountains, strawberries and cream on the lawn, or fresh strawberry shortcake. Above all, strawberries arouse the child in all of us. Not surprisingly they attracted the attention of many early gardeners, as well as one real prodigy, the French botanist Antoine Nicolas Duchesne (1747–1827).

Fragraria vesca and its close relative *F. moschata* are the wild European species that have been gathered wild or cultivated from Roman times, giving rise to various selections over the centuries. *F. moschata*, the ancestor of the 'Moschata' and 'Hautboy' strawberries grown in the early modern period, acquired a poor reputation for fruit set, something that has plagued growers of the fruit for centuries; most strains of this species are dioecious, and so if there is not a good mix of male and female plants, fruiting is bound to be poor.

The opening up of the New World introduced American species to cultivation in Europe. First was *F. virginiana*, known as the 'Virginia,' and next a species found up and down the west coast of both Americas, north and south: *F. chiloensis*, often called at the time the 'Chili.' Early European travelers were impressed by the sheer size of the Chili and suggested that native people had made selections. Despite being known by Europeans for over a century, it was not until 1714 that it was brought to Europe by Amédée François Frézier (1682–1773), an engineer and mathematician who had been sent by the government of France to map the coasts of Chile and Peru and when the opportunity arose, to take a quiet look at Spanish naval facilities. By coincidence, his name is remarkably similar to the French word for the strawberry plant, *fraisier*; Frézier ensured that the fruit was added to the family crest.

Despite his personal care and attention, only five of Frézier's Chilean plants survived the journey back, and politics demanded that they be given to the people who mattered: his cargo master, to whom he was indebted for a variety of favors, not least the water for the strawberry plants; his superior, the Minister of Fortifications; and the director of the Jardin des Plantes in Paris—no question of some basic propagation work first! Not surprisingly, the plant started off its career in Europe with a small gene pool, and given its strong dioecious tendencies, it was most unwilling to fruit. An additional problem was that Frézier had selected plants with fruit, so only female plants had been introduced.

Fruit was produced when plants did occasionally grow hermaphroditic flowers, and the species was slowly spread around the continent's botanic gardens. One that received it was the Botanical Garden of Brest on the west coast of Brittany in France, where the climate no doubt suited this plant from a west coast on the other side of the world. Breton farmers then hit on the secret of success: interplanting the Chili with Virginia or Hautboy forms, which ensured pollination. From 1750 and for the next hundred years, Brittany supplied much of Europe with its best strawberries, appreciated for their combination of flavor and size—even if not as large as the "hen's eggs" Frézier had claimed to have seen in Chile.

In or around 1740, a new strawberry arrived at the Chelsea Physic Gardens, the garden of the Worshipful Society of Apothecaries in London.[17] Its head gardener, Philip Miller, recognized the berry's superior qualities but did not know from where it originated—he had been given plants by "a curious gentleman of Amsterdam" who claimed it had come from Surinam in South America. It was named the 'Ananas,' meaning "pineapple," which at the time was the most highly regarded and most exotic fruit available in Europe. In time it was given its own specific status—*F. x ananassa*.

Because strawberries had separate sexes, observant botanists and gardeners in the early modern period were driven by necessity to confront the fact of plant gender. There is a parallel with corn, whose physically separate male and female flowers presented an opportunity to enquiring minds around this time. It was the young Duchesne, who in 1764 undertook the re-

17. Apothecaries dealt largely in herbal medicines—and were the forerunners of modern pharmacists.

search that led to a clear understanding of what it is that makes a strawberry, and by extension raised the possibility that something similar must happen in other fruit too. Not only did the strawberry finally make the educated of Europe accept the reality of gender in plants, but it also opened the door to the possibility of deliberate crossing of different plant species.

So on July 6, 1764, the seventeen-year-old Duchesne presented King Louis XV of France with a bowl of luscious strawberries. The fruit had been borne by an *F. chiloensis*, which Duchesne had pollinated with a plant of *F. moschata* pollen. Duchesne was honored by having the fruit painted for the Royal Botanic Library and was then given royal encouragement to collect all the known varieties of strawberries in Europe for the royal garden. What followed is one of the first examples of systematic plant collection and evaluation. With the strength of a royal command behind him, Duchesne wrote to professors and farmers, amateur growers and naturalists all over Europe. He collected not just plants, but experiences, including those of the now-elderly Frézier, who was questioned about the conditions prevailing in Chile in which he had found strawberries fifty years previously. A bishop was asked to search the country around a town in the French Alps for a strawberry that supposedly bore fruit twice a year. Even the great Linnaeus was approached, which started a ten-year correspondence and exchanges of plants.

The result was *L'Histoire Naturelle des Fraisiers*, published in Paris in 1766, which described a total of the ten species and eight varieties that Duchesne believed existed. For each he traced its history, detailed its distribution, and outlined its cultivation. He also drew a genealogical tree describing his theories on how the different strawberry taxa were related—which was one of the first times this had been attempted and also made a break with the idea of an unchanging Creation.

Duchesne had been born in 1747 at Versailles, where his father had held the post of superintendent of the royal buildings. His mother had died giving birth to him, and his father devoted himself to the boy's education in the classics, architecture, and natural history. He was particularly fascinated by the plants in the Royal Botanic Garden, the Grand Trianon, and here he was lucky enough to attract the attention of Bernard de Jussieu (1699–1777), the royal botanist, who encouraged him to accompany him in his work, and who taught him much of what he knew about botany and gardening.

It was Jussieu who first put the idea of strawberries having separate sexes in the mind of young Duchesne, an idea he must have borne in mind when he discussed the failure of *F. moschata* plants to flower with the young gardener of a family friend—this experience led him to suspect that the plants were all female. Another experience that got him thinking was the discovery of a form of *F. vesca* in his father's garden with a undivided leaf, as opposed to one being divided into three, as is normal. The plant had originated from an experiment that he and his father had conducted to see how often white berries were produced from red-berried plants. One of the plants had the "mutant" leaves, so the Duchesnes propagated it from runners and then, when it produced fruit, sowed the seed. The vast majority of the seedlings had single leaves—a new variety that appeared to be stable had been created, a fact that flew in the face of the doctrine of the constancy of species.

In the spring of 1764 the Duchesnes obtained some *F. chiloensis* from the king's garden When they flowered, Antoine recognized them as female. Since they had no male plants, he planted it next to an *F. moschata*—the result was the little crop of fruit which ended up being presented to the king.

Duchesne showed how good crops were assured if "pistillate" (i.e., female) plants are mixed with those which were "staminate" (the pollen-bearing males). Experimentation convinced him that *F. chiloensis* could not fertilize itself and that it could not be fertilized by *F. vesca*. A letter to Linnaeus brought the somewhat disappointing response that the great man did not believe him. Confident with his results, however, he went on to undertake more research, which led him to put forward the suggestion that there were laws that governed hybridization between different plants. The idea that closely related species could cross, but that less-closely related ones could not, and certainly not species of different genera, had been advanced by the great French botanist G. L. L. Buffon (1707–1788). Duchesne's work led him to support this idea and to suggest that there were "laws" permitting hybridization between certain kinds and preventing it between others.

Duchesne went on to make crosses between as many strawberries as he could; some of the progeny turned out to be hermaphroditic (i.e., with pollen and ovules in the same flower). Most importantly, he replicated the creation of the 'Ananas,' although this particular theory was not accepted by everyone until the beginning of the twentieth century. He was the first to point out the possible origin of the luscious new variety; in his book he described how "the resemblance of the Ananas to the Frutiller [another name for the Chili] is not that of a father and mother with their child . . . but that of a son with his mother so that I consider the Ananas to be a cross between the Scarlet strawberry [the 'Virginia' and the 'Frutiller']."[18] These hybrids between *F. chiloensis* and *F. virginiana* are the basis of all modern large-fruited

18. Quoted in Wilhelm and Sagan 1974.

strawberries and are given the name *F. ananassa.* The irony here is that two New World species gave rise to a third—one of enormous commercial importance—in the Old World. Strawberries continued to play a key role in new breeding methods (see chapter 5).

SWEET NECESSITY—SUGARCANE AND SUGAR BEET

Trying to make sugar out of beetroot is not something that would immediately occur to most of us. But the eighteenth century was the age of curiosity, and analyzing the sap of roots was the kind of activity in which early scientists excelled. Although he was not the first to notice the sugar content of the roots, the German chemist Andreas Marggraf (1709–1782) discovered in 1747 that the sweetness was sucrose, which when extracted from pulverized beetroots was identical in all properties to the sugar derived from sugarcane.

Sugar has long been something of a human obsession—almost a drug. Once a culture discovers a source of sugar it immediately elevates it and products made from it to luxury status. The Indian and Islamic worlds had access to cane sugar from ancient times, as the sophistication of their sweet cuisines attests. Europeans had to make do with an annual short-lived glut of autumnal fruit, imported dried fruit (for those who were rich), or honey. During the eighteenth century, sugar became the driver for the hideous business of slavery, as European nations sought to grow as much sugarcane as they could in the West Indies, satiating their increasingly sweet teeth at vast human cost.

As the eighteenth century advanced, however, it became apparent that sugar might be got from an alternative, and rather unlikely, source. The story of sugar beet is one of the most dramatic in the whole history of "improvement," as it involved the creation of what was effectively a totally new crop. Beet, descended from *Beta vulgaris*, a European coastal plant, has been used as a green vegetable since ancient times and from medieval times as a cattle feed. Beetroot as a vegetable seems to date back to Roman times, and appeared from the fourteenth century in recipes in England.[19] As a leaf vegetable, we still eat it in the form of leaf beet and Swiss chard and the brightly colored ruby chard, beloved of those who create ornamental vegetable gardens. As a root fodder plant, the beet—or 'mangold,' as the fodder plant is often known—has major advantages for farmers in areas with climates that

19. Campbell 1976.

restricted the amount of leafy winter forage; inevitably selection began to produce beets with bigger roots. The Mennonites, a puritan religious sect of Eastern Europe, traditionalist in many ways but always forward-thinking in agricultural matters, were noted during the seventeenth and eighteenth centuries as feeding their cattle on particularly large-rooted varieties. Later on (see chapter 5), we shall appreciate that we have even more to thank the Mennonites for.

A pupil of Marggraf's, Franz Karl Achard (1753–1821), originally a refugee from France, tried interesting his countrymen in establishing a factory but faced only ridicule. Rescue came from Prussia, his adopted homeland, with a subsidy from King William III to establish a factory at Kunern, in Silesia (now Poland). Achard also collected twenty-six varieties of beet, from which an associate, Freiherr von Koppy, bred the 'White Silesian' beet, from which most of today's sugar beet is descended.

Sugar became one of the casualties of the Napoleonic Wars, with imports to France from the English-governed sugar-producing islands of the West Indies being banned by Napoleon in an effort to damage British interests. On March 25, 1811, probably under pressure from the sugar addicts of the French upper classes, Napoleon issued a decree that set aside eighty thousand acres of land for the production of sugar beet, established factories, set up schools to teach its cultivation, and introduced subsidies for its production. But with the collapse of Napoleon's regime, the bottom fell out of the market; with a sugar content of only 6.2 percent, beet was simply too uneconomical to refine.

Plant breeding eventually came to the rescue. In 1861, Philippe André de Vilmorin (1776–1862) selected a beet with a higher sugar content and yield, the 'Imperial Beet.' His son Louis de Vilmorin (1816–1860) made further advances with the crop through mass selection (see chapter 5) and progeny testing. The idea of the progeny test was borrowed from livestock breeding, where it was commonplace. Essentially it assesses the value of an individual plant's potential in breeding by randomly crossing it with a range of other individuals and then testing its progeny; it is a way of seeing what the plant can provide through heredity. The technique proved particularly useful for beet, especially when allied with the development of a test that was nonlethal to the plant. Plants could be dug up and core samples taken from the root, which could then be tested for sugar content, and then high-scoring plants could be replanted to produce seed. The result of Vilmorin's work was that he increased the sugar content of beets from 10–11 percent to

16–17 percent. Meanwhile, in Germany, breeding was also going on; in order to ensure the genetic isolation of the experimental crops, breeders went around the countryside buying up peasant beehives and dispatching them out of harm's way. In addition, once deprived of honey, the peasants would then have had to buy sugar.

Meanwhile, sugarcane (*Saccharum officianarum*) had undergone a transformation through hybridization between species. Early sugarcane production had relied simply on selection, with propagation from seed made difficult by the plant's photoperiodism. Dutch colonial breeding in Java from 1887 onward made advances in yield and canes were distributed throughout all the possessions of the European empires. An experimental station in Java made an important breakthrough in 1915 when it crossed *S. officianarum* with another species, *S. spontaneum*, incorporating the latter's hardiness and disease tolerance. A British Indian station soon managed another interspecies cross. What was important about these crosses was not so much that they increased yield in the traditional sugar-growing areas, but they made it possible to grow sugar where none grew before.

MORAVIA—LAND OF PROGRESS

The province of Moravia was, until World War I, part of the Hapsburg-ruled Austro-Hungarian Empire; although most of the population spoke Czech, its elite spoke German, which made them part of a world which was developing rapidly, particularly in the fields of science and philosophy. In addition, nineteenth-century Moravia was a forward-looking agricultural society bent on "improvement," its economy underpinned by the production of wool from its sheep and fruit from orchards and vineyards. The breeding of livestock and crops was seen as vitally important for its exports, and the educated members of its gentry were deeply involved in both. Moravia was one of the most dynamic parts of the empire, with an intellectual openness much greater than was usual in Catholic Europe at the time—a legacy of a brief period of a relatively freethinking brand of Protestantism under the reformer Jan Hus (1369–1415). The influence of the liberal thinker John Amos Comenius (1592–1670) should be remembered, too; he was an early example of what is now called an "educationalist," a follower of Francis Bacon, and a believer in the role of natural history in the curriculum.

Moravia's most famous son, Gregor Mendel, was born and lived all his life in Moravia, and that a Czech-speaker should have developed the crystallizing synthesis which has led to modern genetics should come as no surprise.[20] We must assume that Mendel knew of Thomas Andrew Knight's work. In 1816, the journal of the Moravian and Silesian Agricultural Society (founded in 1806), published an account of Knight's work on fruit tree breeding; some of his articles were translated into German, almost certainly reaching the monastery library in Brno where Mendel was a monk. The word "genetic" first appears in the work of the noted breeder of sheep, the Hungarian count E. Festetics (1764–1847), who in 1819 published *Genetische Gesetze* (Genetic laws—although "observations" might be a better description).

Mendel would almost certainly have read about or even visited the park established by the supervisor of agriculture for the Duke of Lichtenstein, T. Wallasheck (1753–1834; later ennobled as T. von Wahlberg) in Lednice, in southern Moravia. Wahlberg had amassed a huge collection of economically valuable plants, much in the spirit of the modern gene bank, the data from which had been collated in the book *Neuste Beobachtungen zur Veredlung des Feldbaues und der Forstwirtschaft* (Recent observations on the improvement of agriculture and forestry), published in Vienna in 1810. It was possibly the first critical evaluation of a large collection of useful plants. Wahlberg regarded the trialing of different plant varieties in a range of climatic and soil conditions as important; he also ensured that the varieties were grown in separate fields to avoid crossfertilization—which he recognized as damaging and an obstacle to their improvement.

Moravia's pioneer in promoting animal and plant breeding was Christian André (1763–1831), a social reformer who argued that humankind needed to understand how the natural world worked in order to better control it—very much in the Baconian tradition. He helped to found a Moravian sheep breeders' association and the Pomological and Oenological Society of Brno. President of the society was leading fruit and vine breeder Jan Sedlácek von Harkenfeld (1760–1827), who was active in making crosses in order to breed new varieties "constant in vegetation and suitable for our climate . . . [and which] would also unite in themselves many superior characteristics."[21]

In 1820 however André was forced to leave the empire because of his liberal politics. About this time, a particularly forward-looking article appeared in the journal of the Pomological and Oenological Society of Brno,

20. See technical note 18: F_1 and F_2 Generations, for a discussion of Mendel's key discovery.

21. Quoted in Orel 1996, http://www.mendelweb.org/MWorel.html.

written by G. C. L. Hempel, secretary of the Pomological Association of Altenburg, near Leipzig in Germany. Shortly before fleeing the repressive Hapsburg dominions, André had asked Hempel for an article on the potential of using artificial methods of fertilization to improve crops. Hempel reviewed Knight's work and went on to suggest that breeders should be able to develop new varieties of fruit not just through hit and miss crosses but by forming an ideal of what was needed, in terms of shape, size, flavor, and so forth, and then aiming toward it. With the current knowledge this was not possible, but at some stage in the future, he suggested, there would arise a new kind of figure who would be able to manipulate the reproduction of crops in such a way as to bring these ideal types to reality.

At the University of Olomouc, the chair of agriculture was held for a while by F. Diebl (1770–1859), who was apparently self-taught and had gained his position through his successes as a farm manager and breeder of new fruit varieties. In 1835, he published a five-volume work on agriculture, the second volume of which discussed the fertilization of flowers and how this might be used to create new varieties. Diebl's origins as a practitioner are reflected in later articles he wrote, in which he stressed the importance of scientists meeting with farmers, particularly in local agricultural societies. Humanity, he believed, had been given dominion over nature by God, and so for him, the gradual improvement of animals and plants through science was entirely natural—as well as a duty.

Mendel would also have known of the work of K. N. Fraas, professor of agricultural botany in Munich, whose *History of Agriculture*, published in Prague in 1852 discussed hybridization along with many other topics. Fraas's book was notable in that it discussed the division between natural scientists and agricultural practitioners. The latter, he said, had for centuries made use of plant forms that many natural scientists did not believe existed, such as fertile hybrids of fruit trees and brassicas. Whereas the scientists supported the idea of the constancy of species, farmers were proving by the crops they grew that there was no such thing. Among the recent examples of selection he cited was that of a school near Würzburg, Germany, where children apparently collected 628,338 apple seeds from 1770 to 1796, from which a nursery selected 26,552 as being "improved."

One of the leading lights of the Moravian and Silesian Agricultural Society was F. C. Napp (1792–1867), an abbot of Mendel's Augustinian monastery at Brno, who took on the restoration of the neglected monastic estate, encouraging villagers to grow improved fruit trees. He is also known to have taken over a nursery that had begun to fail through bad management, moving it onto the monastery grounds. He oversaw practical breeding work, and a number of trials, including one of over a hundred grape varieties, with yields being assessed and wines being made from each variety in order to test quality. In 1840, at a congress of German-speaking farmers and foresters held that year in Brno, Abbot Napp chaired a discussion on fruit tree breeding, at which he supported the idea that new varieties can be made through deliberate fertilization, as opposed to the widely held belief that this was a random process over which the grower had little control. Napp is also said to have written a handbook for village people on how to grow and improve tree fruit. In such an environment, Mendel was not then by any means an isolated man, but rather part of an intellectual mainstream.

Advanced ideas in Moravia's forward-looking society were not uncontested. That there was a struggle between progress and tradition can be seen in the story of a Catholic priest, Father J. Schreiber, who had been in charge of an educational institute established in Kunin, Moravia, in 1796 by an enlightened local aristocrat, the countess Maria Walpurga Truchsess-Zeil. The school, known as the Philantropinum, took in gifted village children free of charge who were then taught useful skills, such as agricultural methods, as well as natural history. André had also been a teacher there. Schreiber imported improved fruit trees from France, which were propagated at the school nursery and distributed among local people. According to the local Catholic hierarchy, he imported more than fruit trees—he also brought back dangerous Lutheran ideas and was accused of teaching more about natural history than religion. He was indicted in 1802 and forced to leave, ending up as parish priest of the village of Vražné, his region of responsibility reaching as far as neighboring Hynčice, which just happened to be the birthplace of Gregor Mendel.

Schreiber also had to face opposition, or at least suspicion, from a conservative peasantry. So in order to distribute new fruit varieties, he and the countess developed a technique that has been used more than once down the ages in order to bring new genes to the countryside: subterfuge. A nursery for trees was established and word put out that these valuable seedlings were under guard, the guards being instructed to make a lot of noise if they heard anybody but not to actually arrest anyone. In a matter of days, all the seedlings had been stolen.

Mendel's monastery in Brno itself had a narrow escape, for in 1854, when the Bishop of Brno, disturbed by the worldly pursuits of the monks and some unorthodox opinions, nearly closed it down.

FROM THE GREENHOUSE TO THE PULPIT—HYBRIDIZATION AND POPULAR ATTITUDES

The concept of hybridizing as a means of improving both crops and flowers steadily gained strength during the nineteenth century. Few accounts or records remain, but horticultural and agricultural literature of the time, at least in Britain, makes frequent references to hybridization. In a letter to the *Gardeners' Chronicle* in 1843, the author writes of his father having raised around ninety hybrids of Cape heaths (*Erica* species) between 1790 and 1841 and having experimented with rhododendrons as well as several genera of small South African bulbs.[22] Rhododendrons, which hybridize easily, were certainly a popular subject for early experimentation. Roses, irises, and various other ornamentals were also widely experimented with (see chapter 13). That the record is indistinct is not perhaps surprising: it was not at all completely clear yet that crossing one species with another was that productive in real improvement rather than an exercise in "curiosity." Religious scruples over meddling with creation may have made some early hybridizers somewhat furtive.

In 1843, John Lindley (1799–1865), the editor of the *Gardeners' Chronicle* and a leader in the worlds of both botany and horticulture, wrote that "as an operation to fill up the leisure hours of the lady gardener and amateur, I do not know anything more pleasing; for there is something akin to creative power in it," followed by a description of the basic techniques that could be used for cross-fertilizing flowers.[23] The next year, however, Lindley felt he had to warn against a hobby which was getting out of hand, using an editorial on hybridization to point out that although "we anticipate through its assistance a change in the whole face of cultivated plants" that many "so-called hybrids are not hybrids at all." In view of hints from a variety of sources that some churchmen were against hybridizing God's creation, he justifies it by referring to the work of the dean of Manchester, William Herbert.[24]

The *Gardeners' Chronicle* also dealt with agricultural matters, and around this time there was a flurry of interest in the matter of improving crops; an editorial from 1845 complains that prizes were being given at shows for wheat seed, "yet it is not possible to tell anything of the quality of the wheat from the seed . . . any more than [someone] could determine the quality of

22. Rollinson 1843.
23. Lindley 1843, 444.
24. Lindley 1844b, 459.

a breed of fowls from an egg."[25] Sir F. A. MacKenzie, sometime president of the Root and Seed Committee of the Highland Agricultural Society, had that year pointed out the absurdity of hundreds of pounds being set aside as prize money for animals but only five pounds for plants, and proposed that the council of the Agricultural Society of England gave higher value prizes for new varieties. Not everyone was enthusiastic about new methods of improving crops; "T.A.F.," writing in the same journal in 1844, reported that he had once heard a landowner addressing a meeting of farmers say that "the powers of the earth are limited like our own, and therefore it was no use expecting greater things of it. . . . [A]ll he wanted to know was how to make a good dunghill, he ridiculed theorists and German philosophers . . . and in all this he was much cheered."[26]

Even not everyone in the plant breeding business thought hybridization a good idea. Alexander Livingston (1821–1898) of Ohio was a grower who specialized in producing round, juicy, tasty, and productive tomatoes, unlike the ribbed, dry and hard fruit that made tomatoes (or "love apples" as they were often called) an also-ran in the vegetable garden. His opinion was that he had "no confidence in hybridizing or crossing as a method of securing new varieties. . . . Like begets like. Rough ones beget rough ones."[27] For him, selection was all; he grew whole fields of tomatoes and endlessly trawled them for chance improvements or grew promising varieties on year after year until they were good enough to name and market. That he was able to do so is probably a reflection of the fact that tomato breeding was in its infancy and genetic variation considerable. At this stage in the evolutionary life of a cultivated plant, such methods are often good enough; although cultivated tomatoes self-fertilize, there is always some crossing, which would generate new variants.

Nineteenth-century objections to Darwinism were only partly on the basis of species evolving or the timescale of evolution being different to that in the Bible; there was a farther and major objection—that of species splitting to create other species. Such a suggestion was seen as blasphemous, because it involved the fragmentation of God's original plan, the sundering of Creation itself. As hybridization between species began to spread during the nineteenth century, it appears that there was a rumble of concern about this too, again largely from religious sources. Documentary evidence for this

25. Anon. 1845, 155.
26. T. A. F. 1844, 788.
27. Livingston 1998, 47.

is hard to find, and evidence for it is largely in the form of gardeners defending themselves against accusations of sacrilege.

Unease at hybridization, originating partly in biblical injunctions against crossing breeds of cattle and against mules, have had a more lasting impact on our attitudes to animals than among plants. "Mongrel" still has negative connotations, and owners of "crossbreed" dogs are still excluded from exclusive dog shows like London's Crufts. Miscegenation was regarded with horror in societies based on racial division, such as the Deep South of the United States. Neither word is used to describe plant varieties today and they have been little used for over a century. Things were clearly different in the nineteenth century, as we gather from Lindley's writing in the *Gardeners' Chronicle* in 1844; he describes how hybridizers were accused of attempting to subvert the whole order of nature by "monstrous practices" but points out natural examples of plants crossing.[28] An editorial in the same journal in 1881 declared that "hybridizing was formerly regarded as a sacrilegious subversion of nature, and those who practiced the art were stigmatized as mischievous intermeddlers in the works of the Creator. Some would have consigned all hybrids to the rubbish heap as being of impure descent. . . . [G]ardeners did not stay their hands in the work of rearing novelties, heedless of the 'confusion' they were causing."[29]

Delegates to the 1899 International Conference on Hybridization and Cross-Breeding in London (see chapter 6) heard M. T. Masters defend the meeting's *raison d'être* in his introductory address:

> Many worthy people objected to the production of hybrids on the grounds that it was an impious interference with the laws of nature. . . . The best answer to this prejudice was supplied by Dean Herbert, whose orthodoxy was beyond suspicion. . . . [H]e succeeded in raising . . . many hybrid narcissi, such as he had seen wild in the Pyrenees, by means of artificial cross-breeding. If such forms could exist in nature, there can be no impropriety in producing them by the art of the gardener.[30]

That William Herbert was a dean of the Church of England reinforced his point. Masters went on to tell, how in the recent past, one firm of nurserymen in Tooting, London, having raised some hybrids of Cape heaths, were wary of selling them as such, exhibiting them instead as new species just ar-

28. Lindley 1844a.
29. Anon. 1881.
30. Quoted in Wilks 1900, 57.

rived from South Africa, and that some nurserymen were afraid to exhibit hybrids at RHS shows "because they might injure the feelings of some over-sensitive religious persons." He also criticized those botanists who objected on the basis that they did not know how to fit hybrids into their system of classification:

> It is indeed altogether surprising that the botanists should have objected to the inconvenience and confusion introduced into their pretty little systems of classification by the introduction of hybrids and mongrels, and that they should object to hybrid species, and much more to hybrid genera. But it would be very unscientific to prefer the interests of our systems to the extension of truth.[31]

SUMMARY

The eighteenth and nineteenth centuries saw the beginnings of a more systematic approach to selecting new crop varieties. Progress in active breeding, in crossing plants to achieve progeny superior to either parent for a particular purpose, had to wait until the sexual basis of plant reproduction had been uncovered. Much deliberate early hybridization was closely linked to this process of discovery, as it was recognized as a very effective technique for learning about reproductive mechanisms and heredity. Others—those more involved with the issues of practical land management and rural economy—recognized its potential as a way of improving varieties. Once the basics of crossbreeding plants and selecting and trialing the progeny became established, it was possible for systematic plant breeding to truly begin. The sugar beet was one of the very earliest examples of this integrated process; it was also a crop closely linked to the development of a more industrial approach to agriculture and to the processing of agricultural products.

Practical plant breeding was a particular interest in Moravia, a part of Europe with a singularly forward-looking elite. So it should come as no surprise that it was here that the key breakthrough was made that opened the gates to truly modern plant breeding—Mendel's understanding of the basics of genetics. That Mendel and his colleagues in Moravia faced resistance to their work, and that experimental breeders in other parts of Europe also appeared to face resistance, indicated that new ideas would take some time to become accepted.

31. Ibid., 58.

FIVE ❦ EMPIRE ❦ GLOBALIZATION IN EARNEST

Anyone who travels in developing countries cannot help but notice the prevalence of bread. It seems as if it does not matter where you are or what the local staple is, in the nearest decent-sized town, somebody has bread for sale, usually wrapped in a clear plastic bag—and usually too expensive for most of the locals. This often puzzles the traveler; why eat bread when wheat does not grow locally and there are other perfectly good and nutritious local sources of carbohydrates? Especially when the bread is often so bad—the poor quality low-gluten flour used in the loaves sold by street traders and most shops means that the stuff often disintegrates as soon as it is cut; in Thailand the thought of eating the soft white triangles sold as sandwiches seems almost an act of madness when a steaming bowl of spicy noodles is also on offer. One researcher reports that among the potato cultivators of the Andes, in whose fields sprout forth a host of diverse varieties, bread is prized as a gift.[1]

What is it that is so special about bread? Wheat may have originated in the Fertile Crescent of the Middle East, but it spread rapidly across Eurasia. When Europeans broke out of their wet and windy corner of the landmass to conquer much of the globe, wheat went with them. Its use to make loaves of bread was very much a European phenomenon: many in Islamic cultures eat it as flatbread; in India as an unleavened slab—the *roti* or *chapatti*; in north China it is more likely to be found as a wrapping for dumplings or for making noodles. Many peasant communities have also eaten it as a porridge or gruel or as a grain-like rice—as in Italian *forro*. The contemporary spread of wheat as bread appears to be one of the most fundamental symbols of globalization, and if the loaf may be likened to a locomotive, there follows a train of biscuits, pies, pizzas, and pastries. Biscuits in particular are the traveler's joy, a local family treat and, one suspects, the foundation stone of many a street trader's business, given their long shelf life and energy-giving irresistibility.

The spread of bread and wheat products may be a cultural symbol of globalization, or it may be to do with the fact that we as a species do find

1. Brush 2004.

it very attractive: compared to the tasteless stodge upon which many cultures depend for carbohydrates, its relatively high protein content and flavor may make it a highly desirable alternative. Wheat flour is also immensely flexible—as has just been noted. Once made, bread and other wheat-based products are very easily transported. The fact that bread is relatively high in protein and even when stale is still edible and nutritious may have contributed to the ability of wheat-eating peoples to endure privations more readily than others. The crew of an eighteenth-century ship may have cursed the rock-hard and worm-infested biscuit they were rationed, but the fact that it did not need cooking and supplied most nutritional needs was an advantage over most other calorie sources. Settler families in wagons crossing hostile territory or soldiers beating a retreat also had reason to thank that they could rely on dry bread rather than anything that needed cooking. Bread may have helped the nations of Europe to dominate the world.

So it is no surprise that wheat has now spread into practically every climate zone where it could possibly do so; the development of local varieties has been the one of the most widely spread of all plant breeding activities over the last two centuries. But there is unease too—to what extent is the desire to eat bread and wheat products simply due to a desire to be "modern" and Western? To what extent does the effort to produce wheat varieties which will grow healthily where none grew before distort plant breeding priorities? This is an issue we shall revisit again in the context of the Green Revolution.

A UNIVERSAL GARDEN—GLOBAL EXCHANGES OF CROPS

Wheat's global spread is simply one of several similar phenomena. Everywhere you go, it seems, there are fields of corn and plots of potatoes and more, as restricted by climate: orchards of apples or vines marching in lines across the landscape. Gardens sprout cabbages, tomatoes, and peppers (*Capsicum* species) almost universally. This global spread of certain crops was facilitated by European power, a process whereby European traders, explorers, soldiers, travelers, and plunderers burst out over the world, beginning in the fifteenth century and reaching its height in the nineteenth. There are three "directions" in which this globalization affected crops: one was the export of classically European crops (such as wheat) to the rest of the globe; another was the importation into the Old World of New World crops, a process which began with alacrity in the sixteenth century; the third was the genetic combination of species from different continents, which sometimes transformed a crop from being a local delicacy to an industry of regional importance.

The nineteenth century is arguably the period in which globalization was at its height—it certainly was in terms of the spread of agricultural systems and crops. Much of this process of globalization was concerned with a small number of European countries transforming the agriculture of their colonies and other countries, nominally independent but dependent upon a European power (Britain's relationship to Argentina is a good example). This transformation was for the benefit of the colonial power and often to the detriment of the colony—and certainly to the detriment of their natural environments. Empire was exploitative and at times destructive, but not wholly so.[2] The colonized often benefited from the introduction of new crops, although in many other ways their lives and communities were disrupted by the reorientation that is at the heart of empire; as economies and political systems cease being locally orientated and turned instead toward the needs of a distant foreign power. One way of looking at the crops that accompanied the administrators, missionaries, and troops of the imperial powers is to see them as imperialists too—those species with genes that enable them to respond well to a wide range of new human-created environments and that are able to provide the goods needed by their cultivators were the ones favored by the spread of empire. These "megacrops" displaced local ones just as tribal peoples were pushed aside by plantations and mines; in Africa, American corn and manioc rode roughshod over finger millet (*Eleusine coracana*), sorghum, and fonio (*Digitaria exilis*), partly because imperial administrators favored their production and partly because their inherent productivity and adaptability made them more attractive to local people. This process is still going on; and those who would argue that imperialism has never really ended might suggest that the imperialism of the megacrops was another aspect of the neocolonial relationship between the West and its colonies.[3] The fact that the imperial period continued until after the middle of the twentieth century is one reason at least why in this chapter we look beyond the nineteenth century.

"Empire" as the term is used here is intended to convey that we are not just talking about the political empires of the European powers but the em-

2. See Ferguson 2003, for an example of a modern balance sheet of the British empire.

3. I refer to the concept of neocolonialism, as first advanced by the Ghanaian Kwame Nkrumah.

pires established by the megacrops—specifically, at how they were bred to be productive in their new homes. In this chapter we will look at wheat and other small grains, which went with Europeans to the newly colonized and settled lands; at potatoes which went the other way; at cotton, a crop with both Old and New World origins, which became a centerpiece of imperial trade; at apples, the classic temperate zone fruit; and at strawberries, whose genetic origins lay on both Atlantic and Pacific shores. Finally we take a brief look at what for the farming and gardening citizens of Europe and North America was one of the most visible signs of vegetable globalization—the seed catalog, with its cornucopia of flowers and vegetables from all the corners of the globe.

As crops were introduced to new soils and climates there was a flurry of selection activity, and as time went on, hybridization, to produce strains and varieties which would ensure success in their new homes; often it was not so much soil and climate that were the greatest problems but the depredations of pests and diseases unfamiliar to both plants and cultivators. Empire stimulated a whole new network of botanical gardens and research stations engaged in trialing new crops for the economic development of the colonies, as well as the foundation of a new science—economic botany. It is important to stress, however, that "official" selection and breeding by colonial authorities is only part of the story. Arguably, the most important work was done by countless small farmers in South America, Africa, and Asia who grew the newly arrived crops, and over the first few generations began the process of making new landraces, adaptable to their conditions and tastes. The process must often have been risky, as crops, perhaps poorly adapted to their new homes, failed, or once harvested, faced suspicion or even disapproval from farmers' families, communities, or customers in the market place. Such innovative traditional farmers are truly among the great unsung heroes of plant breeding.

What is distinctive about the tropical part of the new global network of colonial botanical institutions that developed during the nineteenth century is that it was primarily concerned with the commercial evaluation and propagation of selections among and within species rather than with proactive plant breeding. The best way of finding improved varieties was often through botanical "prospecting" in the wild or in the gardens of native peoples or by making exchanges with other institutions. Most selections were then vegetatively propagated, so breeding as such had a limited role. In some cases, crops of major regional importance started from very limited gene pools; cloves (*Syzigium aromaticum*), for example, are one of several tropical crops that had to be smuggled out by one set of imperialists from under the noses of another—the French managed to collect several hundred young plants from remote parts of the Moluccas in the face of a Dutch embargo. Very few survived, and one tree in the French colony of La Réunion is thought to have been the ancestor of the whole clove industry on this island and on Madagascar.[4]

During the age of empire we see the start of a markedly different relationship between the farmer and their crops. Traditionally, the vast majority of crops would have been consumed close to home, and in the case of those high-value crops exported over longer distances, little distinction was made between varieties. Nineteenth-century empire, however, established political and economic relations that linked producers and processors and consumers over much longer distances. More and more peasant farmers or laborers were growing goods for people they would never meet in lands they would never see or in some cases might not even know about. In many cases they might not even have recognized the goods or the foods made from the raw materials they grew. The converse of this was that consumers in the imperial heartland found themselves consuming more and more food and products about which they knew very little. Tapioca, usually made from the starchy roots of manioc and often known colloquially as "frogs' eyes" in Britain, or "fish eyes and glue" in the United States, is an example.

As empire turned increasing numbers of mostly tropical farmers into producers of goods for export, these farmers and their communities were brought into a cash-based economy for the first time. They were also growing for markets that often had demands and requirements that were very specific, necessitating the cultivation of particular crop varieties. These demands may have been led by particular qualities needed in the final product, but in many cases they were driven by the limitations of the machinery needed to process the product. As the industrial revolution brought the power of the machine increasingly to bear on the links between producers and consumers, so did the machine become to resemble a master in its own right. Often, as is illustrated particularly by the story of cotton, the demands of the machine and the desires of the producers conflicted.

The age of empire saw the growth of a more systematic approach to plant breeding, with the beginnings of systematic selection and an increasingly proactive approach to hybridization—the latter reflecting the fact that a number of individuals began to understand basic genetic principles. The

4. Wit 1976.

nineteenth century saw, to use Darwin's terminology, the beginning of methodical breeding, where a goal is set and the breeder tries to work toward it.

Darwin's huge contribution to the life sciences was very much more than the theory of evolution for which he is most famous. He was not a plant breeder, but his work on pollination and fertilization mechanisms in plants provided an essential basis for much breeding work. Of vital importance for the development of plant breeding as a scientifically led practical discipline was his exploration of self- and cross-fertilization, much of which appeared in his *The Effects of Cross and Self Fertilization in the Vegetable Kingdom* (1876). He also wrote extensively on plant domestication and breeding and the work of breeders in the years preceding him, particularly in *The Variation of Animals and Plants under Domestication* (1868). He drew upon the work of earlier students of botany, such as Sprengel, and on breeders, such as Knight, synthesizing their work into a much more comprehensive and incisive understanding of plant reproduction. Whereas many early breeders, such as Knight and Rimpau (who is discussed later; see page 110), were *technologists* (working through trial and error and simple experiments in order to work out practical ways of solving practical problems), Darwin, like Mendel, was a *scientist*, seeking to understand the underlying mechanisms.

FROM ONE HEDGEROW TO MANY FIELDS—FINDING AND DISTRIBUTING NEW CEREAL VARIETIES

One day during one of the final years of the eighteenth century, Mr. Woods, a farmer in Sussex in southern England was walking over his fields when "he met with a single plant of wheat growing in a hedge. This plant contained thirty fair ears, in which were found fourteen hundred corns. These, Mr. Woods planted the ensuing year, with the greatest attention, in a wheat field: the crop from these fourteen hundred corns produced eight pounds and a half of seed, which he planted the same year; and the produce amounted to forty-eight gallons."[5] So began the story of 'Chidham' wheat, one of the most widely grown types over the whole of Britain between 1800 and 1880.

Chidham was recognized by its finder for its large-sized ears, which produced a quality crop. Previously such good varieties might not have traveled far, but in a time of rapidly improving communications and a growing interest in "improvement," they were more likely to spread. Productive and re-

5. Quoted in Chidham and Hambrook Parish Council 2006.

liable landraces began to lose their connection with their home regions as gentlemen farmers wrote to each other about new finds, displayed them with pride at meetings of agricultural societies or sold them to dealers in seed.

The story of 'Forward' wheat is another good example of the informal processes by which new varieties were found and distributed. In June 30, 1794, the *American Mercury* published an article entitled "Forward, an Account of a New Species of Wheat,"[6] describing a new, hard winter wheat that matured some fifteen to twenty days earlier than any other and had a large plump grain whose hardness reduced the risk of premature germination before harvest. It had originated seven years earlier—a Mr. Isbill of Caroline County, Virginia, had observed a single ear in a field of mixed wheat. Over the next few years Isbill bulked it up, sharing it with neighbors so that by 1794, several thousand bushels were being produced on several farms in the county.

As an early wheat, Forward allowed farmers further north in New England to harvest in late summer rather than autumn and then thresh and mill before winter, whereas before they had to wait for watermills to mill during periodic thaws in the bitter weather; the result was that Forward allowed farmers to realize gains six months earlier than before. Forward took off rapidly in New England after an enterprising farmer had started selling it in Connecticut.[7]

The fact that seeds of new varieties were being sold rather than being exchanged between neighbors was a sign that seed was increasingly becoming a commodity to be traded and to be dealt in by specialists—seed merchants. Part of their importance lies in the fact that in changing times, new crops and new varieties had to be sought, which created an opening for traders. There was also a widespread belief that seed from some areas was better than others, probably a result of a link being erroneously made between plant health and seed health. Crops were less prone to disease in some areas than others; consequently, it was believed that seed grown there would be better quality. Particular areas began to become associated with particular crops, probably because of a handful of good and honest growers; in England, London was known for leek seed, Deptford its onions, Battersea its cabbage and Windsor its peas and beans (*Vicia faba*); heirloom vegetable specialists still sell a bean variety called 'Windsor,' which dates back to the early eighteenth century at least. Seed merchants almost invariably only traded—they

6. See Destler 1968.

7. Ibid.

bought in seed from growers or other dealers, or directly off a ship at the dockside. They and their customers had only the word of their suppliers—reputation was everything.

That important new crop varieties were found through the combination of sheer luck that brings together a new hybrid or new mutation and an observant farmer (who also needed to be entrepreneurial and well-connected) illustrates that there was little or no concept of active improvement beyond the acres of a very few inspired individuals. As the nineteenth century wore on, the number of such individuals steadily increased, and as improving communications and media made it more likely that the work of innovative growers spread, more and more growers began to take an active interest in looking closely at their crops. Various methods were tried in order to pick out better forms; as with the methods used by traditional peoples, these were sometimes appropriate, sometimes not. A key breakthrough was the realization that perhaps the best way to improve a crop was not by picking out the product but by looking at the plant that produced it. This was first realized by growers of cereals but gradually spread to other crops; in the antebellum South, Henry W. Vick became a successful seed dealer and cotton breeder through such an imaginative leap.

Whereas other planters picked out their cotton seed for planting by separating out the best-looking heads of cotton, Vick sent his most trusted slaves out into the fields before the harvest to pick out the largest and most productive plants and collect their seed. During the early 1840s, he spent winter evenings sorting through cotton plants, and one day he realized that "I plainly saw that what I had supposed to be a homogenous stock of cotton seed, consisted in fact of ten or a dozen different varieties." This discovery led him to select out and grow the best one, which in its first year yielded exactly one hundred seeds. Named '100 seed,' it went on to become one of the most widely grown cottons of the late antebellum South.[8]

To gentleman farmers intent on improving their land, the relative difficulty of making much progress with field crops must have been frustrating: cattle and other livestock had begun to have pedigrees as part of an obvious improvement in their quality. Whilst apples and other tree fruit could be grafted and endless progeny distributed around the land so that eventually everyone from the lord with trees painstakingly trained to kitchen garden walls to the laborer with a tiny garden could grow the same quality varieties. It was a relatively straightforward business to put a good bull with a

8. Quoted in J. H. Moore 1956, 100.

good heifer, but how do you begin to improve wheat, oats and other seed-grown crops?

Patrick Shirreff of Hopetoun, Scotland, was one of the first to try, as recorded in his brief memoir, *Improvement of the Cereals and an Essay on the Wheat-Fly* (1873). His experiments began in 1819 with a series of *pure line* selections of wheat and oats; he later moved on to hybridization. He was not the first to produce a hybrid wheat—this appears to have been accomplished by a Mr. Maund of Bromsgrove in Worcestershire, who showed some at a meeting of the English Agricultural Society in 1846. Maund was quoted as saying that in hybridizing, the interests of the farmer, the miller, and the baker needed to be considered, "added to which the caprice of the public."[9]

Shirreff began his work on wheat with "sports"—exceptional individuals that stood out from the mass. He noticed that if their seed was sown and grown on, it tended to be uniform, a fact he seemed to accept without investigating further; he was a practical man, seemingly uninterested in exploring the science. So, without realizing it, he discovered the technique of pure-line breeding.[10] However, he was frustrated by the fact that exceptional plants were rare; he only managed to find enough to bring four into commerce during forty years, but these included three that were successful for much of the nineteenth century: 'Mungoswells' wheat and two oats, 'Shirreff' and 'Hopetoun.'

Making only limited progress with selection, Shirreff increasingly turned to hybridization, suggesting that "before commencing to cross, consider what properties the new variety is wished to inherit; and fix upon such kinds as they possess in the highest degree the desired properties."[11] He went on to describe a process that involved two people manipulating the flowers of wheat to transfer pollen from one parent plant to another. One fruit of his labors was 'Shirreff's Squarehead' wheat, produced in 1882, widely grown and used for hybridization work all over Europe over the next few decades.

A key early wheat breeder was John Le Couteur (1794–1875), who farmed on Jersey, a British-administered island off the French coast. He had made a collection of wheat varieties, exceeding 150, but it seems as if his interest in improving wheat arose following the visit of La Gasca, a professor from Madrid University who had researched the range of wheat varieties available in Europe. La Gasca commented on the range of different characteris-

9. Quoted in Anon. 1846.

10. See technical note 19: Mass Selection, Pedigree, and Pure-Line Breeding.

11. Quoted in De Vries, 1907, 113.

tics visible in each one of his host's plots of wheat selections, in one case distinguishing over twenty distinct forms in one field of one variety. After his departure, Le Couteur saved the grains from each of the forms and sowed them separately.[12] Le Couteur selected what he reckoned to be the best ones and bulked them up to sell; of the seven he launched commercially, one—'Bellevue de Talavera'—was particularly uniform and continued to be grown in northern France until well into the twentieth century. This was possibly of Spanish origin and noted for its strong (i.e., high gluten) flour. Le Couteur did not, as far as we know, engage in hybridization, but Darwin suspected that some of his selections were crosses, as he found them "incorrigibly sportive."[13]

Another British farmer, F. F. Hallett of Brighton, also began to work in the mid-nineteenth century, but rather than picking out individuals and then developing a new strain directly from their progeny, he picked out the best kernels from the best heads of wheat, sowed them, and then repeated exactly the same process the following year, a process he called *pedigree-culture.* With Hallett's approach the role of the seed raiser was vital, as he could always claim that he was improving the stock; with Le Couteur and Shirreff's strains, once a farmer had bought the strain, he could save the seed-corn every year himself—benefiting from the strongly self-pollinating character of wheat and barley. The grain fields of early-nineteenth-century Europe and North America would have looked very different from how they do today; landraces are by their nature heterogeneous, so individual plants in a field would have been very obviously different in height, color, and overall form. As landraces began to give way to new named varieties selected from just one or a few similar promising individuals, greater uniformity would have been the result, and farmers if they were asked, would have been able to give a name to the variety they were growing. The sale of named varieties—indeed, the awareness that there were such things as distinct varieties that could be distributed nationally (or even internationally)—was a major change. In fact, the trade in seed and the naming of varieties went hand in hand; good seed strains had to acquire names if they were to be sold. In France, the appearance of fields would have gone through an even more dramatic change, for here the *meteil* system was often used, with mixtures of wheat and rye or barley and oats being sown together, much as crop mix-

12. Darwin 1875.
13. Ibid., 353.

tures can still be found in some developing countries or in forage mixtures anywhere.

In Britain and North America, the beginning of the naming and the widespread trading in distinct varieties can be very roughly dated to the 1830s, in France and Germany, somewhat later. In France during the mid-century, Ukrainian wheat began to be imported; it rapidly crossed with native French varieties resulting in new populations noted as being good for baking, adaptable, and with good rust resistance. Some of these were developed and marketed as distinct named varieties by entrepreneurial landowners, including 'd'Aquitaine,' which became particularly successful, and 'Noé,' named after the Marquis de Noé, who distributed it—this latter variety was noted for thriving in the hot dry south.[14]

BREEDING A DYNASTY—THE VILMORIN FAMILY

During the 1830s, the French seedsman Philippe-André de Vilmorin (1776–1862) grew some wild carrots, made a selection of the best, sowed their seed, and so on; within three years he had a crop that was pretty much as good as domestic carrots. This was a surprise to him, as it challenged the belief at the time that many years were necessary to turn a wild plant into a worthwhile crop. Such an experiment was typical of the man and of his son Louis—two members of a remarkable dynasty founded in 1743 by Pierre d'Andrieux de Vilmorin (1713–1780), the "royal botanist" of the French court and a collaborator with Duchesne. Another member of the family, Philippe-Victoire de Vilmorin (1745–1804) had been a close friend of Duchesne; there is little doubt that the family's relationship with Duchesne enabled them to make the progress they did in breeding.

The Vilmorin attitude to seed production combined scientific experimentation with good business sense—the result was a company which played a major role in plant breeding. From 1840 to 1920, it was the most important seed company in the world, especially for cereals. It is still one of the world's largest transnational companies dealing in seed, although it is no longer a family firm.

Philippe André, as with many other enlightened growers at the time, realized that there was a clear distinction between improving crops through good husbandry and improving the crop itself. It is probably true to say that

14. Vilmorin-Andrieux 1880.

many farmers and growers throughout history knew this, but it was only in the post-Enlightenment period that it came to be clearly articulated, and most importantly, acted upon. He made "improvement through the seed" central to his production process, and in doing so made the decisive break with the widespread contemporary belief that seed was best grown in particular areas and with the belief that healthy plants of necessity produced healthy seed.

Philippe André's son, Louis, undertook crucial work on heredity and selection during the 1850s; his main contribution to plant breeding was his method of pedigree breeding—genealogical selection. In its details it was very similar to what Hallett had been doing, and what happened later in Svalöf, Sweden (see chapter 6). Building on the now widely accepted use of pedigrees in cattle breeding, he was primarily interested in stabilizing lineages through simply selecting the best of each generation and repeating over several generations. Louis formulated a set of principles that he wrote up in a series of articles, which were widely circulated in France and translated into English. This approach was then systematically used by the family company for variety improvement. The core concept was to isolate good characters, with the ability to transmit them being described as *être bon étalon* (to be a good stallion), an expression borrowed from animal breeding; "it is only by acting on unique individuals that it is possible to establish a nobility in plants, that is to say a series of individuals in which the individual qualities are transmitted without alteration from generation to generation."[15]

Pedigree breeding enabled Vilmorin and other practitioners to stabilize the characters they desired in a variety. However, they had to have worthwhile characters in the first place, and in order to achieve this, they needed to increase the level of variation. Louis achieved this by periodically carrying out the very opposite of pedigree selection—random selection in various directions over several generations without regard to selecting for particular characters or even for selecting the most extreme characters; good ones were then selected and then new lineages set up. Improved sugar beet was one of his greatest achievements; he recognized that the progeny of good lines could vary enormously, but that they could also be homogenous, and he made it a priority to find lines with the latter property (see chapter 4).

Louis's untimely death at the age of 44 cut short a life which, for all we know, may have resulted in him coming to the same conclusions as Mendel.

15. Quoted in Gayon and Zallen 1998, 247.

His son Henri (1843–1899) headed the company from 1866 to 1899 but was more interested in business than science, building the company up so that it became one of world leaders in seed production until the First World War. Unlike his father, he believed in acquired characteristics[16] and rejected statistical approaches to studying heredity, both of which must have held back progress. What he did do, however, was make use of hybridization to improve wheat, beginning in 1873, inspired apparently by Shirreff's work and the acquaintance of botanist Charles Naudin (1815–1899), who had undertaken research on hybrids and heredity. His guiding principle was that crosses between races provided a much greater variability than spontaneous variation within them. Hybridization fulfilled the same function as the random breeding practiced by Louis—the generation of new characters—but did so more efficiently. The approach worked only because wheat tends to self-pollinate; the same method would not have worked with cross-pollinating plants as the process of pedigree selection following hybridization would have resulted in gradual deterioration through "inbreeding depression" (see chapter 10).

New wheat varieties were the great success of the Vilmorin company. In 1883, the company started distribution of seed from the first modern French wheat, 'Dattel,' the result of a cross between a French and a British wheat. This was followed by others, so that by 1918, the traditional French wheat landraces, *blé de pays*, occupied only half the wheat lands of the nation. Vilmorin wheats were exported to Argentina, Chile, Russia, and the Maghreb (Morocco, Algeria, and Tunisia), and were predominant in France until the 1950s.

ONWARDS AND UPWARDS—THE "GERMAN METHOD"

Whereas the Vilmorin company aimed at producing a variety which was stable and predictable enough to be called a "final product," Hallett's slow and gradual selection was a continual process, an ongoing improvement of a crop variety. This was the approach adopted by many German breeders, too, who worked with a variant of mass selection, producing "elite" strains, and every year selecting the best to grow on for the next year's selection; indeed, this became known as the "German method." One result of this approach, which suited breeders and seed merchants, was that a seed strain would

16. See technical note 22: Lamarck.

never stay the same but would be constantly improved. A farmer might buy seed of a particular strain, but in a few years' time, a new batch would be an almost guaranteed improvement. One particular rye, 'Schlanstedt,' was produced this way by a German breeder, Wilhelm Rimpau (1842–1903), during the 1860s and was a great commercial success, ousting many other ryes from the fields of north Germany and France. Rimpau kept his elite strain isolated from other ryes to avoid contamination—as unlike wheat, rye cross-pollinates. Such contamination, followed by deterioration of the crop, inevitably happened to the vast majority of the crops grown in subsequent years by the farmers who bought Rimpau seed and then resowed from their own crop. For the first time, farmers were being led away from the traditional practice of saving their own seed and becoming dependent upon outsiders.

Rimpau also worked with wheat from 1875, making crosses as well as selections; his work emphasizing the combining of German hardiness and bread making quality with American and English productivity. His first commercial wheat variety was the hybrid 'Rimpaus früher Bastard' derived from crossing the British Squarehead with an American winter variety; immensely successful commercially, it was grown in Germany until the early 1940s.[17] It was eventually jettisoned, along with some 90 percent of the country's national seed list, by the Nazi regime, who established as totalitarian a grip on the varieties which could be used by farmers as they did on other spheres of life.[18]

Rimpau observed regular patterns in characteristics among the progeny of crosses and observed phenotypes but did not breed all the F_2 generation and so was unable to classify their genotypes.[19] He noted the relative uniformity of the F_1 generation and the segregation of the F_2 but did not seek to investigate this farther. Rimpau, like those others who came within a hair's breadth of discovering what Mendel did, was first and foremost a practical breeder, not a pure scientist, so growing on a whole range of plants whose phenotypes were less desirable would have seemed a waste of time. Mendel, in his monastic fastness, working under no commercial pressures, with a mission to understand the workings of the Creator through nature, had the time and the inclination to do this.

17. Porsche and Taylor 2001.

18. Pistorius and Van Wijk 1999.

19. Explanations of *genotype* and *phenotype* are given later (see technical note 23: Phenotype and Genotype).

WEAVING THE BREAD BASKET OF THE WORLD

The day in the 1840s when Mrs. Fife shooed the family cow away from a patch of wheat in Ontario, Canada, turned out to be a momentous one. The animal had already eaten two out of the three shoots on the plant, one of several grown from a sample of a shipment which had arrived in 1842 from the port of Danzig (now Gdańsk), Poland. The survivor was grown on by Mr. David Fife and produced the first crop of 'Red Fife,' which went on to become one of the most successful Canadian wheat varieties.

One of the major features of the nineteenth century was the spread of European settlers across the globe; in particular, anywhere with a climate not too dissimilar to home would be seen as ripe for settlement. Of all the regions which were settled by Europeans and their crops, that of the Midwest of North America underwent perhaps the most thorough transformation, with the greatest global consequences. From being vast expanses of grassland and savannah, dotted with herds of buffalo many thousands strong, it went in a comparatively short period of time to being a region of corn and wheat. From here, much of the United States is fed and much is exported; and in times of famine, Midwestern-grown US food aid finds its way to many remote corners of the globe. While the moister eastern region became the Corn Belt, the drier western part—the Great Plains, together with a great swathe of Canada—has become the most productive breadbasket in the world.

The transformation of the great open middle of the North American continent may have occurred rapidly, but it was not a foregone conclusion, particularly where wheat was concerned; it was plant breeding that enabled the Great Plains and Canada to become preeminent as wheat lands.

To be sure of its potential as a wheat-growing country, Canada needed varieties that could cope first and foremost with a short growing season, shorter than any in Europe. This was achieved through the efforts of a family team, William Saunders and his sons. William Saunders (1836–1914) was a pharmacist by training who had emigrated to Canada in 1848; he made a career producing medicines from herbal extracts and exposing fraudulent practices among the snake-oil salesmen of untested remedies who were such a feature of life in North America and Britain in the late nineteenth century. Following an old British tradition, he was an obsessive hobby breeder of gooseberries (see chapter 4), and had an apparently fanatical belief in the value of crossbreeding plants. In 1886 a commission of the Parliament of

Canada granted him funding to start the Dominion Experimental Farms. Despite having no experience of wheat, he plunged into what was initially a very unsystematic program of wheat breeding; sending off for samples of wheat seed from all over world and showing his son Percy how to cross-pollinate wheat before sending him off to work with the crop on farms across the west. However, the large number of plants needing to be grown on rather overwhelmed them, and the crossing program had to give way to simply growing on batches of wheat for enough years to produce a sufficient quantity for evaluation purposes.

Saunders was particularly interested in Russian wheats for their early-ripening qualities. One was 'Ladoga,' a hard red wheat from Lake Ladoga near St. Petersburg, which had been distributed to a number of experimental farms. In those days there were no small-batch milling tests, and hundreds of bushels were needed to evaluate quality. Since it took several years for a new variety to be bulked up enough to produce this quantity, evaluation was inevitably a very slow process. Years of expectation must have led to many bitter disappointments; once the Royal Flour Mills finally got its millers and bakers a wagonload of Ladoga, for which there were high hopes, but it turned out to be weak (i.e., low in gluten) and made coarse yellow bread.[20]

Another son, Dr. Charles Saunders (1867–1937), initially followed a career in music, leading choirs and teaching at a college in Toronto, but he returned to work with his father on wheat breeding, and in 1903 was appointed Dominion Cerealist. He was a lot more discriminating than his father and had the idea—almost a revolutionary one at the time—of crossing an established variety with others from which he wanted to import particular characteristics. One of his first series of crosses was between 'Red Fife' and varieties obtained from Russia and northern India by his father; of these, 'Marquis' resulted from crossing Red Fife with the Indian 'Hard Red Calcutta.' Saunders had selected a single head in 1903, after six or seven generations from the original cross; the plants had consistently ripened a week earlier than Red Fife. The new wheat he dubbed 'Markham.' Saunders, however, lacked a mill or even an oven to bake bread and had to rely on the time-honored, if not exactly conclusive, method of testing for gluten content—chewing the grains to see if they formed an elastic mass. By 1906, he had obtained a small mill and oven and was confident that the variety he was bulking up was worth it. His records of the period indicate that he made a

20. See technical note 20: Gluten.

note of the volume of water added to the flour to make a dough and paid a great deal of attention to the condition of the dough at each kneading, that he had a system for assessing the quality of the resulting loaf's texture and crust, and that he drew cross sections of the loaves. Bread quality was clearly something which for him was as important as the earliness of the crop and its hardiness.

The year 1906 saw Charles Saunders with two selections from 'Markham': 'Marquis' "A" and "B"; by the next year he had settled on "B," and 'Marquis' wheat was born and launched commercially in 1909. Marquis—high quality, early maturing, nonshattering, and high yielding—became the cornerstone of wheat production in western Canada and increasingly set the standard for bread wheat quality globally. Forty years later, it was still making up 70 percent of the Canadian spring wheat harvest.[21] The success of 'Marquis' must have had a great deal to do with Charles Saunders's pioneering scientific rigor in testing and evaluating the bread he made from his wheats. But he was never only an agronomist; after a period of illness he left for France in 1922 to read French literature at the Sorbonne, returning to Toronto in 1928 to teach French *and* agronomy.

Farther south, in the Great Plains region of the United States, Russian wheats were also making an impact. One name in particular is associated with their introduction and early breeding—Mark Carleton. Carleton's reputation is as a rough-hewn and plain-speaking man, who, in the words of one colleague, "went at everything like he was driving cattle." He was socially difficult, obsessive, humorless—the unfortunate characteristics of many visionaries.[22] Born 1876, in Cloud County, Kansas, he knew poverty as a child and it is said that he "grew up with black stem rust." He got a teaching job at Wichita University and then a post at the Kansas Experiment Station at Manhattan, where he discovered that rusts were not indiscriminate but were very particular about which grain varieties they infect. Clearly, this had major implications for breeding. He developed a philosophy that reflected the harsh environment of the plains—"it isn't what a wheat yields in the best years—it's how it stands the worst ones."[23] Carleton was the kind of man who regarded the devastation of his research plots by early winter weather as a blessing in disguise; he recognized that the future lay in what survived.

Like Saunders and many other breeders of this period, he sent for wheats

21. DePauw and Hunt 2001; Buller 1919; Scowcroft 2003.

22. Quoted in Paulsen 2003.

23. Quoted in De Kruif 1928, 9.

from all over the world, but few of them even began to cope with the plains, its droughts, and early frosts. In 1894 he got a job with the USDA, and over the next few years he traveled all over the Great Plains studying rust infestations, especially the dreaded black stem rust, but became deeply frustrated at having no ready answers for farmers whose livelihoods were being ruined by epidemics of the disease. His conclusion was that the wheat varieties being raised on the plains were inadequate, that he had to find some new sources that were frost-resistant, drought-resistant, *and* rust-resistant.

Carleton became obsessed with Russia as a result of friendships with Russian Mennonites, religious refugees who had the reputation of being the farmers who were most in tune with the land and climate of the Kansas plains. From Russia they had brought over a tough wheat dubbed 'Turkey Wheat' by those already settled. Turkey Wheat was tough in more ways than one—cold-hardy, but also hard in texture; the millers did not like it, as it needed more time and power to render into flour; they had been initially reluctant to mill Red Fife and other spring wheats, too. In widespread cultivation, though, Turkey began to deteriorate in quality, and it lodged (fell over) badly. It was suggested by some that the soil or climate of Kansas was causing this, but others held that the plants were not breeding true, and that without any selection, the variety was reverting to inferior ancestral types. Carrying out selection work, they thought, was beyond the abilities of the average farmer.[24]

Carleton's belief was that he should go back to the source of the Mennonite wheats. He fell in love with the idea of Russia, learned the language, and studied its geography and everything he could find out about its agronomy. In July 1898, in the opposite direction of the waves of immigrants pouring into the United States, he set out for his chosen land as a special agent of the new Office of Seed and Plant Introduction of the USDA. He found the Russian plains remarkably like home and came back with what was agreed to be several worthwhile hard red winter wheats. Successful from the word go, these were the parents of the 'papoose' wheats, varieties with Indian names, released just after World War II and which did much to raise productivity at that crucial time.

In 1900, after another Russian trip, he had brought back 'Kharkhov,' which spread very rapidly as an autumn-sown winter wheat; by 1914, half of the national hard red wheat grown was down to this variety. Other Russian wheats he had come back with earlier—'Arnautka,' 'Kubanka,' and 'Per-

24. Malin 1944.

odka'—began to look promising, the latter two of which the North Dakota Research Station found to out-yield all others. Indeed, after a rust outbreak in 1900, 'Kubanka' produced thirty bushels per acre, compared to two to eight per acre for standard varieties.

Conflicts arose, however, as Carleton was now chief cerealist of the USDA (a position to which he had been promoted in 1900); he was expected to direct the research agenda toward what the milling and food industries wanted. Carleton had other ideas; he saw the potential of 'Kubanka,' which he had brought back from the remote central Asian region of Kirghizia (now Kyrgyzstan). Vigorous, hardy, and rust-resistant, 'Kubanka' looked like the ideal start for a new American wheat, but it was hard to mill, and its flour did not suit all tastes. It got nicknamed "goose wheat," even "bastard wheat," suitable only to be exported to make macaroni.

'Kubanka' was a durum wheat, high in protein and therefore nutritious, used for high-quality Russian bread; but it was not suitable for the machinery then used by American millers—it slowed down the milling process considerably. The millers had for some time been denouncing the recent trend towards breakfast cereal (as opposed to bread or porridge), and in particular, the "graham," or whole-grain, flour used in many of these products. Carleton spent the next decade of his career trying to get the new wheat accepted; he waged a personal propaganda campaign that seriously damaged his professional reputation. At his behest, the USDA worked with bakers on baking experiments; test loaves were also sent out to famous cookery writers and others thought to be influential on the domestic front. In 1903, in an interview with the journal *North Western Miller*, Carleton announced that baking tests had shown durum flour blended with soft flours made perfectly good bread, leading him to reject the "macaroni wheat" label. But Carleton also tried to promote macaroni as the new national dish, criticizing cooks in hotels who did not know how to cook it, filling government scientific bulletins with recipes for semolina pudding, semolina soufflés, and fritters—not surprisingly, he became known as the "macaroni messiah." The editor of the *North Western Miller*, William Edgar, would have none of it, denouncing the USDA in the August 1903 issue for Carleton's observation that durum wheat would overtake other wheat as "effrontery, terrifically appalling . . . in defiance of the opinion of thoughtful wheat raisers and flour makers," and going on to denounce "professors and theoretical farmers," calling scientists who were promoting its high nutritional value as "half-baked."[25]

25. Margreaves 1968, 217.

The journal's hostility carried on for years, not dying away until the end of the First World War.

Farmers on the Great Plains grew hard wheat because it flourished in drought, even though the millers would pay so little for it. Durum wheat prices fell after 1903, and in August of that year, North Dakota farmers denounced the millers and the "wheat ring," and in a meeting gave a vote of thanks to the USDA for their plant introductions. In 1904, black stem rust hit—'Kubanka' survived while others blackened and rotted. Durum slowly increased in popularity as a result; in 1905, twenty million bushels were milled. In April 1908, the North Dakota Durum Wheat League was started, and governor John Burke announced a Durum Day devoted to advertising the crop. Milling technology began to change; a bleaching process was invented so that darker wheats could be made acceptable to a public which equated whiteness with quality. The value of hard Russian wheats was finally established, and from now they could make an impact on the gene pool of American wheats.

In 1916, Carleton published a book outlining his plant breeding methodology and philosophy, *The Small Grains*. In it he summed up the three goals of the contemporary breeder:

1. to introduce new varieties from other regions and assess them
2. to select better individuals from heterogenous populations
3. to cross these individuals and make selections from the progeny

Carleton's career, unfortunately, was cut short by personal misfortune. His daughter contracted a terminal illness, another child died, and he got involved in some wheat-farming speculations in Texas to cover medical expenses; worst of all, he borrowed money from grain dealers of the wrong political party, leading to the USDA sacking him for conflict of interest. He ended up working in South America, trying to repay his debts to old colleagues and salvage his personal honor. He died of malaria in Peru in 1925 at the age of 59.

Australia also became a wheat-growing land, with breeder William Farrer (1845–1906) making such an impact on the agriculture of the country that his face ended up on the two-dollar bill—the only plant breeder so far to be thus remembered. His initial work had been on the rust that had done great damage in the 1880s. In 1901 he released 'Federation,' which was derived from stock descended from 'Red Fife' and Indian varieties; it became the mainstay of Australian farming and also did well in the far-western U.S. states.

At the core of the spread of the wheat across the globe was a redistribution; its cultivation shifted away from the moist maritime areas it had traditionally been grown in to drier continental ones with a shorter growing season—the Ukraine, central Asia, and the Great Plains—or drier but not colder climes—South Africa, Australia, and Argentina. The new breadbaskets were land rich but labor poor, their exploitation only possible because of the railways which made grain exports possible. Actual world yields rose 17 percent between 1886–1890 and 1926–1930; this was overwhelmingly down to breeding.[26] The varieties that made this extraordinary achievement possible were not from the traditional wheatlands of western Europe, but from eastern Europe, Russia, central Asia, and India. Wheat in North America arguably gained more from Russian, central Asian, and Indian wheats than from western European. This redistribution of the crop's geographical variants laid the basis for farther dramatic developments in wheat productivity during the twentieth century.

POVERTY AND THE POTATO—
A NEW CROP IS ADAPTED SLOWLY TO A NEW HOME

The story of the potato and corn is an example of an effect opposite to that we have seen in wheat—that of staple crops of conquered and now firmly marginalized peoples taken up by the European powers and distributed in the opposite direction, from the New World to the Old. Both corn and the potato are remarkably productive crops, added to which they are relatively easy to grow and process, factors which have made them attractive to small or poor farmers the world over. Corn began a steady march through the smallholder plots of Europe and Africa soon after its introduction in the late sixteenth century and had become a staple over much of northern Italy by the seventeenth century. Potatoes, however, had to overcome considerable prejudice in Europe before they became popular, let alone staples.

Potatoes had more than prejudice to overcome. The story of the potato in cultivation is one of the clearest indications of the importance of adaptation to day length—photoperiodism. The first potatoes introduced to Europe in the sixteenth century were members of the 'Andigena' group from the Altiplano of the Andes, between the equator and the Tropic of Capricorn; they were adapted to relatively little change in day length, and in

26. Olmstead and Rhode 2006.

higher latitudes did not begin to form tubers until day length was around twelve hours. In Europe, with its long summer days, this meant that tuber formation did not begin until the autumn, when frost was likely to damage the plants.

Potatoes remained exotic curiosities in botanical collections until botanists had swapped enough seed amongst themselves for variation to throw up some strains which begin to form tubers at longer day lengths. By the late eighteenth century they had become a feasible crop, but one that was regarded with great suspicion by many Europeans as potentially poisonous, vulgar, or belonging to the devil. The first populations who developed an enthusiasm for the alien tuber tended to live in those areas where a mild climate reduced the likelihood of late frosts—such as Ireland, where it became a lifesaver for a peasantry whose numbers were rapidly increasing but who were forced by oppressive English landlords to live on tiny patches of land. Occasional new introductions from South America, such as the Chilean stock from latitudes (and therefore day lengths) closer to North America and Europe, led to new varieties of commercial importance, such as the 'Rough Purple Chile,' introduced into the United States around 1850. Following several blight epidemics during the 1840s, introductions of new varieties were particularly sought after; those from the US-Panama consulate 1851 proved especially valuable for their relative disease resistance.

There was a widespread belief in the eighteenth and nineteenth centuries that potato varieties "degenerated" over time—we now know this is due to a steady accumulation of virus diseases, but at the time it seemed as if this was yet another example of a lack of moral fiber in the plant itself—which added to the moral opprobrium of a crop that for many was condemned merely by not being mentioned in the Bible. In some places, potatoes were regarded with real fear, with an attitude not dissimilar to today's widespread worries over GM food. When famine struck Naples in 1770, many people would not touch the potatoes sent as food aid, believing them to be poisonous and disease spreading. Growing from seed produced virus-free plants, and given the potato's strong tendency to cross-pollinate, this was the main source of new varieties in Europe during this period; however, because of the day length problem, flowering was usually late and seed set poor, so limiting propagation by seed as a route to improvement.

The great tragedy of the Irish famine of 1845–1849 is well-known. The famine was the result of overdependence on one crop. The introduction of the potato to Europe without many of the diseases with which the species would be in equilibrium back home in South America is a potent example of the dangers of human-driven evolution of species outside their native land. European potato varieties evolved and were bred in the absence of potato blight, so their resistance was not tested and was effectively lost. It was not until much later, with the introduction of further potato varieties from South America, that there was any hope of breeding in blight resistance (see chapter 11).

It is also worth pointing out though that much of the Irish peasantry suffered regular malnutrition during the spring and early summer when potato supplies ran out or deteriorated—early maturing potatoes were not bred until after the famine. Indeed, very few potato varieties were grown in Ireland. Most common was the 'Lumper,' originally from Scotland, where it was grown for animal food: high yielding, low quality, reliable, and good in poor soil. Even the Irish said that pigs would eat it last of all. As a Napoleonic War ditty goes, "Our gentry who fed on turtle and wine, must now on wet lumpers and salt herring dine."[27]

Until more genetic material was introduced from South America from the 1850s onwards, would-be potato breeders were very limited in their ability to improve the crop. The ever-inventive T. A. Knight developed a method of preventing tuber formation by lifting the plant's roots out of the ground, which appeared to stimulate flowering and plentiful seed set; he proposed crossing early-maturing varieties with productive kinds to produce new potato ones; unfortunately, he did not undertake any crossings himself, and the method did not appear to have been taken up by anybody else.[28] When more varieties became available for breeding, the strongly outbreeding potato began an orgy of proliferation, aided and abetted by breeders who now recognized that not only was this a crop that was remunerative but that it also left a great deal to be desired. Anxious to put off rivals, breeders were fond of inventing spurious origins for their varieties—stories of gifts of potatoes from sailors being a suitably mysterious enough origin for many.

One breeder who was open about his sources was the man who bred the first early-maturing potatoes. The Reverend Chauncey Goodrich of Utica, New York, who had given up his calling to serve humanity through gardening and plant breeding, obtained some seed from plants that had been part of the American consular introduction of potatoes from Chile in 1851. Within ten years, Goodrich had launched the first true early-maturing potato—'Garnet Chile.' This variety was the progenitor of many more, includ-

27. Quoted in Zuckerman 1998, 140.

28. Knight 1807.

ing the most successful American potato of all time, the 'Russet Burbank' (see chapter 8).

SEEDS OF CONFLICT—VARIETY SELECTION AND IMPERIAL RULE

Much of the contact between imperial powers and the farming populations they ruled was commercial; in many cases the entry of imperial powers into existing societies stimulated—or forced—a transition from tradition and community self-sufficiency to the hurly-burly of the commercial world. This was particularly the case where pressure was put onto farmers to grow crops or varieties of crops that the imperial power wanted, but that were not wanted by native inhabitants. A good example is provided by the Japanese annexation of Taiwan in 1895, an event that was partly driven by a need for access to more agricultural land and was given a major boost by the Rice Riots of 1918, when a rise in prices, caused by shortages, set off rioting in Japan. Taiwanese rice (part of the *indica* rather than the *japonica* group) was not acceptable to the Japanese, however, especially the many red rices, which they particularly despised. Research into improving rice began soon after the invasion; a survey completed in 1910 showing there to be 1,197 rice varieties on the island. At the same time, a Native Rice Improvement Project was started. In a startling example of "the tyranny of taste," the growing of native red rice was suppressed by the police, with the result that in the following decade nearly 70 percent of native varieties were lost—in many ways a disaster, as native rices were notably drought-tolerant. The Japanese introduced their own rice to Taiwan, which grew extremely well, but many Taiwanese farmers continued to grow their own *indica* varieties, most frequently in the off-season.

What in the end happened in Taiwan was actually reconciliation, both genetic and political. Great success was achieved with hybrid rice varieties that were bred from both native Taiwanese and Japanese rice—the *ponlai* rices. These were particularly responsive to both fertilizer and irrigation. Methods of cultivation were somewhat different, but once they had adapted, Taiwanese farmers were able to make substantial gains. The eventual implications of the success of the Japanese intervention in rice breeding and cultivation in Taiwan however was less harmonious; arguably it contributed to the Second World War. With plentiful cheap rice flowing into the country from Taiwan and from its other colony in Korea, Japan experienced what Britain had experienced following the repeal of the Corn Laws in 1846—agricultural depression. Growing and prolonged rural disorder helped fuel farther foreign adventurism (including the invasion of Manchuria in 1931) and the rise of militaristic nationalism.

Another island where the issue of variety selection led to revolt was Mauritius, here, though, events hastened institutional and political change. Mauritius is a small island in the Indian Ocean, colonized initially by the French but then by the British after it was seized in the Napoleonic Wars of 1805–1815. Its checkered history left it with a highly stratified society: a tiny elite with sizeable estates, a mass of small farmers, mostly of Indian origin, and slaves or ex-slaves. Sugarcane was far and away the most important crop, but the elite and the small producers had very different priorities, which crystallized over the choice of variety to use. During the late nineteenth century, large-scale producers organized themselves into a Chamber of Agriculture, which effectively became a second government and, needless to say, dominated the direction of agricultural policy and the research agendas of the Royal Botanic Gardens and later the island's Department of Agriculture. Small-scale producers relied heavily on a group of cane varieties known as 'Uba,' which were easy to grow in a wide variety of conditions, including those less than favored, but had a high fiber content and a relatively low sugar content, which made them unpopular with the processing factories, particularly since suppliers of cane were paid by weight. The Uba group was effectively a landrace, having been selected by the small farmers themselves, not from the "standard" *Saccharum officinarum* but from the more resilient but lower-quality *S. sinense*. Originally the plant had arrived in Mauritius from Durban, Natal, to be trialed by the Chamber of Agriculture; it derived its name from the only letters visible on the rather battered crate the young plants had arrived from Durban in: u-b-a. During the 1920s there were other importations of cane, including natural hybrids between *S. officinarum* and *S. spontaneum*, one of which came by the name of 'Uba Marot'; further hybridization with existing *S. sinense* clones probably happened too.

By 1936, sugar prices had dropped to an all-time low, and the processing factories began to offer lower prices for Uba varieties. Many small farmers were in debt and needed to borrow yet more money in order to replant their fields with cane varieties more acceptable to the processors. Strikes and violence erupted over the island, and one particular factory and estate, Union Flacq, was the target of arson on its fields and an attack that resulted in the deaths of six protesters—as is usual, those in a position of power were better armed and more ruthless. The appointment of a new governor in 1938 helped calm the situation, followed by a commission of enquiry, and political measures which began to draw nonelite groups into the political process. Agricultural extension services were a leading target for reform, with the commission making the failure of the Department of Agriculture to maintain contact with small farmers and recognize their needs a particular point of criticism. The department launched a list of approved cane varieties, which effectively banned Uba, and embarked on an "uba replacement scheme." By 1939, the department was recommending the use of a variety known as 'M134/32' as a stopgap measure while research continued and demonstration plots were established across the island to promote greater fertilizer use.

COTTON, SPINNER OF MANY TROUBLES

Of all the crops of empire, it was cotton which caused the most conflict, and arguably still does today—as is witnessed by the ongoing struggles over GM cotton in India. Cotton, a crop that at least until recent times involved a huge amount of labor in its growing and processing, has always had a great tendency to drive economic and social relations; we should not forget its role in reviving slavery in the antebellum American South. It is also distinctly pest-prone, and issues over pest control have often become politicized. On the one hand there are pressures toward growing pest-resistant varieties and on the other toward using large quantities of often highly toxic pesticides; during the early 1990s, around 55 percent of the total pesticide use in India was on cotton, the result of a massive shift away from native, often pest-resistant, Indian *desi* varieties toward tetraploid nonresistant American-origin ones.[29]

Much of the interest of cotton as a crop goes back to it being the product of several species of the genus *Gossypium*. Old World diploid cottons (*Gossypium arboreum* and *G. herbaceum*) traditionally grown in Asia (since as far back as 5000 BCE) were largely displaced by Spanish and Portuguese introductions of New World tetraploids during the early colonial period (*G. hirsutum* and *G. barbardense*). From the eighteenth century these were in turn displaced by three independently derived types: 'Upland' (derived from *G. hirsutum*), 'Sea Island,' and 'Egyptian' (both from *G. barbardense)*.

Originally there were seven races of *G. hirsutum*, of which one, the Mexican *latifolium*, was introduced into the United States and became known as Upland. There it became long-day tolerant, with genes for short-day re-

29. Rao 1991.

sponse being selected against and dying out. All Upland cottons in cultivation originated in the U.S. cotton belt. Upland production in the United States skyrocketed after the invention of Whitney's saw gin in 1793, which revolutionized the labor-intensive process of separating the fibers of cotton from their attendant seeds. From the United States, Upland also spread rapidly throughout the tropics during the period of the Civil War. Another variety, Sea Island, arose somewhere on the coast of Georgia and the Carolinas during the late eighteenth century and was very rapidly taken up for its very high-quality long staple—in distinction to the shorter staple of the original Upland. A key fact was that machines worked better with long staple material.

In the wake of Napoleon's short-lived occupation of Egypt (1789–1799), a French textile engineer, Louis Alexis Jumel (1785–1823), was invited to Egypt in 1817 to set up a new spinning and weaving mill. Apparently through sheer good fortune, he discovered an unfamiliar cotton in a garden in Cairo, a bush bearing a fiber superior in length and strength to any other currently being cultivated in the region; despite being such good quality it was only apparently used by a few local women. By 1821 Jumel had propagated enough, to produce two thousand bales. The progressive ruler of Egypt, Muhammad ʿAlī Pasha (1769–1849), encouraged Jumel and helped to fund his research, recognizing the value of cotton as a replacement for wheat, which until then had been Egypt's main export. In an early example of a "top-down" or state-led development project, ʿAlī ordered the new cotton (called 'Jumel' or 'Mako') to be grown on a large scale, with major irrigation projects launched to support it. Some hundred to one hundred and fifty thousand *feddans* (a *feddan* is more or less equivalent to an acre) were set aside for cotton; credit and seed were offered to farmers, gins were manufactured and distributed to villages, and experts from elsewhere in the Arab world sent for and distributed around the villages to show peasants how to grow and process the crop. By 1823, exports started and attracted some of the highest prices in Europe.

A law of 1830 (*Laihat ziraʾat al-fallah*; law on the cultivation of cotton) specified how cotton was to be grown, and outlined the duties of government officials in overseeing this. *Laihat ziraʾat al-fallah* was an example of horticultural specifications made law, dictating how seed was to be grown, plants selected, how the cotton should be picked, when plants should be discarded, and so forth. The new crop went down well with peasants, as they earned good money from it—just as well, as most were having to grow it under threat of penalties. ʿAlī's plan was similar in intent to that of the Green

Revolution of the 1950s and 1960s—to jump-start development on a basis of agricultural production, particularly industrialization and universal education; ʿAlī is rightly remembered as one of the most forward-looking leaders the Arab world has ever had. The new cotton may have launched Egypt's reputation as being among the world's foremost producers of fine cotton (a position it still holds), but ʿAlī failed to establish a cotton processing industry; the atmosphere was too dry for spinning and weaving and the use of prisoners rather than free labor greatly reduced the efficiency of the factories. High tariff barriers put up by European powers anxious to stifle competition were also implicated.

In India, generally regarded as the home of cotton in the Old World, there was great variety: Dacca (in present-day Bangladesh) had the very finest cottons, woven from the short-staple *Gossypium arboreum*; they were apparently so fine that when wet with morning dew they actually disappeared. Elsewhere the annual *G. indicum*, derived from hybridization between *G. arboreum* and the originally Persian *G. herbaceum*, also produced good quality material, while the north Indian *G. bengalense* produced a lower quality cloth from a coarser staple. Until the arrival of the British, there was little doubt that India produced the world's finest cotton goods. India's British rulers, however, set out to destroy this industry and replace it with industrially woven cotton from the mills of Manchester.

In the earlier years of the British Raj, native Indian *desi* varieties were branded inferior by the British and cotton farmers were highly taxed, so that there was a strong encouragement to grow high-volume, low-quality varieties, even if farmers were given any choice, which often they were not. Indigenous techniques were shunted aside for the development of a centralized system of ginning and transport, all geared to the needs of the Lancashire mills. Local production was discouraged by a tax on spinning wheels and looms. Much cotton that got to England was adulterated with mud, added by overtaxed and disgruntled farmers, to increase the weight. Even this strengthened the British mill owners' case for long-staple cottons—long fibers can be more easily made to engage with a machine if they break because of adulteration.

Traditionally in India, the length of the cotton staple was not important. However, the processing machinery used throughout the industry internationally had been developed to deal with the long-staple American cotton; the machines did not work well with the short-staple Indian cottons. Indian varieties had been selected over centuries to withstand the vagaries of climate and for pest resistance; many landraces had been developed that coped with local climate and soil conditions. Given that the aim of the British in India was to feed British industry by using India as a source of raw material and as a market for finished goods, it was inevitable that long-staple varieties would be promoted at the expense of the traditional Indian ones.

Side by side with the destruction of the Indian cotton industry was British encouragement of *bengalense* types, which gradually displaced *indicums* during the nineteenth century. American cottons were imported but did not adapt well; many planters recorded accounts of devastating insect attacks. In a good example of nondeliberate breeding, however, some of the original US Upland strains did survive among the *desi* cottons, developing the leaf and stem hairs that protected them against insect predation. From these a new race of Indian Upland cotton was developed during the early twentieth century.

The more perceptive among the British could see what was going wrong. James Dunsmore, a retired cotton merchant, wrote letters to the government between 1823 and 1827 on the subject of the differences between American and Indian cottons, arguing that Indian cotton was superior and that spinning machinery had been invented by people who knew only American cotton: "the manufacturers have no views beyond immediate gain and in general want no change in the material because it would occasion a heavy outlay in new machinery"—his letters remained unacknowledged. Gray, a planter in Madras, noted that "the indigenous plant of India will not thrive well on any land except denominated cotton soil, while on the same soil the plants of the western world inevitably fail." Another planter, Shamrow, wrote that "the government lavishly wasted money on experimental farms . . . upon the chimerical suggestions of would-be amateur experimentalists and ignorant imbeciles." George Watt, a botanist and advisor to the government argued that prevailing trade policies meant that the importance of indigenous varieties was deliberately underplayed, that the "the labors of centuries thrown away," all for the "attainment of high yield of a worthless staple."[30]

The cotton commissioner for Berar and Central provinces, Rivett-Carnac, appointed in 1864, made it his main task to introduce Egyptian and New Orleans cotton. The first monsoon unfortunately wiped out all the new varieties, and so he turned to improving local varieties instead, popularizing in particular the 'Hingunghat' variety and arranging exhibitions for farmers and merchants. At a trade fair in 1869, 'Khandish Hingunghat' was declared the best out of three hundred varieties for its cleanness, strength, length,

30. All quotes concerning Indian cotton are from Shambu Prasad 1999, 5, 6.

brightness, silkiness, and evenness. Eventually even the British manufacturers began to favor *desi* cottons; printers in particular found that Indian cottons took better, as they had more essential oils in their fiber.

Eventually planting policy had to change; during the last half of the nineteenth century, Indian cotton production was increasingly dominated by native varieties. Colonial officials began to make attempts at either the collection and preservation of *desi* varieties or their improvement. Many varieties however were lost, including those that provided the raw material for the renowned Dacca fabrics. After the First World War, cotton improvement and quality control moved up a few notches; crop improvement began to be undertaken by the Indian Central Cotton Committee, which oversaw breeding work at government research stations and the distribution of seed through pedigree supply schemes. The Cotton Transport Act restricted the distribution of lower quality cottons, and distinct new regional varieties began to appear.

As the twentieth century advanced, a variety of opinions began to be advanced about the best way forward for the breeding of cotton. Interestingly, after more than a century of British rule in India, it could no longer be assumed that Indians would promote native Indian varieties or that Britons would be against them. At the First Conference of Scientific Workers on Cotton in India in 1937, an Indian scientist, Ayyar Ramanatha, put forward the suggestion that US cottons should be favored over *desi* varieties, whereas the British J. B. Hutchinson argued that the potential of the *desis* had not been fully exploited, and the American types could not be developed much farther.

The fight against the discrimination against native cottons and taxation on a native textile industry became—not surprisingly—a central core of the Indian freedom movement's campaign against the British. Mahatma Gandhi's wearing of homespun *khadi* cloth, made from short-staple *desi* cotton, in the form of the traditional peasant *dhoti*, and his promotion of spinning and weaving as a home industry were not just romanticism (although there was that too) but part of a well-thought-out and coherent political and economic strategy. By stressing *khadi*, Gandhi recognized the problems faced by Indian small-scale producers and signaled his support for a fundamental reorientation of the industry. *Khadi* became a movement, which still exists today, linking fashion statement with politics; by wearing *khadi*, one signals one's support for a gentle Indian nationalism and sympathy for Gandhian doctrines.

In the 1930s, during the height of the freedom struggle, the *khadi* movement was as much a technical response to the situation of the Indian cotton industry as a political and social one. The *khadi* movement aimed at replacing centralized manufacture with localized industry in accordance with the Gandhian aim of eliminating Indian rural poverty through small-scale and craft-based industry. American long-staple cotton was simply unsuited to *khadi* manufacture (the seeds of varieties of American varieties were too soft, and got crushed in hand gins), whereas *desi* cotton was ideal for home production. The freedom movement set up its own laboratory to test cotton at its Satyagraha Ashram—a center for a range of progressive experimentation and associated with the poet Rabindranath Tagore. An All-India *Khadi* Board (later called AISA, the All India Spinners' Association) was established, with a technical department headed by Gandhi's nephew, Maganlal Gandhi. As well as promoting research into village-level processing and spinning, the *khadi* board supported research into cotton breeding, too, particularly in the promotion of the perennial *Gossypium arboreum*, largely through seed exchanges. *G. arboreum*, being perennial, was easier for farmers to manage, quite apart from its short, thick, and strong fibers being so ideal for small-scale *khadi* production; its level of pest and disease resistance was also good. During conferences organized by the Indian National Congress (the pro-independence movement) there were exhibitions of *khadi*, and Congress workers were encouraged to bring in specimens of cotton, along with details of where they were grown: the soil, climate, and so forth. Maganlal Gandhi argued for science to be applied to an Indian cultural and geographical context, and the ashram laboratory was apparently well equipped with modern textile testing apparatus. Maganlal Gandhi's untimely death in 1928 deprived the freedom movement of one of its most forward-thinking members.

After independence, it might have been expected that patriotism would drive policy towards *desi* production. Not at all, as the government headed by Jawaharlal Nehru, the first prime minister (1947–1964), chose a model of development more influenced by state socialism of the Soviet kind than a decentralized Gandhian one. With the loss of much of the Punjab to Pakistan, from where much cotton came, Indian government policy was aimed at growing enough of its own to reduce imports, with an emphasis on high yielding varieties, which responded to intensive use of fertilizer. The AISA under Dadabhai Naik continued cotton research and vigorously contested the government policy of supporting big mills; it set up a committee to look into "whole cloth." The industry standard for quality definition at the time stressed fiber strength and a "tear test"; instead Naik suggested a "wear" test as being more relevant to how cotton goods are actually used. Krishnadas

Gandhi, the secretary of AISA, tried to get the Indian Central Cotton Committee to research indigenous varieties, arguing that the most durable cloth of the last twenty years had been from *desi* and not from long-staple varieties, which was why his organization was involved in supporting research into pure-line *desi* selection. In the end the milling industry won; they needed long-staple cotton in order to work efficiently, and with a new generation of insecticides, the lack of resistance amongst *G. hirsutum* strains became much less problematic, enabling them to be grown in areas where production had been impossible before. *Desi* and *khadi* were again marginalized—although their popularity (for example, among politicians wanting to stress their roots) never went away.[31] Today they are increasing in popularity through the growing use of organically grown cotton—usually *desi* varieties.

AN APPLE A DAY—THE RISE OF THE FRUIT INDUSTRY

One of the most notable developments during the age of empire was that of an enormous increase in the eating, growing, processing, and trading of fruit. For nearly all of human history, fruit—at least outside the tropics—has played a minor role in nutrition; its short season and the difficulty of keeping or preserving it for any length of time has condemned fruit eating to the brief bonanza of its harvest time. There has been a long history in Europe of selecting apples for their eating or cooking qualities, but on the whole the main emphasis in terms of selection was on those fruits that could be turned into alcohol or dried—and therefore traded over long distances. The grape, first grown in the Caucasia region, was ideal for fermentation, and the plant's naturally high level of diversity resulted in its being developed by the peoples of the Mediterranean basin and west Asia into a great many varieties. The selection of apples for cider and pears for perry also resulted in much local variety selection.[32] It is worth noting that all these fruits tend to produce bud sports, so that there are two roads to varietal diversification: the making of selections from seedlings, and the propagation of chance mutations.[33]

By the seventeenth century, as the populations of western Europe began

31. The author himself was wearing *khadi* the day this period was being researched in the library at Kerala Agricultural University.

32. "Cider" here is understood to refer to fermented apple juice, i.e., "hard cider" in American parlance.

33. See technical note 21: Sports.

to experience a gradual rise in wealth, and given the importance of food as a marker of social status, it was inevitable that there was an increase in fruit eating and growing; the burgeoning gardening literature of this and the next century is particularly rich in material on fruit growing. Fruit's real arrival as an important part of diet and trade occurred during the nineteenth century. Not only were many more people able to afford a more varied diet, but improving transportation—notably the railway and cargo ship—made it possible to shift large quantities of fruit over comparatively long distances. Advances in preserving technologies also made it possible and worthwhile to grow fruit in one place and ship it to another without risk of spoilage. For the first time, whole districts could give themselves over to fruit production, and given the attractive nature of orchards and the healthy and tasty nature of the product, regional identities often became linked to fruit growing. At home, the increased availability and cheapness of both sugar and glass during the nineteenth century enabled housewives to preserve their own fruit through bottling or in the form of jam. So, it should come as no surprise that the breeding of new fruit varieties became of enormous interest from around the middle of the nineteenth century. A particular feature of fruit breeding was its internationalism, in particular, the transfer of seeds and propagating material from one side of the Atlantic to another.

Tree fruits are generally cross-pollinating, which slowed down the development and propagation of new varieties. The Chinese and the Romans were able to graft apples, pears, and vines, so "fixing" the genes of desirable varieties, which could not be propagated through other means of vegetative propagation. Mangoes (*Mangifera indica*), however, were only propagated by the sowing of their stones, which made it almost impossible to ensure the continuation of good clones, at least until the Portuguese introduced grafting in the sixteenth century.[34] The fact that much tree fruit seed when sown produces such a range of sour or inedible fruit was the focus of a great deal of interest to the increasing numbers of experimental gardeners from the seventeenth century onwards.

The celebrated Belgian chemist, Dr. Jean-Baptiste Van Mons (1765–1842), spent a great deal of his life in the work of ameliorating fruits, particularly the pear. The American garden and landscape writer Andrew Jackson Downing recorded that, in 1823, Van Mons's nurseries had over two thousand seedlings of merit, of which four hundred and five were named. Van Mons was

34. L. B. Singh 1976.

influential with a theory that, although incorrect, took a long time to be disproved. He believed that seed from older trees was more likely to revert to producing wild and therefore useless fruit, so he developed a system of sowing seed from young trees, and then sowing the seed from these as soon as it was produced, claiming also that such a succession of sowing resulted in improvements every generation and that in each generation the trees would bear fruit younger. Apples, he said, needed four generations to produce a good variety from an inferior sort, pears five, stone fruits three. There was also a limit, he held, beyond which a fruit could not be improved. He also subscribed to an idea we would now term Lamarckian, believing that it was necessary to "subdue or enfeeble the original coarse luxuriance of the tree"; in order to do this, fruit was to be selected before it was ripe, allowed to rot "in order to refine or render less wild and harsh the next generation"; cutting off of tap roots and pruning also helped, he claimed.[35]

That Van Mons achieved such an impressive reputation during the nineteenth century is perhaps down to his sheer productiveness—which seems to be a pattern in plant breeding history; someone who produces some good varieties and a great many tolerable ones and is evangelical about promoting them is often able to obscure erroneous technique or frequent failure: Ivan Vladimirovich Michurin and Luther Burbank could be seen as being in this mold. Van Mons, like Burbank, clearly believed in propagating on a large scale; with so many individuals to choose from, he then had a good chance of finding something worthwhile. He is quoted as saying "to sow, to re-sow, to sow again, to sow perpetually, in short to do nothing but sow, this is the practice to be pursued; and in short is the whole secret of the art I have employed."[36] Various attempts were made to prove or disprove him, but by the end of the century, Knight's discussion of crossbreeding had been held up instead as not only a much better way of producing new varieties but also somehow a more accurate description of reality. At the time it was generally held that the crossbreeding of tree fruit was first practiced in Holland, with the branches of different trees being held close to each other to encourage cross-fertilization.[37] By the early years of the twentieth century, Van Mons's work had been thoroughly rejected; "he placed himself in the unfortunate position of conceiving a theory and setting out to prove it," in

35. See technical note 22: Lamarck. Quoted in Downing 1869, 6.

36. Quoted in Webber and Bessey 1899, 469.

37. There is comparatively little on Van Mons; my sources were Talbot 1882; Gourley 1922; and Warder 1867.

the words of one writer who continued to recognize his place in history; "however in doing so, he greatly stimulated the science of plant-breeding; and although his theory was without foundation, the net result of his work is a landmark in the progress of the origination of new varieties of fruits."[38]

Much of the progress made with improving fruit during the eighteenth and early nineteenth centuries was not made, however, through self-conscious breeding programs like that of Van Mons, but through sharp-eyed gentleman farmers or landowners taking advantage of chance discoveries: seedlings that cropped up in tenants' gardens or sports from established fruit trees.

The biggest breakthrough in apple breeding—and arguably with certain other fruit, too—was the taking of apples to the New Worlds of North America and Australia. Things, however, did not start well; European apple varieties often died or never produced in the more severe climates they encountered in North America. Knowledge of grafting was limited among the settlers, so there was widespread propagation by seed, augmented by the spontaneous germination of pips from frontier orchards, resulting initially in a lot of poor-quality and inedible fruit. However, it appears that much early American apple growing was for cider production, where the qualities of sweetness, durability, and good appearance were much less important than they were for the dessert table or cooking. Thoreau's comment on such apples—"sour enough to set a squirrel's teeth on edge and make a jay scream"—could probably have been made about a great many nineteenth-century American apples.[39]

No story of American apple growing can be made without mention of one very well known supplier of cider apples—John Chapman, better known as "Johnny Appleseed" (1774–1845). He grew apples from seed, establishing nurseries that he would then leave in the hands of local farmers, before moving on to follow the next batch of pioneers, usually with a boat- or wagonload of seedlings. In this way he set up a chain of nurseries from western Pennsylvania to Ohio, and when he died he left 1,200 acres of orchards. Clearly a drifter, he never established a home but appeared to have had good business sense and organizational skills. He was a vegetarian and regarded it a sin to fell a tree or ride a horse, and apparently he preferred the company of Indians or children. He was renowned as a philanthropist who worked at peace with the Indians, as a healer, and as a (rather unorthodox)

38. Gourley 1922, 312.
39. Quoted in Pollan, 2001, 9.

evangelist. Chapman would have nothing to do with grafted trees; it is reported that he had said, "that [grafting] is only a device of man . . . and it is wicked to cut up trees that way. The correct method is to select good seeds and plant them in good ground and God only can improve the apple."[40]

The mass growing of apples from seed, by Chapman and others, unleashed an avalanche of genetic diversity and recombination of genes, with a whole range of new varieties the result—seven thousand by the end of the century.[41] From an early stage, the traffic in varieties began to move back across the Atlantic; today many American apples are popular in Europe, for example, 'Jonathan,' 'Delicious,' and 'American Beauty' varieties;. Benjamin Franklin noted that the 'Newtown Pippin' apple found in Flushing, New York, had already spread to Europe by 1781.

Taste and appearance were inevitably important to fruit breeders, but so too were carrying quality, longevity, and size—small fruit took longer to pick and had a higher proportion of unwanted pip and skin to flesh. Given that many tree fruits were introduced into regions at the edges of their viability, a key issue in breeding has always been hardiness and time of flowering. North American breeders, particularly those in the northern Midwest with its viciously cold winters, have had a history of being interested in fruit varieties or species imported from Russia. Professor Joseph L. Budd (1835–1904) of the Iowa Agricultural College visited Russia in 1882 and came back with a sizeable haul of fruit varieties and spent much of the rest of his life breeding new fruits for the plains.[42] The use of genes from locally native species has also been tried; Niels E. Hansen (1850–1900) at South Dakota State College of Agriculture and Mechanics worked with native Midwestern sand cherries, *Prunus besseyi* and *P. munsoniana*, in an effort to produce a new race of plums (*Prunus domestica*) in the early twentieth century.

Little serious attention was given to apple breeding in the US until the latter half of the nineteenth century, when there was an interest in new varieties which has been described as the 'Great Apple Rush,' with country people on the lookout for new varieties that might make it big. The key event which brought about the change from mass planting of seedlings to the selection of quality varieties seems to have been the discovery of the 'Wealthy' apple in Minnesota. Frustrated by the inability of any apple to survive the bitter winters of his new home, pioneer nurseryman Peter Gideon

40. Quoted in ibid., 16.
41. Fowler and Lower 2005.
42. Hedrick 1911.

(1820–1899) tried breeding with apples of Russian origin but found that they produced poor quality fruit that was susceptible to fire blight. Eventually, during the 1860s, he did come up an apple that not only survived the winter but tasted good and kept well—until May. He called it 'Wealthy' after his wife's name. Like much of Gideon's life, the details of how he bred the Wealthy are obscure; in later years he claimed he and his family were on the point of destitution and he was on the point of leaving the state, when, in his words:

> I resolved to leave Minnesota and with this resolve I went to bed after coming to this decision and was meditating as to where to go and how to get there, when, to my surprise, a strange man from the spirit world stood by my bedside and told me to write to a certain man in Bangor, Maine and I would get scions and seed from which I could grow trees that would stand in Minnesota.[43]

Gideon went on to produce many other apples, but none came to be as highly regarded as 'Wealthy,' which in actual fact, as research by Professor Budd in time revealed, did turn out to have Russian parentage. The apple made his reputation and got him an official position as manager of the Minnesota State Experimental Fruit Farm, established right next to his own property on the shores of Lake Minnetonka. Such was the fame of the nursery and state farm that families would come on steamer outings to view them. Gideon was not an easy man, however, and he alienated friends and colleagues with his deeply held and sometimes eccentric opinions, which he seemed unable to separate from his work as a nurseryman and breeder; reports to the Minnesota Horticultural Society would be peppered with his views on horse racing (against), theories on "animal magnetism" (for), and spelling reform (for).

From this point on, there was no stopping ambitious Americans who were looking and hoping for new apple varieties. In an age defined by gold rushes and a raw entrepreneurial energy, even a humble apple could be a source of riches. Apple names often reflected the hope and expectation of their finders: 'Westfield Seek-No-Further,' 'Twenty Ounce Pippin,' 'Bread and Cheese,' or 'Paradise Winter.' One such apple that only just made it was a seedling that kept on appearing in the ground of an orchard in Peru, Iowa, during the 1860s; it was in between the rows and therefore in the way. Despite the efforts of the owner Jesse Hiatt to hack it down, it kept on re-

43. Quoted in Curran 2004, 160.

sprouting. Finally taking its continual reappearance as a sign—"if thee must live, thee may"—Hiatt transplanted it and then discovered one of the best flavored apples he had ever had. He called it 'Hawkeye' and sent four off to the Stark Brothers Nurseries in Louisiana, Missouri, who ran an annual competition for new varieties. It won and was renamed the 'Delicious.' However, the name of the entrant was lost and the nursery had to wait another year and hope that it was re-entered—which it was. Its good appearance, distinctive flavor, good shipping qualities, and relative disease resistance ensured a great future.

Paul Stark of Stark brothers had a reputation as an "aggressive pomologist" and would go to almost any lengths to get a good new variety. In 1914 he set off on a journey by narrow-gauge railway and horseback to find the source of an apple that he had been sent by one Anderson Mullins in Clay County, West Virginia. Mullins had noticed a precocious seedling with large yellow apples in his garden in 1905 and that it bore well in nine years, even when others failed. Stark realized its potential from its exceptional flavor and how well it had kept on its journey. It is thought that it had originated from a cross between 'Golden Reinette' and 'Grimes Golden' and was possibly of Johnny Appleseed origin. Stark paid Mullins $5,000 for the tree and a nine-hundred-square-foot tract of land around it, which he had protected with a woven-steel cage plus a burglar alarm. For thirty years the company sent Anderson's nephew Bewel Mullins $100 a year to look after the tree and keep records. The ancestral 'Golden Delicious' tree proved very adaptable and was used in many breeding programs; it expired in 1958, unlike the 'Delicious' tree, which has recovered from severe winters on more than one occasion and is still alive.

THE STRAWBERRY—A SOFT AND JUICY STORY CONTINUES

Apples have had a long history in cultivation, for while their widespread trading and transport is relatively recent, their keeping qualities have always ensured that country people and the wealthy have found it worthwhile to continually select new varieties. The same could not be said of soft fruit, which until the invention of food preservation technology could not be kept for more than a few days. Interest in soft fruit as delicacies for the tables of the elite became serious in the eighteenth century and expanded exponentially in the nineteenth. The reasons included improved transport and preservation and—at least in the case of the two most important, the strawberry and the raspberry—the fact that breeding had brought about huge changes in what these plants were capable of.

In the case of the raspberry (a complex of hybrids of *Rubus*), a major breakthrough was the occurrence of natural hybrids between European and American raspberries in gardens in both the United States and Europe where different species were being grown together; the hybrids combined the eating qualities of the European *Rubus idaeus* with the drought and cold tolerance of the American R. *strigosus* and *R. occidentalis*. Deliberate breeding in raspberries has gone on to contribute genes for disease resistance and drought from other American species, such as *R. parvifolius*, *R. biflorus*, and *R. kuntzeanus*.[44]

A key stage in the development of plant breeding is the realization that a plant may be improved by going back to basics, looking over the garden wall or trial plot fence and going back to nature for new genes, perhaps even completely remaking a crop or horticultural plant from wild relatives. In chapter 4 we saw how the strawberry played a crucial role in the development of modern ideas about flowering plant reproduction. It also played a pioneering role in the use of genetic material from the wild and in systematic hybridization.

The mouth-watering prospect of getting two strawberry crops a year stimulated some of the earliest and most widespread attempts at systematic hybridization: in Germany, France, and England and its American colonies. Genes for this delightful prospect came from the Californian strawberries, a race of *F. chiloensis*, introduced to Europe in the middle of the nineteenth century (see chapter 4), which, after its crossing with European strawberries, produced a number of hybrids. Of these, 'Vicomtesse Héricart de Thury' was the first widely cultivated variety to produce two full crops a year. One of its parents was 'Elton' bred in England in 1819 by T. A. Knight; he reported that his experimental strawberry garden included over four hundred hybrid lines in 1818—surely one of the earliest examples of extensive variety trialing.

Strawberries crossed relatively easily; indeed, the fact that until the mid-nineteenth century different male and female plants (of necessity genetically different) needed to be crossed to produce a decent harvest must have been a major factor in encouraging hybridization. Furthermore, the short life of strawberry plants encouraged growers to raise them from seed, which in

44. Jennings 1976.

turn encouraged more genetic mixing and variation. Very little of this hybridization, though, appeared to be planned; most growers mixed together their favorite varieties and sowed whatever seed was produced. As small plants which matured quickly, they were easy to trial, and there was little incentive for anything as time-consuming as hand pollination. A good plant breeder at this time was basically someone who was prepared to grow a lot of plants and who had keen powers of observation. Many growers named their varieties, which soon resulted in profusion, and no doubt confusion. In the United States, the Prince family of Long Island were pioneer hybridizers, establishing a nursery in 1830, which later became a center for horticultural and botanical research—the Linnaean Botanic Garden. The Princes, father and sons, imported all the strawberry varieties they could lay their hands on; one son, William Prince, wrote one of the first American gardening books (*Short Treatise on Horticulture*, 1828); another, William Robert, grew over ninety varieties, many of which he himself had originated.[45]

The old problem of dioecious plants reducing pollination and hence fruit set acted as a block on large-scale strawberry production, but by mid-nineteenth century the more forward-looking growers understood the need for a hermaphrodite, and hence self-fertilizing variety. It was James Wilson of Albany, New York, who broke through this important barrier. He knew what he was looking for, and yet did not plan any crosses, simply planting together seedlings of a number of what he considered the best available varieties—'Ross Phoenix,' 'Hovey,' and 'Black Prince'—and let them get on with it, collecting the seed and making selections from the progeny. In 1852 he launched the first true hermaphrodite strawberry, 'Wilson's Albany,' often just known as 'Wilson.' The plant took off with a vengeance, as it eliminated the need to ensure that different male and female varieties were intercropped; good crops of the juicy red fruit were a certainty. Now anybody could grow strawberries, and they did; agricultural historians have even described the period between 1859 and 1870 in the United States as "the strawberry fever" years.

'Wilson' was not without its critics, and looking back at it today, we could perhaps say that it was the first mass-market-named vegetable or fruit variety, with all the accusations directed at it that have been aimed at mass-market varieties ever since—at 'Golden Delicious' or 'Red Delicious' apples, for example—widely derided as tasteless. 'Wilson' had all the attributes of a perfect crop for growers, especially large-scale ones, and few of the char-

45. Additional material on the nursery may be found in DeWan 2007.

acteristics that made it a favorite on anyone's table. The fruit looked good (firm, well-shaped, and dark red), it cropped dependably and did so for many years, it was easy to pick, and it appeared indifferent to soil conditions or the peculiarities of local climate. But "indifferent" is how it seemed to many consumers—tart with a poor flavor, the forerunner of what many consumers would say has been one of the overriding features of modern food production: the concentration on varieties that are easy to grow, easy to pack, and have a long shelf-life, but taste of . . . nothing.

The wild has always been an important source of fresh genes; once it was realized that many cultivated plants were composites of related races or species from a variety of origins, it made sense to deliberately introduce them as a source of variety. Albert Etter (1872–1938), an early-twentieth century strawberry breeder in California, turned to the wild strawberry (*F. chiloensis*) that grew along the beach at the nearby Cap Mendocino. With it as a parent he raised several varieties, including 'Ettersburg 80,' which he produced around 1910 and described as being "as resistant to drought as a young cactus"; its value was recognized by growers all over the world, even if no one really believed him.[46] A few years later, it appears that he made a variety with a genetic makeup completely different than the 'Ananas': 'Ettersburg 121,' derived from the European Alpine strawberry (*F. vesca*) and a *F. chiloensis* from the beach. Etter paid close attention to usage in his breeding; this was a time when California was establishing its fruit canning industry. He classified new varieties on the basis of a whole list of characteristics, including suitability for shortcake, cake topping, jam making, and "canned fruit for the sick room."

Etter's attention to such details was part of a whole new precision in plant breeding that was clearly related to the increasing specialization of function in agriculture and commercial horticulture. By this time, not only had the profession of "plant breeder" been recognized, but breeding was becoming increasingly separated from that of planting stock production—California's wizard breeder, Luther Burbank, grew very little on himself, preferring to sell newly developed varieties onto nurserymen who propagated and sold them to both commercial fruit growers and the public. Those who grew the crops were increasingly no longer involved in producing or selecting new varieties; as a consequence those involved in plant breeding needed to formalize and articulate the means by which they assessed the value of new varieties. How well they did so was a reflection of how closely they understood

46. Quoted in Wilhelm and Sagan 1974.

the needs of the growers. Productive, the breeders certainly were—by 1925, one publication listed 1,362 strawberry cultivars in the United States.[47]

A FREE-FOR-ALL—THE ARRIVAL OF THE SEED CATALOG

Looking through late-nineteenth-century or early-twentieth-century seed catalogs or the journals written for gardeners or farmers, one thing becomes obvious immediately: the enormous number of new plant varieties and the enthusiasm with which they were being promoted. Looking through letters from readers, it is also apparent that many of the new varieties were not quite satisfactory. The seed and rapidly expanding nursery trade, feeding on growing wealth and a desire to experiment and improve, was a dirty business. The selling of a product that would take the buyer at the very least months to assess was a field that was almost an invitation to fraud. New varieties could be promoted way beyond their worth, old ones given new names, or seed could be adulterated; in Britain it was such common practice to add killed mustard seed to higher-value but identical-looking cabbage or cauliflower seed that special seed-killing machines were advertised for sale.[48] There were no regulations, no official trials, no growers' associations to discipline their members. It should come as no surprise that the later decades of the nineteenth century began to see a growing level of both government regulation of the seed trade and of self-regulation by the industry itself. Until this became established, which took decades, the business was a free-for-all.

In a world where rapidly expanding communications made it possible for people to find out about and try new plants, and where there was, for the first time in human history, a widespread understanding that plants could be improved and developed and changed, it was not surprising that there was a lively market in novelties. "New" plants could be found anywhere on a spectrum from the genuinely scrumptious and prolific to the totally fraudulent. In the middle there might be those new varieties sold in all honesty by the hopeful and the naive—like the unnamed farmer who found an odd-looking cotton in his field in Alabama sometime in the 1830s; the bolls grew directly from the main stem rather than side stems. Growing on the seed the plant produced, the man in question began to sell the seed as 'Twin' or

47. Fowler and Lower 2005.

48. Earley Local History Group 2006.

'Okra' cotton; in 1837 the seed was selling for 59¢ *each*; in a few years, however, everyone realized it was worthless—better neither in yield or quality than others.

A good new variety could have an international and immensely long-lived impact; its name could even enter the dictionary—like the 'Williams' Bon Chrétien' pear. Writing in 1816 in the *Transactions of the Horticultural Society of London*, the eminent botanist William Hooker discussed a pear seedling that had been found some twenty years previously in the garden of a Mr. Aiton in Berkshire, England. Aiton had grown it on out of curiosity and liking it, gave it to a local nurseryman by the name of Williams—who, as was usual at the time, named it after himself, hoping to promote his business.[49] It was so successful that it is now grown all over the world and sets the standard for the ultimate in juicy and tasty pears, the kind that dribble juice all over your chin as you bite into them. What is more, in Eastern Europe, where almost any fruit is liable to be fermented and distilled into a potent firewater, it is the mainstay for the production of a pear-based spirit; in Slovak this is actually known as 'Williamsovka.' Introduced to the American colonies, it lost its name, so Enoch Bartlett of Massachusetts started to sell it under his own name, and as the 'Bartlett,' it went on to become a great canning pear.[50]

More common was the marketing of established varieties as "new" ones, of good new varieties being grown on by somebody else and then rebranded, or seedsmen needing to make their way in the world growing on minimally different varieties as "new," with their names attached to the product—the more that could be marketed the more their name would be seen. There was often a poor understanding of what value in a new variety meant, like the Okra cotton, and new varieties were often promoted on the basis of curiosity value rather than being better quality or easier to grow. It was the failure of so much that was being sold that lay behind the quiet fury of so many of the letters to the magazines. A good new variety might, after a few years, be found bearing as many names "as there are persons who desire to make money by selling the seed," in the words of an 1855 writer.[51] As an example, 'Banana Cotton,' originating in Warren County, Mississippi, soon showed up all over the South as 'Boyd's Cluster,' 'Boyd's Prolific,' 'Wash-

49. Hooker 1816.
50. Roach 1985.
51. Quoted in J. H. Moore 1956, 102.

burne's Olive,' 'Hogan,' 'Hebron's Banana,' 'Mitchell's Pomegranate,' and many more.[52]

From the late nineteenth century on, seed companies began to play an increasingly important, if not dominant, role in breeding non-cereal crops and a major role in producing varieties for market gardening and for private growers. The production of new cereals was a somewhat different matter—the fact that they were so vitally important for national food supplies and involved large-scale and long-term work made it more likely that they would be the concern of government. There were exceptions though, one being the family firm of Gartons of Warrington, Lancashire, in the north of England. Their production of cereals—oats in particular—was appreciated as internationally important during the latter quarter of the nineteenth century and the early years of the twentieth; Mark Carleton visited them in 1898 and was reportedly astonished at their work. Garton varieties were widely exported throughout the British Empire (then by far and away the world's largest political unit) and the United States. That private companies could be so effective in breeding cereal grains indicated that there was no link of necessity between their improvement and the publicly funded research that was to so dominate this sector over so much of the next century.

SUMMARY

The development of agriculture worldwide from this point on was also linked to the introduction of new crops, as the Old and New Worlds undertook a vast exchange of cultivated species. Europeans may have been the agency for this massive genetic transfer, but much of the initial breeding was carried out by non-European farmers, who turned exotic and alien new crops into reliable landraces. Many crops new to Europeans were cultivated in botanical gardens and other institutes very often in their overseas colonial possessions; primary selection work was carried out. A similar process occurred in the late nineteenth and early twentieth centuries as Americans got to work on new crops, but on their own territories. Colonial intervention in Asia, Africa, and the Pacific often involved using native populations to grow crops which primarily benefited the populations of the occupying power; plant breeding priorities were inevitably distorted, sometimes fuelling the resentment of subject peoples.

Among the greatest beneficiaries of this global exchanges were the cereal

52. J. H. Moore 1956.

crops: corn, wheat, barley, and rice. Wheat in particular, the grain of choice for the empire-building cultures of Europe (and perhaps one of the reasons for their historical success), became the subject for much early experimentation in improvement, first in Europe, and then in the United States and Canada. Its success in the New World kept prices low and helped feed the urban populations of Europe. The potato also achieved great success in Europe and Asia, but only after major disease problems were overcome. The successes of wheat and potato cultivation helped reduce the price of staple food for many, and linked with a rising standard of living for many populations, other food crops grew in popularity, particularly fruit. This, and the invention of new transport and preservation technologies encouraged a great expansion in fruit growing and breeding.

SIX ❦ BREAKTHROUGH ❦ GREGOR MENDEL

"They'll revert, you know," says the flower show customer to her companion. "After a few years they'll go back to yellow, it always happens." Popular wisdom—or misconception—is always interesting to compare to reality but also to try to understand scientifically. The human race seems to veer between *essentialism*—believing that certain characteristics are built in when they are not, as with much racist and eugenic thinking—and the opposite thinking that characteristics can easily be changed when in fact they are genetically determined. That one plant variety can transmute into another is an example of the latter way of thinking, linked perhaps to the belief which persisted until Pasteur, that life, in the form of worms or vermin, is spontaneously generated by decaying matter. It is not uncommon for a plant such as a polyanthus or a columbine to be bought, planted, and enjoyed for a few years, and then . . . somewhat mysteriously, a few years later, there it is, but a different colour. What usually happens is that the plants set seed, and their seedlings replace the parent plant (often a short-lived species anyway), creating the illusion of transmutation.

This is the nearest many get to seeing Mendel's concept of segregation actually happening in practice, as genes are recombined in reproduction, and colors or other characteristics not visible in the original plant are expressed in its seedlings. The general tendency is for plants to "revert," because the genes that determine their appearance as wild plants are the dominant ones—so polyanthus (*Primula* Pruhonicensis[1]) become yellow, like the wild ancestral cowslips (*P. veris*) or primroses (*P. vulgaris*), and columbines blue, like the wild *Aquilegia vulgaris*.

The story of Gregor Mendel is well-known. Between 1856 and 1863, he conducted extensive trials with pea plants, growing over twenty-eight thousand of them, which led him to draw up his famous Laws of Inheritance. It was the genius of Mendel to recognize traits to study that are obvious, not influenced by the environment, with clear cut descriptive categories, and the under control of one (as yet to be discovered) gene. A great many char-

1. The scientific name reflects the complexity and confusion surrounding the genetic origins of what are commonly known as *polyanthus primulas*.

acteristics are controlled by several genes, often resulting in quantitatively expressed characters in the phenotype that do not result in such clear phenotypes, or are likely to be influenced by the environment.[2] Mendel avoided these; it was this ability to recognize the simple and the immutable that led Mendel to his discovery. Mendel's laws do not tell the whole story of inheritance, but they describe accurately the bare outlines with remarkable clarity.

We know only so much about Mendel's work—many of his notes were destroyed after his death by his fellow monks, possibly out of ignorance but also possibly because of the desire to suppress material regarded as dangerous by a conservative religious hierarchy. A few notes survive of some practically orientated plant breeding by Mendel that involved a planned program of apple improvement aimed at combining good eating characteristics with resistance to cold winters. He is also known to have exhibited vegetables as early as 1859 as a member of the Horticultural Section of the Moravian and Silesian Agricultural Society.[3]

As we have seen, many involved in breeding work, such as T. A. Knight and Wilhelm Rimpau, almost reached the same conclusions as Mendel did. However, the fact that he was working in an environment without commercial pressures enabled him to concentrate on the pure science and grasp an essential truth that eluded others who were more focused on breeding better plants. Initially little notice was taken of his work by contemporaries; his paper "Experiments in Plant Hybridization," read at two meetings of the Natural History Society of Brno (then generally known by its German name of Brünn) in 1865, was only cited three times over the next thirty-five years. However, in 1900, in a remarkable example of the zeitgeist at work, Mendel was rediscovered independently by three researchers: the Dutch botanist Hugo de Vries (1848–1935), the German botanist and geneticist Carl Correns (1864–1935), and the Austrian agronomist Erich von Tschermak-Seysenegg (1962). The effect was truly galvanizing; from that time on, the considerable number of people scattered over several countries who were working on inheritance and animal and plant breeding had a focus, a theory to test, experimental trials to replicate, and a set of principles to argue over.

Mendelian genetics were not immediately taken up by all involved in inheritance studies or plant breeding. Indeed, it could be said to have taken several decades before it was accepted by everyone working in the field.

2. See technical note 23: Phenotype and Genotype.

3. Orel and Vávra 1968.

The fact is that inheritance is a highly complex field, and the interactions of genes often obscure the underlying patterns. Those who were not inclined to believe in Mendel could find plenty of evidence to support their skepticism. Many before Mendel were also making clear progress with a variety of approaches to breeding work, including hybridization, and some would continue to do so after his work was publicized, but without reference to him. It is clear that the reactions to Mendel's work were quite different in different countries. Nowhere was the reaction more wholehearted than in the United States.

SOWING THE PRAIRIES AND THE PLAINS—PLANT BREEDING IN THE LATE-NINETEENTH-CENTURY UNITED STATES

It should come as no surprise that plant breeding should have occupied such an important place in the young United States. As the nation's enormous range of environments, many of them very testing and unfamiliar to people of European origin, were being filled with new farms and farmers, huge demands were made on existing crop varieties. Cereals were needed that survived drought and unfamiliar fungal diseases, as well as fruit that set reliably in a short growing season and vegetables that survived a long rough ride to distant markets. As we have already seen with apples, varieties from "old Europe" were often not up to the mark. In addition there were new crops to try, unfamiliar ones that might survive harsh conditions or flourish in climates verging on the tropical.

From the 1860s on, agricultural research began to be given land, finance, and political support at levels arguably more generous than has happened anywhere else in history. The year 1862 saw the foundation of the USDA, and the Morrill Land-Grant Colleges Act passed control over certain federal lands to colleges so that they might teach agriculture and certain other land-based disciplines. Then in 1888 the Hatch Act funded an agricultural experimental station at each land grant university. In 1890, a second Morrill Act set up colleges, primarily for black students in the South. The new institutions resulted in the practical techniques of agriculture and horticulture being studied alongside basic sciences as botany, entomology, and chemistry. Teaching and research, all too often separate, were thus integrated.

For much of this period it was farmers themselves who undertook plant selection work, picking out the best to sow next year, as farmers have always done. From the mid-century on, a growing network of county agricultural fairs provided an opportunity for farmers to distribute good crop strains more widely. Farm journals and almanacs also popularized the idea of improved varieties and offered advice to their readers on how to go about making seed selections.

Wheat production during the 1870s was particularly good, to the point where prices became depressed—one factor among several that led many to think of diversification; if new crops could be found these could reduce imports and provide new ones for export. Federal administrators and politicians looked to science-based agriculture to help achieve this. Scientific expertise as a result began to be seen as part of what made an agricultural expert—itself a new concept—scientifically led interventions in farming being what such experts did. Hybridization of crops was one such intervention.

A key figure in the agricultural and horticultural history of the period is Liberty Hyde Bailey (1858–1954), who was born on a farm in South Haven, Michigan. Bailey's father was famous for his orchards—he once exhibited 320 varieties of apple, all grown on his land, at the local South Haven and Casco Pomological Society. The younger Bailey started to graft at age ten, and as a teenager he became so skilled that he was employed all over the neighborhood. As a young man he read Darwin's *On the Origin of Species*, which made a deep impact. He entered Michigan Agricultural College (later Michigan State University), where his genius for plant study was soon recognized by William James Beal (1833–1924), a professor who had been a student of Asa Gray (1810–1888), a key early proponent of Darwinism and a pioneer in the laboratory method of teaching botany. Beal had himself been involved in testing the yield of open- and cross-pollinated corn and the systematic trialing of potatoes.

Bailey's career took him to Michigan State Agricultural College in 1885 and then Cornell University in 1903 as dean of the Department of Agriculture. Bailey was clearly a polymath, one of those figures able to immerse himself in several disciplines at once and see the links between them. He was instrumental in developing agriculture as an applied science, at the heart of which was his passion for education. Bailey developed a system that integrated students and faculty members into practical fieldwork and research based on sound scientific principles. Gone was the European division between the high and mighty professor and the students who sat at his feet. Gone too was the idea of agriculture and horticulture as crafts, to be learned through years of practice. Instead, students learned about basic scientific principles and undertook experimental work; with a sound grasp of principles, they could then approach new problems in their professional lives in a rational and systematic way.

Plant breeding was at the core of Bailey's work; indeed, it can be said that he invented the term "plant breeding" and created the discipline and identity of plant breeding as a profession.[4] Students were taught the essentials of breeding, including hybridization, within a context of learning about classification and nomenclature. Bailey's own work in breeding began in 1886, with a study of cucurbits, members of the *Cucurbitaceae*, a family on which he became an authority and remained fascinated with all his life. For him, melons (*Cucumis melo*), pumpkins and squashes (derived from *Cucurbita pepo* and *C. moschata*), and gourds (various members of the *Cucurbitaceae*) combined economic value with being good—and beautiful—experimental subjects. His experimental work and contact with growers all over the United States led him to believe that breeders could not be expected to be generalists but instead must focus on particular varieties for different soil and climate zones. This became a central tenet in plant breeding for more than half of the twentieth century, only being overturned by Norman Borlaug's shuttle-breeding of wheat in the 1950s (see chapter 12). In order to have the range of potential needed for different environments, a crop must, he argued, maintain a wide and diverse level of variation. He was particularly critical of workers in some experimental stations who were trying to reduce the number of varieties. He also believed that fruit and vegetable varieties had to be constantly improved—"I look upon new varieties as so many starting points for still farther development, not as final or permanent things in themselves . . . . [T]he amelioration of the vegetable kingdom is a slow unfolding of the new out of the old, through the simple and quiet agencies which man employs in cultivation and selection."[5]

In 1896, Bailey addressed the Philosophical Club of Cornell with his arguments that the evidence of deliberate breeding was for him the strongest proof of evolution. "I assume you all believe in evolution . . . the proof which appeals to me most strongly is the fact that gardeners and breeders have it in their power to make new forms and that they have been making them since man began to deal with plants and animals."[6] He went on to argue that plant breeding was a field of enormous promise, as the origins of most new forms

4. Perkins 1997.

5. Quoted in Rodgers 1949, 128.

6. Quoted in ibid, 258.

lay in random and chance discoveries, which indicated that the advent of deliberate breeding would deliver much more.

Bailey was a prolific writer, producing over seven hundred articles and more than sixty books, ranging from elementary school textbooks to volumes of poetry. Toward the end of his life, he began to turn more and more to the kind of homespun philosophical musings typical of many American figures of this period; he was clearly influenced by the currently fashionable idea of transcendentalism, a belief in the importance of a personal spirituality free from the framework of organized religion. He outlined his ideas in a number of books, of which *The Holy Earth* (1915) was the most important. In many cases, his ideas, poetically and passionately put, presage those of modern environmentalism. Science and evolutionary theory, he believed, did not threaten religion—but then, his spirituality was a pantheistic approach to nature and God; for him, God was revealed through nature and nature study.

Bailey's identification of plant breeding as a distinct discipline ensured that he took a systematic approach, one linked to outlining its principles and methodology and the importance of imparting these principles to students and colleagues. In 1891, at the age of 33, he published *Philosophy of Crossing Plants*, and three years later the first book-length work in English on the subject, *Plant-Breeding, Being Five Lectures upon the Amelioration of Domestic Plants*, which brought together material from his lectures and several previous publications; it went through edition after edition and was translated into a great many languages, including Chinese and Japanese. In the third edition, published in 1902 he incorporated a discussion of Mendel. A passage in his preface to the fifth edition, published in 1915, is revealing:

> To one coming out of a plant-growing relationship, the masterful works of Darwin had introduced order, and the forms of cultivated plants had been made worthy of serious study. . . . All these writings were fascinating to read. How to produce new forms of vegetation seized some of us with irresistible power.[7]

This is a statement that reveals a truly Baconian vision of humanity's relationship with nature. To those who see having power over nature as being inherently troubling, this might be read as the statement of a megalomaniac threatening global domination over the ecology of the planet, but for those who saw themselves as having responsibility for feeding and clothing the

7. Quoted in Perkins 1997, 54.

earth's growing population, and, more keenly felt at the time, the vastly increased numbers living in cities, this was a brave and confident clarion call of intent.

A key element in the development of U.S. agriculture in the late nineteenth and early twentieth centuries was the introduction of new species or new varieties from abroad. In 1897, the USDA set up the Office of Seed and Plant Introduction to introduce as wide a range as possible of potential crop plants for all regions of the United States. As well as this office sourcing material directly, consulates were instructed to send in plant material. Expeditions were also mounted—nearly fifty in the first two decades of the twentieth century. Seeds and propagating material were distributed among colleges and experimental stations and to farmers and growers among the general public, including varieties of hops (*Humulus lupulus*) and barley from Bavaria, edible-podded peas from France, radishes (*Raphanus sativus*) from Japan, figs from Turkey, oats from Sweden, hemp (*Cannabis sativa* subsp. *sativa* var. *sativa*) from Egypt, and much, much more.

Much of the USDA introduction program was part of the attempt to develop crops that had never before been grown in North America. As the United States was settled from coast to coast, new climate zones were opened up. Growing unfamiliar crops was, initially at least, a very hit-and-miss business; success depended on trialing and identifying promising varieties and then improving them. Rice was being planted in Louisiana; flax in the northern states; and citrus and other Mediterranean climate fruit, such as loquats (*Eriobotrya japonica*), avocados (*Persea americana*), and figs, in California and Florida. In many cases introduction was only the first stage—a strong tendency was for American and Canadian growers to push the limits of what was possible, so apples, plums and cherries[8] were grown as far north as possible; citrus and pineapple growing in Florida was also a pretty marginal activity due to occasional, but potentially very damaging, frosts. For many crops, breeding had to follow on from introduction almost immediately for any cultivation to be worthwhile.

Niels E. Hansen (see chapter 5) was one of those who believed that introduced genes from similar climate zones were vital for breeding new crops for the northern parts of the United States. In 1895 he arrived at the South Dakota State College of Agriculture and Mechanics at Brookings and set

8. "Cherry" covers a multitude of species. It is generally used as a common name for any small-fruited member of the genus *Prunus*. The common sweet cherry is derived largely from *P. avium*.

upon growing hardier crops for the plains. He joined what for a few decades became a regular pilgrimage of American agriculturalists and breeders trekking across Russia to collect hardy grains and fruit varieties. In 1897, he built the first known greenhouse for fruit breeding.

At the other end of the climatic scale, Walter T. Swingle (1871–1952), originally from Kansas where he had worked on cereal diseases, took a job developing a tract of land near what is now Miami in Florida. Swingle was renowned as a linguist, which enabled him to go on several expeditions around the Mediterranean studying the cultivation of citrus, dates, figs, and olives. On the Miami plot and at a laboratory set up in 1892 by the USDA at Eustis, he developed a range of citrus hybrids: the tangelo, citrange, and limequat; he also worked on pineapples and grapes. On one level it was a change from his previous place of work, but on another level he had the same enemy—frost. In Kansas he had once had a great reverse when severe weather froze the progeny of over two hundred wheat crosses; in Florida he suffered a similar experience one winter. Though originally he had been breeding for disease resistance, from that time on, cold resistance was the priority.

That plant breeding was central to agriculture in the United at this time and was advancing very rapidly is evident from the yearbooks of the USDA; for three years running, 1897–1899, there was an entry on progress in plant breeding. These yearbook articles are not just a fascinating window onto an exciting time in breeding, with the full value of hybridization just being realized; they are also evidence of a period when the young study of heredity was turning into genetics. All three articles were at least partly written by Herbert J. Webber (1865–1946). Webber had originally been picked by Swingle to work with him on citrus at Eustis, but in 1899 he was summoned back to Washington D.C. to take charge of a new laboratory of plant breeding. Webber was also noted as a botanist; he had made a considerable impact through his work on cycad reproduction, which forced botanists to rethink much received wisdom on plant evolution; so radical was his discovery that some colleagues suggested that his work in plant breeding in the Florida heat had addled his brain.

Now in charge of the USDA plant breeding section, Swingle's main areas of investigation were wheat (breeding for earlier maturity, drought resistance, resistance to smut, increased protein content, and yield), cotton with longer fibers, and citrus. A letter sent on August 31, 1901 to Albert F. Woods at the Bureau of Plant Industry about his work with cotton captures the mood of the successful plant breeder.

Eureka Woods. I have two hybrids. First bolls opened today which if we can reproduce true in the third generation will beat any upland varieties known. Big boll like Upland, small black seed, an abundant lint equal in quality to Egyptian or low grade Sea Island. Just the type I have been working for but which I feared was unattainable. Wish I could show you these cottons. You could see more cotton by coming to Columbia now than you could probably see anywhere else in the world.[9]

Hybridization, noted Swingle and Webber in 1897's yearbook, had many advantages, but for these to be made the most of, useful characteristics had to be fixed; they drew on Darwin to suggest that this could be done through self-pollinating or breeding with very similar plants. Among the frequent advantages of hybrid plants they listed were an increase in size, growth rate, and/or yield. But they went onto suggest that hybridization's greatest use would be in combining characteristics from different species. In the world of ornamental gardening, a way forward had been shown by well-publicized work in improving the hardiness of ornamental rhododendrons by crossing the highly decorative Himalayan *R. arboreum* with the less showy but hardy *R. catawbiense* of the Appalachians (see chapter 13). In Florida, Swingle had just started to do the same with citrus, making a difficult transgeneric cross between citrus and the tough and frost-hardy *Poncirus trifoliata*. In a few years, this work would bear fruit with the citranges and would be repeated several times over the next few decades as growers throughout the world attempted to grow citrus on the edge of its natural zone of hardiness.[10] Vine rootstocks were being produced which were resistant to the virulence of the insect pathogen phylloxera, and by using the American native *Vitis berlandieri*, were able to grow on iron-deficient soils. Tobacco was being improved through crossing Havana tobaccos with others. Even the chemistry of plants could be changed; Webber and Swingle suggested that American breeders had much to learn from a German grower who had bred potatoes with an increased starch content.

The 1898 yearbook showed Webber grappling with the problems of fixing hybrid traits, drawing upon the work of Henri de Vilmorin in suggesting that while hybridization greatly increased the range of possibilities, the real technical difficulties came at the second stage. A key conceptual breakthrough for practice was his recognition of the importance of setting goals

9. Quoted in Rodgers 1949, 267.
10. Cameron and Soost 1976.

for improvement programs with a definite notion of the ideal types wanted from a breeding program.

Webber did not always get it right, as in discussing the improvement of Sea Island cotton by selecting early-maturing forms, he repeats erroneous advice that was widely believed during the nineteenth century, that seed taken from immature seed heads would produce earlier fruiting plants. He also believed that plants could be improved by selection of vegetative material, even quoting research conducted by the USDA's Division of Vegetable Physiology and Pathology with violets, which appeared to show increased productivity and disease resistance over several years of selection using cuttings. Thorns, he held, could also be removed from citrus by continual vegetative selection.

One particularly interesting section in the yearbook deals with growers who showed the links between wild plants and cultivated forms. A British grower called Buckman, Webber tells readers, took wild parsnips and over several generations produced a variety good enough to farm commercially. Eventually this was taken up by the Sutton Seed Company, which released it as 'The Student'; it was so good it remained widely available for nearly a century.[11]

BRITONS UNBENDING—THE 1899 CONFERENCE AND THE INTRODUCTION OF MENDEL

The year 1899, just one year before Mendel's work was rediscovered, goes down as historic in its own right, for this was when London was the venue for the First International Conference on Hybridization and Cross-Breeding. Organized by the RHS, it was an immensely British affair. The conference report makes clear that this was one of the grand events of the kind at which Victorian England excelled. Bands played, luncheons and dinners were eaten, speeches made, toasts proposed and drunk to. The report even tells readers what music the bands played and gives the menus. A particular feature was the display of hybrid plants—exhibiting plants is something the RHS has always done—and still does—well. The plants shown were a tribute to the expertise of British breeders—and those foreign delegates who were able to bring plants or samples with them. There were a great many orchids, reflecting the obsessive interest wealthy gardeners had for these still novel and exotic plants. Rhododendrons featured strongly, along with

11. H. J. Webber 1898.

roses and clematis. There were some surprises among the hybrids too: ferns, bromeliads,[12] and nepenthes, a genus of tropical pitcher plant rarely grown today. Remarkably few fruits or vegetables made an appearance, possibly because the event was held in July. As the conference was hosted by the RHS, plants were judged and awards given, the list of winners being dominated by the names of aristocratic and wealthy garden owners—men with large estates and teams of gardeners.

Clearly a good time was had by all; one unnamed foreign professor said that the dinner on the second day reminded him of student days, and that he "did not know the English could unbend so far"; another stated that "as long as I live I shall remember that dinner."[13] Toasts were proposed by Herbert Webber, Hugo de Vries, Henri de Vilmorin, and Walter Swingle; the last honored Fairchild and his pinks in the first few sentences of his toast. The three-day proceedings concluded at Waddon House, near Croydon on the outskirts of London, in a marquee in sweltering heat, before special trains to London returned the delegates to the capital, seen off by another military band.

In Britain, the key figure in the introduction of Mendel was biologist William Bateson (1861–1926). In 1900 he was on his way from Cambridge to London to deliver a lecture to the RHS when he read about Mendel in a German journal; he promptly set about revising his notes in order to be able to incorporate the discovery into his lecture. That same year he published an article in the society's journal entitled *G. Mendel: Experiments in Plant Hybridization*, which had the effect of launching interest in Mendel across the English-speaking world.[14] He then caused considerable controversy by giving a lecture on Mendelian genetics at Oxford—from this point on, he was cut adrift from other, mostly older, biologists who did not believe him or Mendel.

In 1902 Bateson published a book, *A Defence of Mendel's Principles of Heredity*, and locked horns with two sceptics: Walter F. R. Weldon (1860–1906) and Karl Pearson (1857–1936) in the journal *Biometrika*. Biometricians were involved in precise statistical descriptions of populations and much com-

12. Bromeliads are members of *Bromeliaceae*, the pineapple family, and have had something of a small cult following among horticulturalists since the late nineteenth century.

13. Quoted in Wilks 1900, 40.

14. Mendel 1901. Bateson wrote the foreword for this edition.

plex mathematics, but their work was very largely descriptive and they had little understanding of experimental research—illustrated most clearly by the erroneous belief that there was no difference between sexual and asexual propagation in inheritance. Building on Darwin's work on natural selection, they believed in a *continuous* spectrum of hereditary variation—that most variations were very slight. Misunderstanding Mendel's work when it was first published, they assumed that because he discussed heredity in terms of *discontinuous* characters, there must be a conflict between Mendel and Darwin. This erroneous belief held up a synthesis of Darwin, Mendel, and the biometricians for some fifteen years.[15]

When *Biometrika* stopped publishing Bateson's work he had to have it privately printed instead. Bateson and Weldon, who had been at St. John's College, Cambridge, together, then found a particular focus for their disagreements—the origin of the then-popular ornamental daisy relative, the cineraria (now classified as *Pericallis x hybrida*), whose huge dense heads of daisy flowers adorned many an Edwardian conservatory. Weldon, drawing on Darwin, claimed that the cultivated varieties must be the result of accumulation of many small variations over many years, starting from a single original species. Bateson's theory was that hybridization between several species had occurred, resulting in a much greater level of change. The disagreement between Bateson and Weldon gathered pace with letters to journals, an accusation by Bateson that Weldon had accused him of dishonesty, and then a final falling out—never were they to be friends again.

Bateson set out to prove he and Mendel were right by designing a series of experiments, assisted by one of the group of younger people who were gathering around him and the exciting new theories—Miss E. R. Saunders. Saunders went on to become a noted authority on plant genetics herself, publishing work on stocks (*Matthiola incana*) in 1928. Several other young women worked with Bateson—at a time when women had only just gained the right to university education, were still fighting for the right to use university libraries, and were certainly not expected to engage in controversial research, especially if it concerned reproduction. One of the older scientists

15. A point made by Provine 1971. There was a definite link between early biometricians and eugenics (see chapter 7); Pearson in particular became a famous "social Darwinist," who once declared that "When wars cease mankind will no longer progress for there will be nothing left to check the fertility of inferior stock" (Ferguson 2003, 264).

critical of Bateson once reduced a tea party hostess to tears when he publicly upbraided a woman guest for discussing the crossbreeding of small animals.[16]

Following in the footsteps of Mendel, Knight, and others who had used members of the pea family, Bateson carried out extensive trials with sweet peas (*Lathyrus odorata*). These were initially grown in the family garden, despite the appeals of Mrs. Bateson, who insisted that they needed all the space to grow vegetables to feed the family. When yet more land was needed, Bateson managed to get some space on the university farm. At harvest time, everyone, including Mrs. Bateson, joined in to pull down the peas and record their characteristics.

In August 1904, Bateson presented his results to his colleagues at a lecture, based on work with poultry, sweet peas, and stocks. It was one of those lectures, so immensely gratifying to the lecturer, where the turnout was so good that all the windowsills were taken and latecomers had to stand. Bateson won over many, but not all; Weldon continued to attack him and his colleagues, and Bateson was free to conduct research on his own terms; he did not gain proper funding until 1910, when he took up the directorship of the John Innes Institute. From that time on, he worked toward not just proving the validity of Mendelism but also became actively engaged in running an independent institution concerned with plant science.

In contrast, American interest in Mendel was considerable and immediate—and it came from those involved not in "pure" research but from those involved in practical breeding. Straight off the mark, in 1901, the USDA's *Experimental Station Record*, an information clearinghouse for the experiment stations, published a detailed synopsis of Mendel's work based on Bateson's article for the RHS. It quoted Bateson's claim that Mendel's laws were "worthy to rank with those that laid the foundation of the atomic laws of chemistry."[17] The USDA also helped popularize Mendelism through its Graduate School of Agriculture, inaugurated in July 1902.

Some immediately saw the commercial possibilities, with Willet M. Hays of the University of Minnesota suggesting at a conference that Mendel's laws amounted to "hundreds of millions of dollars' worth of added annual income with but little added expenditure."[18] Bailey was quick to recognize the value of Mendel, proclaiming in an article in 1903 that "we can no lon-

16. Russell 1966.

17. Quoted in Paul and Kimmelmann 1988.

18. Quoted in ibid.

ger be satisfied with mere 'trials' in hybridising [*sic*]: we must plant the work with great care, have definite ideals, 'work to a line,' and make accurate and statistical studies. . . . [T]he breeder . . . must have very specific and definite knowledge of what makes the plant valuable and what its shortcomings are."[19] However, Bailey, the man of experience, also spoke, warning that "the wildest prophecies have been made in respect to the application of Mendel's laws to the practice of plant breeding."[20] Mendelism, he thought, offered explanation and the ability to plan breeding programs. But it did not offer shortcuts or ways around traditional wisdom. Nor would rules take the place of intuition; breeding, Bailey thought, could not and would not be reduced entirely to mathematics. Many plant breeders today would still echo these thoughts.

Experimental evidence, much of it directly linked to variety improvement, soon began to pile up in support of Mendel. An example is a paper by an American researcher on wheat, "Quantitative Studies on the Transmission of Parental Characteristics to Hybrid Offspring," which reported that the study had begun in 1899, but when completed had been entirely in accordance with Mendel.[21] Of the various implications of the acceptance of Mendel's work, one was the beginning of the end of the widespread belief that male and female parents controlled different progeny characteristics; a typical example being Professor Ludwig Wittmack (1839–1929) of Berlin, whose contribution to the 1899 conference in London was titled "On the Particular Influence of Each Parent in Hybrids." The paper discussed how his work with such disparate plant groups as cereals and bromeliads "tends to show that the mother plants usually determines the habit, that is the vegetative portion; while the father has the greater influence upon the floral parts."[22]

A NEW WORD FOR A NEW CONCEPT— "GENETICS" AND THE 1902 AND 1906 CONFERENCES

On October 3, 1902, Bateson wrote excitedly to his wife, "At the train yesterday, many of the party arrived with their 'Mendel's Principles' in their hands! It has been 'Mendel, Mendel all the way,' and I think a boom is beginning at last. There is talk of an International Association of Breeders of

19. Quoted in Rodgers 1949, 279.
20. Quoted in Paul and Kimmelmann 1988.
21. Spillman 1902.
22. Wittmack 1900.

Plants and Animals and I am glad to be right in the swim."[23] The occasion was the Second International Conference on Plant Breeding and Hybridization, held in New York. Bateson, giving the first paper, explained why the practical plant breeder found hybridization so promising in the light of Mendelism: "He will be able to do what he wants to do instead of merely what happens to turn up. The period of confusion is . . . passing away, and we have at length a basis from which to attack that mystery."[24] Bailey's contribution was entitled "A Medley of Pumpkins," reflecting his long fascination with the highly variable genus *cucurbita*. Luther Burbank's speech was "Principles of Plant Breeding," and there were lectures on disease resistance, rose breeding, the improvement of strawberries, corn, oats, cotton, and much else.

The New York conference sounds a much more sober affair than the British conferences that preceded and followed it. There was no mention of marquees filled with floral exhibits, prizes awarded or medals pinned on chests, nor cadets from West Point coming down to play music to delegates. We can only assume that the food was good, for the proceedings report does not list the menus. A report of the conference in the journal of the RHS reported that the level of state involvement for plant breeding in the United States was "greater than anything comparable in Europe." It also reported that "it was evident that the audience was keenly interested in the subjects discussed. To those who are accustomed to the sometimes apathetic reception such communications may receive in London, the close attention and the animated discussions which frequently followed the papers were most stimulating. . . . Those who came did not do so out of compliment or sense of duty, but strictly with a view to business."[25] Two reasons may account for this. Firstly, many of the British audience were gentlemen amateurs who were there for reasons of status and were probably no more interested in plant breeding than in any other of the numerous other pursuits of this famously generalist class. Additionally, many of the audience at the London conference were probably suffering the aftereffects of a jolly good lunch.

It was 1906 when the seal was set on the use of the word "genetics" to describe the new discipline of the scientific study of heredity. This was when the Third International Conference on Genetics, organized by the RHS in London (or, to give it its full title, Hybridization [The Cross-Breeding of Genera or Species]). The report of this conference, like the 1899 one, was written by

23. Quoted in Paul and Kimmelman 1988.

24. Quoted in Kloppenburg 1988, 69.

25. Quoted in RHS 1902, 1060.

its secretary, the Reverend W. Wilks (1843–1923), a minister of the Church of England and the creator of the Shirley poppies (see chapter 13). This must have been one of the last occasions that a minister's name appeared in such a position; such men had been among the most indefatigable of that great breed of Victorian Briton, the amateur naturalist. Science was now on the verge of taking over; the study of the natural world and its manipulation was now more and more the role of graduate specialists, full time and salaried.

Historically, it had been the breeding of animals that had attracted more attention and effort than the breeding of plants—cows and sheep had had their pedigrees written down and discussed by gentleman farmers long before wheat or peas had. So it is interesting to note here that it was plant breeding that launched the concept of genetics as a formal science and played such an important role in elucidating early principles. The inaugural address was given by Bateson, which is worth quoting at length, as it conveys clearly and colorfully the sense of excitement and confidence of the young science of what he, later on in his speech, suggested be called "genetics."

> It is just seven years since, on the hottest day of a very hot summer, the first conference . . . assembled at Chiswick . . . that definite results might come from that beginning we naturally hoped, but of those who endured the heat of that stifling marquee, or inspected the plants exhibited in that tropical vinery, not one, I suppose, anticipated that in less than a decade we should have such extraordinary progress to record. The predominant note of our deliberations in 1899 was mystery. In 1906 we speak less of mystery than of order.
>
> When formerly we looked at a series of plants produced by hybridization we perceived little but bewildering complexity. We knew well enough that behind that complexity, order and system were concealed. Glimpses indeed of pervading order were from time to time obtained, but they were transient and uncertain. As casual prospectors we picked up occasional stray nuggets in the sand, but we had not located the reef, nor had we any machinery for working it if discovered.
>
> Then came the revelations of Mendel's clue, with all the manifold advances in knowledge to which it has led. The most Protean assemblage of hybrid derivatives no longer menaces us as a hopeless enigma. We are sure that even the multitudinous shapes of the cucurbits, or the polychromatic hues of orchids—though they may range from one end of the spectrum to the other—would yield to our analysis. Methods for grappling even with these higher problems have been devised. The immedi-

ate difficulties are chiefly of extension and application. Thus the study of hybridization and plant breeding, from being a speculative pastime to be pursued without apparatus or technical equipment in the hope that something would turn up, has become a developed science, destined, as we believe, not merely to add new regions to man's knowledge and power, but also to absorb and modify profoundly large tracts of the older sciences.[26]

As with the previous London conference, this was a grand affair, beginning on Monday, July 30, with an informal *conversazione* around exhibits arranged on a series of tables in the RHS hall. The official report records that a military band played in the background while delegates had a light meal of "sandwiches, cakes, chocolat [*sic*]" éclairs, fancy biscuits, ices, and a variety of French and German wines.[27] Among the exhibits, R. H. Locke of Caius College, Cambridge, had a display of peas illustrating Mendel; Miss M. Wheldale of Newnham College, Cambridge, showed cross-fertilized antirrhinums; Rowland Biffen of Cambridge displayed wheats and barleys; and the director of Public Gardens of the British colony of Jamaica showcased hybrid pineapples (*Ananas comosus*). Another exhibit was of peas from Messrs. Arthur Sutton of Reading—this was a reminder of the great importance commercial seedsmen were in the development of many vegetable and flower varieties.

No RHS event is complete without an awarding of medals; here the prestigious Veitch Memorial was awarded to Bateson, Wilhelm Johannsen, Professor Wittmack, and Maurice de Vilmorin. Among the recipients of the Silver-Gilt Banksian medal were Biffen and Miss Saunders—a photograph shows her in a jacket with a starched collar and a loose tie. Bateson's words were echoed by Niels Hansen of the USDA: "what was formerly a chaos of empiricism is now becoming a one of the exact sciences. . . . No longer is heredity a jungle." The report adds that "during the delivery of the professor's speech one of the heaviest thunderstorms in living memory raged."[28]

One of the American speakers at the 1906 London conference was the botanist and fungal disease specialist Erwin Smith, a close friend of Liberty Hyde Bailey, who lectured on the plant breeding then being carried out by the USDA, which he described as focusing on four issues: disease resistance,

26. Quoted in Wilks 1907, 90–91.
27. Quoted in ibid., 29.
28. Quoted in ibid., 72.

cold hardiness, resistance to drought and alkali soils, and greater productivity. His discussion of cold hardiness largely focused on the subtropical crops that were still so young in the United States: citrus, dates, and pineapples. Date cultivation in particular was still a very young industry; in Arizona and California, the trees had to contend with colder conditions than they would normally thrive in, and there was a widespread issue of very alkaline soils in semiarid regions. Facts and figures poured forth, all of vastly increased productivity, of resistance to disease, of higher quality, of lands previously thought impossible to cultivate now bearing harvests. Even the yield of the humble violet was not beneath his sights, for it too now bore ninety flowers per plant as opposed to the fifty before.

Indeed, the USDA and the network of experimental stations run by the various states led the world in the systematic breeding of new varieties, and its representatives made frequent appearances at a growing number of international conferences devoted to the subject. Such events, now the mainstay of many an academic's life, were themselves very much a novel feature, the product of a genuinely globalizing world.

Now firmly launched as a discipline, plant breeding soon began to acquire teachers and students. One of the first university courses in plant breeding was established by Rollins Adams Emerson (1873–1947), brother-in-law of Webber, at his old school—the University of Nebraska—in 1899. In 1906 Congress passed the Adams Act, which gave a great boost to genetics research as it directed funding to solving theoretical problems at experimental stations and other institutions (before, the stations had only been working on particular practical problems). Research into the laws of inheritance was among the topics the act was aimed at funding; often this research fed directly into plant breeding practice. In 1911, Arthur W. Gilbert, professor of plant breeding at Cornell, reported in an address to the National Corn Exposition in Columbia, Ohio, that over a thousand students were following courses on plant breeding.

It is always possible to tell when a group of people following a particular path have chosen to become a profession—they form an association or guild. This happened to plant breeders in the United States in 1903 with the foundation of the American Breeders Association. The germ of the idea was taken home from the first hybridization conference in 1899. All the leading US plant breeders were members; early journals show their concerns to be overwhelmingly practical, but with a great interest in the association of characters, the prediction of performance through understanding the linkage of phenotypical characters to genotype, and the first discussions of biometry. Tensions in the new profession soon arose, which in some ways are still with us today—scientists from the USDA and the land-grant colleges on one hand wanted to research plant breeding and genetics in depth, whereas the members representing commercial organization wanted practical answers to problems.

LIGHT FROM THE NORTH—SCANDINAVIAN PROGRESS IN CROP GENETICS

Mendel and his laws seized most of the limelight at the turn of the last century, not just for the conciseness and explanatory power of his ideas; he was also a very romantic figure—a rather mysterious man of God cloistered away in a far-off country, whose work was not appreciated in his time, whose notes had for the most part been burnt by his order after his death. Truth is often stranger than fiction. But there were other figures doing immensely important work around this time, less glamorous people producing less dramatic results but very important all the same.

Just as Moravia may have seemed peripheral to the English-speaking world, so too might have Sweden. But it too had a strong and progressive farming community, and a great tradition of botanical research—this was after all, the country of Linnaeus. The best work with cereal crops was being undertaken by a company set up to sell seed—the Agricultural Experiment Station at Svalöf in southern Sweden was not supported by government and had no educational or purely scientific remit, which makes it an unusual beast for its time, when systematic research tended to happen only in universities or when undertaken by wealthy gentleman amateurs. The driving force was Birger Welinder, an estate owner at Svälof who in 1886 founded the Southern Swedish Association for the Cultivation and Improvement of Seed. The association was responsible for breeding, and the General Swedish Seed Company, established in 1891, took on the task of the production and marketing. The Swedish state supplied some grant aid, which led to an increasing level of state control; in 1913 the rules were changed so that the government appointed the majority of the board.

Svalöf arguably became the world center for the development of genetics in the early twentieth century, as the station combined fundamental work in genetics with the solution of practical problems. Svalöf acted as a hothouse whereby practically orientated plant breeding forced the pace of fundamental research into the nature of heredity. It was also an institution that was also very successful in combining the private and the public. It was more centralized than the US state-based system and more organized than the British network, which at the time relied heavily on gentleman amateurs and an idiosyncratic university system. Much plant breeding work had, up until the turn of the twentieth century, been conducted with a large element of trial and error; clearly much effort and time got wasted as worthless plants were grown on. One of Svalöf's great achievements was to reduce this through a very economical use of resources. Unglamorous it may have been, but methodology was perhaps the institution's most important contribution.

The basis of the Svalöf methodology was the work done by the German Kurt von Rümker (1859–1940), who, in 1888, while still a relatively young docent at Göttingen University wrote the first book on plant breeding in German, *Instructions for Cereal Breeding on a Scientific and Practical Basis*; the next year, he delivered the first lecture course on the subject in Germany. His "methodical selection" included both pedigree breeding and mass selection. In his view, heredity was constantly changing and intensely varied; for him the difference between seeds in one ear was as interesting as between the seeds of different plants.

Based on Rümker's work, record keeping became Svalöf's secret weapon—record keeping like no one had done before, with a huge number of plant characteristics being recorded meticulously. "Almost incredible" is how the normally sober de Vries described it after one of his visits. The bookkeeping seems even more astonishing when we learn the primary genetic reason for the success of Svalöf and its most prominent director, Nils Hjalmar Nilsson (1856–1925), was that individual plants were the basis for all their experimental strains.

Nilsson was one of the first to put into effect a belief in the accurate measurement of the end results of plant breeding. To be effective, breeding requires detailed measuring of the characteristics of the produce, as Charles Saunders in Canada had done through paying such close attention to the bread made from various wheats. The institute's first director had been the German Thomas Bruun von Neergaard, an agricultural engineer by training who developed instruments for measuring, weighing, and sorting cereal grains: "measure, number, and weigh" could have become the institute's motto. Neergaard also developed a classification system for making sense of the multitude of forms to be found in the cultivated races of cereals, an idea which Nilsson took a stage further with an elaborate system based on what he described as "botanical characters" often of no interest commercially but that he believed could be correlated to commercially important characteristics such as yield, hardiness, rust resistance, and so forth.

Swedish winters are long, cold, and, perhaps worst of all, dark. There is rather more incentive to working patiently through vast piles of data than in many other countries. During the winter of 1891–1892, Nilsson spent much time poring over the results of trials with cereals without making much progress. He and Neergaard had been working with the "German method" of mass selection with 'Squarehead' wheat and 'Chevalier' barley from 1886; their breeding program picked out and purified selected botanical characteristics, which where then tested for their value in field conditions. But there was no effect on uniformity—in particular Nilsson was frustrated by the inability to find a barley with consistently stiff straw; a sense of impatience even appeared in academic papers he wrote at the time.

Nilsson then recalled that many good new varieties had originated from farmers or peasants noticing chance mutations. Every now and again Nilsson found records of populations descended from such individual plants, which were so distinctive that they had not been fitted into the Svalöf classification system and had been kept in separate plots—it was these, he found, that had an exceptional uniformity. It then dawned on him that perhaps the best approach would be to start with the individual plant.

Realizing that grain samples inevitably yielded heterogenous progeny, Nilsson then kept each ear or pod separate, sowing seed from two thousand samples (mostly cereals, some peas) one year and then growing on their progeny, with the result that the variation was hugely reduced. Of his oats, of 422 sown, 25 showed variation, 397 were completely constant. For the most part, having created unvarying stock he found that uniformity was the rule, and it was transmitted to the next generation. Such uniformity was possible only if the progeny of each individual was kept separate; in other words, the heterogeneity of most contemporary cereal strains was because they were mixtures of different forms. Purity could be assured once a selection had been made, but only if the stock was kept free from pollen from others. De Vries, in summarizing Nilsson's work, pointed out that "ordinary varieties of cereals are built up of hundreds of elementary forms, which with few exceptions have hitherto escaped observation."[29] It was only later that Nilsson discovered that Vilmorin had reached the same conclusion some years earlier.

29. De Vries 1907, 90.

Not far from Svalöf, just over the body of water known as the Øresund, the Dane Wilhelm Johannsen (1857–1927) was also working on pure-line selection and in 1903 came to a conclusion similar to Nilsson's. Johannsen started out as a chemist, working for the brewer Carlsberg on barley, moving later to botany and in particular plant physiology. Much influenced by Vilmorin's pedigree breeding method, and disenchanted with Darwinism, Johannsen was very much a "pure" scientist, which may account for his relative lack of interest in the more instrumentalist approach taken at Svalöf, whose personnel were first and foremost practical breeders. Johannsen spent the last few years of the nineteenth century growing beans, specifically, the self-fertilizing bean *Phaseolus vulgaris* (known as 'Brown Princess'). He grew the progeny of each individual bean, weighed the beans they produced, and repeated the exercise the year after. The average weight of the beans remained stable, showing that this was an inhereted characteristic, and that each bean plant and its produce could be seen as part of a "pure line," which once created could not be developed any further. He also showed that any variation in the weights of his purebred beans was the result not of their genes but in the conditions in which the plants grew; this led him to make one of the most important conceptual clarifications in the history of genetics, that between genotype and phenotype. For good measure he also suggested that the basic units, or particles, of heredity be called "genes." It was his theory of pure lines that explained why breeders must start with one plant and not many; but if breeders wanted a wide range of variation they had to cross different parents. Johannsen's bean counting had confirmed experimentally what Vilmorin and Shirreff had realized practically: that (with inbreeding crops) selecting a particular line out of the mass of lines available in a land race leads to consistency.

Johannsen recognized that self-fertilizing pure lines were very different to situations where plants were cross-pollinating—such populations were, he thought, mixtures of many different types. His leap of imagination was to state that although most populations do not consist of pure lines, the study of these pure lines was the place to make a start—only when they were understood could the more complex case of cross-fertilizing lines, or hybridization, be considered.

H. Nilsson-Ehle (1873–1949), a later worker at Svalöf who served as director from 1925 to 1939, developed a program of hybridization based on Mendelian genetics between 1900 and 1910. By 1906 he was confident enough to promote hybridization as a vital part of the institute's work. He set off on this track after looking at the problems of how to improve winter wheat;

there were some promising new races but all had serious flaws, so the idea of combining strengths through hybridization was obvious. By 1908 he had combined Mendelism and pure-line theory, showing how the pure lines, or, "pedigree varieties," which can be isolated by pedigree selection, represent special cases of the hereditary variability produced by hybridization. From this it followed that all hereditary differences up to this level consist only of various combinations of elementary properties, all of which are subjected to segregation and recombination according to the laws of Mendel. The new characteristics displayed by mutations were, he held, like any other characteristics—they could be resorted and recombined with other characteristics in the next generation.

It was once assumed that hereditary variation was continuous both in the expression of characteristics and over time, from one generation to another, and the technique of mass selection was founded on this assumption. Mendel's breakthrough was to show that it was not, that inheritance followed a predictable pattern of segregation and recombination, with a mathematical ratio, Johannsen's breakthrough was that populations consisted of a number of pure lines—at least in the case of inbreeding species. Nilsson-Ehle showed how the technique of developing pure lines, followed by the controlled crossing of these lines, would give the breeder much more precise control over the outcome. Mass selection, such as Rimpau's much esteemed "German method," belonged to the past.

Nilsson-Ehle also made another breakthrough. Much of the opposition or skepticism about Mendel's work concerned the very obvious fact that many traits were not inherited through neat 3:1 ratios; in many cases, characters were clearly blending from one generation to another. Nilsson-Ehle's masterstroke was to show how some characters are quantitative, with a whole range of values, and that these are the result of several different genes at work, each one of which behaves in true Mendelian fashion—the results in the phenotype display a continuous variation, which is the result of the interactions between multiple genes. This work with continuous variation played an important part in the development of population genetics (see chapter 11). As is so often the case however, the zeitgeist was at work—in the United States, Edward Murray East did similar work during 1910–1918, showing how Mendelian theory could account for continuous variation.

Nilsson-Ehle showed how it was possible, through an appropriate selection of parents, to produce individuals which displayed levels of variation outside the range of those characters shown by the parents. Working on a crop which was very important in Sweden, and indeed throughout northern

Europe, he showed that oats could be produced that were either more or less susceptible to rust than either parent. There was, he said, a distinction between breeding which involved a *combination* of characters, and *transgression* breeding where there was an intensification of characters beyond what was present in the parents.

For some years there was controversy between Nilsson and Nilsson-Ehle; the former defending the original "Svalöf method" of individual selection, believing that in many crops that there was enough natural variation with which to work without resorting to hybridization, the latter that planned hybridization between plants with desired characters was necessary. Nilsson eventually came round to share his colleagues point of view; for prior to 1915, planned hybridization had not achieved much, but after this the breeding of most new cereal varieties involved hybridization.[30]

The success of the Svalöf station's elegant dovetailing of theory and practice can be appreciated by the 40 percent increase in grain yields during the first fifty years of the institute's existence, of which about half has been put down to the effect of new varieties. However, this varied hugely between regions: southerly fertile Skåne achieved a 45 percent increase; northerly Värmland, with meager soils and a short growing season, only 5 percent. Such differences are often seen in the years after an advance in plant breeding, for example during the Green Revolution, as new crops bring about major increases in yield in favoured regions but often only small ones on marginal land.

SUMMARY

Mendel's brilliance revealed the underlying workings of heredity, but his work took some time to be revealed, and even more to become accepted as a guide to the underlying principles of breeding. His work received the most enthusiastic reception in the United States, where there was a greater interest in applying science to agriculture and horticulture than anywhere else, with a high level of government support for agricultural research. Science was seen as vital for the task of improving and diversifying agriculture, an agriculture which was having to face the task of finding crops and crop varieties which would thrive in a wide range of unfamiliar environments. In particular, the United States was ready for the idea of hybridization more than anywhere else.

The concept of genetics was launched at a series of international confer-

30. Norton 1902.

ences around the turn of the twentieth century, and with it an appreciation of the massive potential of hybridization. There was a strong sense of fog dispersing, opening up a view over a whole new landscape of possibilities. The understanding of Darwin's Theory of Evolution and the uncovering of the basic principles of heredity was intimately linked to plant breeding. Breeding was not only an important experimental tool but an obvious application of new theories.

Plant breeding could now become targeted and deliberate. For real progress to be made, however, there needed to be both a better understanding of mechanisms and improved methodologies. Advances in both came from Scandinavia, in particular from the research institute at Svalöf in Sweden, where meticulous record keeping and a recognition of the potential importance of all plant characters in breeding work enabled the quantity of data to be assembled to gain a better appreciation of how heredity worked. In particular there was a recognition of the importance of pure-line breeding in inbreeding crops, which enabled varieties to be produced with the consistent reproduction of particular traits. The combination of the idea of the pure line with Mendelian principles and an elucidation of continuous inheritance made a powerful new tool for crop improvement. Breeders were now able to develop pure lines and to cross them, making a decisive break with the traditional practice of mass-selection. The era of landraces was coming to an end, to be replaced by consistent and predictable varieties.

SEVEN ❦ GERMINATION ❦ MENDELISM AND PLANT BREEDING IN THE EARLY TWENTIETH CENTURY

The picture of 'Sutton's Perfection Marrowfat' peas completely dominates page 15 of the 1910 Sutton and Sons Ltd. Seed Catalogue for amateur gardeners. The company appeared to have had a policy of printing life-size pictures of its crops wherever possible—black and white photography had advanced sufficiently to produce high-quality images that, despite a lack of color, still created an impression of luscious plenty. Early-twentieth-century seed catalogs were often opulent productions—the promise of the vegetable garden followed by the free-flowering abundance of the flower pages. The sheer range is impressive, often with a wider selection of vegetable crop varieties available than today—the same is true of many annual flower seed species, too. Photographs of trial beds or of greenhouses where plants are grown for seed were frequent—much more so than they are in catalogs today; it was as if the breeding and production process was something in itself to be proud of. In the 1922 catalog, there are pictures of men in white coats in laboratories surrounded by glass jars, and a strategically placed microscope. The contrast with the nostalgic tone taken in many seed catalogs of today is striking.

It is clear that selection has been the main breeding process—"saved from the richest and most varied flowers grown in our immense collection" is a fairly typical phrase. Where plants are hybrids, this is often pointed out, the word "hybrid" appearing to have a definite cachet of superiority; in any case, the word is often used in the variety name itself. In many cases, varieties offered are "strains," implying a lack of genetic purity, but consistent enough in appearance. Occasionally customers are warned that the odd inconsistency is to be expected, such as a few singles in a crop of doubles.

There is no mention of Mendel, but one cannot help but suspect that behind such phrases as "after years of continual selection and crossing," the company breeders have perhaps started to make use of the time-saving predictive power provided for by the monk's theories. In many cases though, a Mendelian approach may have been superfluous, as so much could be achieved through a few seasons worth of random crossing followed by rigorous selection.

Mendelism captured the imagination of so many because it provided both an explanation for observations made during experiments and a guide to planning future experiments. But Mendelian theory did not find the same response everywhere. In the United States it was taken up quickly, except by a few mavericks like Luther Burbank (see chapter 8). In Europe, Mendelism had a more mixed reception, with enthusiasm in some quarters and indifference in others—occasionally even opposition and hostility. Much of this chapter is spent looking at the varied and cautious response to Mendelism in Europe. Inevitably too, the story of early-twentieth-century breeding will revolve around the different pace at which Mendelian methods were accepted.

The view in the United States was that plant breeding could be reduced to rules based on Mendel—scientists could produce not just new crops and flowers but more crucially, abstractions and general principles, which could then be applied many times over. There was a parallel here to many other developments happening during the first two decades of the twentieth century in manufacturing industry, where complex tasks were being broken down, and their constituent parts re-assembled, as in Henry Ford's mass-production of motorcars, or the "scientific management" of Frederick Taylor.

The reason Americans were more enthusiastic in their response to Mendel than Europeans can perhaps be sought in geography—the United States was trying to adapt a wide range of crops to the conditions prevailing in its territory, so genetics was a very obviously useful tool. The European empires, however, had a major investment in plantations in their colonies, growing crops that were not only often vegetatively propagated but were also growing in optimum conditions. For them, the old-fashioned art of plant hunting and introduction—trawling peasant plots, jungles, and bush for possibly useful variations of crop varieties— was still a commercially valuable one.

Mendelian genetics was seized upon in America not just because it made sense but also because it was a way for scientists to carve out a place for themselves and enhance their status. Mendelian theories were important in professionalizing breeding creating a new class of specialists separate from farmers and amateur breeders. From now on, scientists and professional breeders appeared to be keen to distance themselves from the latter group, and to stress that only they had the requisite knowledge to improve plants. Specifically, they wanted Mendelian methods to be seen as superior to the process of selection which many farmers, of corn in particular, used, and used with considerable success. At least in the earliest years though, American breeders tended to use Mendelism to *interpret* rather than guide and plan breeding programs. It was only when breeders and researchers with corn began to work with Mendelian methods that the new genetic paradigm really began to make an impact.

In Britain, however, this split between Mendel-led scientific breeding and a less professional approach took very much longer to open up. In a more traditional and less–socially mobile society, scientists, landowning farmers, and gentleman amateurs were usually part of the same social class, with family, school, and university links. Scientists were under less pressure to define themselves, and there was room for more nonconformity.

TWILIGHT OF THE GENTLEMAN AMATEUR—MENDEL IN BRITAIN

In the summer of 1922, a group of grateful English farmers from the wide fertile fields of the eastern county of Essex presented Rowland Biffen with a silver cup in thanks and commemoration for new varieties of wheat—'Little Joss' and 'Yeoman.' The farmers must have been among the first real beneficiaries of Mendelism, for these varieties, released in 1916, were ideal for the wide fields of East Anglia—Britain's leading arable area. Yields were improved, as were plant strength and rust resistance. Yeoman in particular had great standing power, resisting lodging in weather that would flatten many other wheats; it was also the first British wheat that could be used for quality bread making without mixing in with imported wheats. One of its parents was 'Red Fife' (see chapter 5).

While Bateson had been instrumental in introducing the concept of Mendelian genetics into Britain, it was Rowland Harry Biffen (1874–1949) who first made practical use of the new science. The background to the productive new wheats had been a long period of agricultural depression. In 1846, the Corn Laws were repealed—the result of a long campaign by representatives of Britain's rising middle and working classes. Tariff barriers on wheat imports came down, much to the delight of those for whom the price of bread was the single most important figure in their lives, and the lives of their employers—the new factory owners. No longer protected by a wall of high tariffs, British wheat was displaced in the marketplace by much cheaper and higher quality North American grain. In an early example of the contradictory effects of "globalization," cheaper food benefited one part of the population, but brought pain to others—a spirit of depression soon fell over the countryside.

With this depression continuing into the twentieth century, calls for more government effort to support farming grew. Prime Minister Lloyd

George (1916–1922) had a deep concern for the land and farming, in particular for how it could be made to be of benefit for all the population—his Liberal Party even had a "Song of the Land" in its repertoire. Lloyd George himself was a great believer in the power of science and its ability to contribute to creating greater equality in society. Action followed words with the establishment of the School of Agriculture at Cambridge University in 1910 and in 1912 the Plant Breeding Institute, with Biffen its first director.

Lloyd George's attempts at social engineering marked a sea change in British politics and society, for they were an attempt at simultaneously challenging the power of the old aristocratic elite and at maintaining British power through applying the power of the state to reform and research. Up until now, many scientists had preferred self-reliance to seeking state funding; one of the first institutions to engage in Mendelian genetics was privately funded, the John Innes Institute, of which Bateson was made head in 1910. Bateson had tried—and failed—to get funds from his own university and from various nongovernmental bodies, such as the RHS. There was, however, pressure from breeders and seed companies on government to invest in breeding work at this time, probably because of the problems they had in protecting their rights over a product which anyone could legally reproduce and sell.

Bateson's being appointed to head the John Innes Institute was recognition of the Mendelian approach at the highest level; he set about redirecting its program towards genetics, initiating work which combined pure science and practicality. In terms of practical work, fruit—very much an expanding industry in Britain at the time—was of particular importance, with much effort put into looking at breeding compatibility (or otherwise) among different apples, pears, cherries, and plums.

Rowland Biffen studied natural sciences at Cambridge University but only became interested in the application of science during an 1897–1898 expedition to South America to look at the sources of rubber (*Hevea brasiliensis*). Here he realized the importance of science in making the most of Britain's tropical empire; he was apparently very taken aback to discover that the native peoples of Brazil had a great knowledge of rubber and what could be done with it, and yet his own country had no science of plantation agriculture. On his return he took up a lectureship in botany at the new Department of Agriculture at Cambridge, where he started to work with wheat. Siding with Bateson in the disputes over Mendel, he was not a man to get actively involved in academic arguments, preferring to get on with the job. A keen amateur grower of roses and alpine plants, he was described as having "the eye of an artist"[1] for the kind of plant he wanted to create. In his retirement he bred garden flowers, simply for fun, never launching them commercially; he had a particular passion for the florist's auricula (hybrids of *Primula auricula*). His last, posthumous, publication was a study based on the pedigrees he kept of his auriculas. Intuition was clearly a large part of Biffen's success as a breeder, but the artist's eye is not always a good guide—he loved the Swedish 'Iron' wheat for the beauty of its ear, and used it as a parent in many crosses. Many years and tens of thousands of crosses later, he had to admit it did not make a good parent. In his spare time, he was a keen amateur watercolorist, and he left money for a new gallery in Cambridge's Fitzwilliam Museum.

At heart a hands-on researcher, Biffen was a teacher only reluctantly, yet he was apparently brilliant; eschewing the wall charts, lantern slides, and models that were the contemporary equivalent of PowerPoint presentations, he covered blackboards with precise little sketches. His practical, instrumental approach to his work was not appreciated by everybody; there is a story reported by the American botanist Edgar Anderson about when the visiting Russian Nikolai Vavilov switched from studying with Biffen to Bateson, and when asked why, explained that "Beeffeeen, yess, he is a gude man, a fery gude man, but you see, he has no phee-low-so-phee."[2]

Biffen was all too aware of the decline in British agriculture; as he saw it, the Industrial Revolution needed to be brought to British agriculture, along with a rational scientific approach, for which there was no better starting point than the principles of Gregor Mendel. Mass selection had gone as far as it could—it could go no further because it combined too many variables, with numerous plants of different genetic makeups obscuring what was important. What Biffen knew to be important was the identification of genetic factors that could be combined. Bateson and Biffen thought of plants as being composites of hereditary factors; these factors were the foundation on which they worked, they could be combined in an infinite number of different ways in a manner which was both was mathematical and predictable. There was an analogy between Mendelian genetics and the rapidly developing science of chemistry—genetic factors as atoms which could be built up into as many different compounds as could be conceived.

In 1905 Biffen published his first results with wheat, describing how more than a dozen characteristics—morphological, histological, and constitu-

1. Engledow 1950, 20
2. Quoted in Anderson 1967, 77 (spelling in the original).

tional—were determined by genes, not just plant form but also response to growing conditions and baking quality. In 1910 his first variety was released, 'Little Joss,' which was resistant to yellow rust. Susceptibility to the disease, he had discovered, was inherited as a single gene according to Mendelian laws, which made it one of the very first fruits of scientific plant breeding.

It was on the back of his success with Little Joss that Biffen became the first director of the Plant Breeding Institute. Research went into established crops as well novel ones—of the latter, some were successful, such as sugar beet, others became outmoded, such as willow (for basket making), while others were failures: flax, hemp and . . . tobacco!

The creation of strong links between genetic research and plant breeding enhanced the professionalization and institutionalization of breeding, but possibly weakened the links between breeders and farmers. This was not what every institution wanted. The attitude at Cambridge was that scientists could improve agriculture without the active involvement of the farming community—at the University of Reading, however, there was another ethos—that of supporting the farming community and always staying close to it.

John Percival (1863–1949) at Reading was very skeptical of the value of Mendel, believing taxonomy and systematics to be more important in plant breeding. Sterile interspecies hybrids were a major problem for breeders, and since sterility was linked to the degree of relationship between species, Percival and other skeptics put a great deal of emphasis on taxonomy in an attempt to predict which species could produce viable seed. Both Biffen and Percival were products of the "new botany" developed in the United States and England in the late nineteenth century, where Bailey and others sought to make the science as relevant as possible to practical issues and to engage with agriculture and forestry. They were, however, from very different social backgrounds; whereas Biffen was from the traditional gentry class who dominated academic life at the old universities of Oxford and Cambridge, Percival was from a more humble background—a north-country independent farming family, and Reading was a "new" university (founded in 1860) with strong institutional links to the farming community.

Percival directly challenged Biffen and Mendelian-based breeding. For Percival, wheat was an immensely complex organism for which a purely Mendelian approach was too reductionist; he rejected the analogy with chemistry. He particularly stressed the great complexity of wheat in Britain and saw it as important to link the origins of different wheat varieties with their characteristics and their taxonomic status. For him, farmers could play a positive role in breeding if they could more accurately assess their needs and pick out "sports" with potential. This difference in views is paralleled today by radical differences between breeders, for example, the Riccharia affair in India could be seen in this light (see chapter 14).

Different approaches to practical breeding became particularly apparent over the issue of gluten in wheat. Strength (high gluten) in wheat could only be gained at the expense of yield, and millers were not prepared to pay a premium for higher gluten wheat—there are similarities with Mark Carleton's battles with the millers in the United States (see chapter 5). Percival recognized that the interests of millers and farmers could diverge, which had not occurred to Biffen. Given the closeness of his university to the farming community, it was natural that Percival was going to side with the growers, but also, he was more likely to look at breeding problems within a wider context. Today we would probably argue that Biffen was "right" in a technical sense but also appreciate that he was working in the proverbial academic ivory tower and possibly out of touch with the real world of farming and commerce.

Plant breeding, once an obscure hobby for gentlemen, was now seen as essential for the survival of the countryside, even of the nation itself. Britain's scientists had still not completely taken over, however. One figure of the old school of the gentleman-amateur stands out during this era: Edwin Sloper Beaven (1857–1941). He too was another Mendel skeptic.

From a relatively prosperous background, Beaven went to a minor public school,[3] where his tutor took the attitude that students should find out things for themselves rather than be simply taught facts, a daringly progressive idea for Victorian England. This resulted in his father withdrawing him at the age of thirteen. At the age of twenty-one he joined a brewery in Warminster in southwest England, which started a lifelong fascination with barley. He was particularly interested in how barley varieties respond differently to different soils. By the 1890s, he was able to hire two assistants to help him conduct experiments in his four-acre garden, which became home to a huge collection of barley varieties. An enthusiastic correspondent, he was in touch with Danish and German botanists, as well as many early British crop scientists working in universities and institutions.

Using selection, Beaven bred two varieties that were very popular: 'Plumage' and 'Archer.' Around 1900, inspired by Biffen, he began crossing barleys and analyzing the results with Thomas Wood, a professor of agriculture at

3. Somewhat perversely, British "public" schools are in fact private, elite institutions.

Cambridge who was engaged in pioneering work on using statistics in crop breeding. Beaven's autobiography showed his attitude to Mendel as being mixed; on one hand he gives him credit for helping him focus on the need for separate characters in breeding work, but on the other, he had his own breeding system, which Mendel's added nothing to. The key factors of yield and malting quality were, he recognized, extremely complex and controlled by multiple factors—and he did not see how the Mendelian system actually helped. Geneticists, he thought, could be dangerously and impractically reductionist. Differences of opinion, though, did not stop him collaborating with Mendelians; he become a leading light in the British Seed Corn Association, which he had founded with Biffen. That he had no scientific background was not held against him by those working in science in 1922; the academics granted him their ultimate seal of approval when the University of Cambridge gave him an honorary degree. Beaven was a classic figure of the now-vanishing English Victorian era: practical, largely self-taught, independent. He was active as a Sunday school teacher, a member of the Young Men's Improvement Society, critical of scientific materialism, and of the opinion that it was the role of landowners and others in society to promote agricultural progress, not government.

Beaven's skeptical attitude to Mendel was not untypical in Britain—after the initial rush of enthusiasm there was widespread disillusionment and disappointment. At the Imperial Botanic Conference in 1924, Frank Engledow (1890–1985), professor of agriculture at Cambridge, claimed in a lecture entitled "The Economic Possibilities of Plant Breeding" that breeding had contributed an increase in output of only 3 percent to 5 percent. Plant breeding was still, he thought, an art rather than a science, and took the view that Mendel had brought about too many "wild hopes"; he thought "the genetic base of phenotypical characteristics were too complex for science to understand."[4]

As late as the 1930s it was still possible to say that amateur breeders like Beaven had done more for breeding than scientists. In 1938 the secretary of the Agricultural Research Council had stated that "the scientific man in this country has a very hard row to hoe, because the amateur breeder, the non-research-institute breeder, has made such remarkable progress that the scientific man does not find much room for direct improvement; that is due to the work of men like Dr Beaven."[5]

4. Quoted in Roll-Hansen 1997.

5. Quoted in Palladino 1996, 424.

Another example of the ambiguous British attitude to Mendel, and to the science of genetics generally, is the curious figure of Sir George Stapledon (1882–1960). Brought up by a free-thinking mother and admirer of Darwin, Stapledon maintained a lifelong balancing act between science and semi-mystical holistic ideas—an act not unusual in either Britain or Germany in the 1920s; it was during this period that the organic movement, with its mixture of the empirical and the mystical, began. Appointed in 1919 as the first director of the Welsh Plant Breeding Station, Stapledon's field was the hitherto very neglected one of forage grasses, and his passion was the regeneration of the British countryside, especially its infertile and poor hill farming areas. Credited with playing a major role in stopping the decline into poverty and dereliction of upland Britain, he had a reputation as a real rebel who had no time for committees or academic rules and procedures; he was committed to using scientific progress for the common good, but at the same time he became increasingly skeptical about science as he got older.

Forage crops are notoriously difficult to work with because of the physical problems of controlled pollination amongst vast numbers of plants and the need to be able to bulk up seed to the very high quantities needed for commercially viable new varieties. Stapledon's impatience with attempts at total rigor can therefore be seen as a positive advantage in a situation where a precise and systematic scientist may not have been able to make much headway. An example was his 'S21' cocksfoot grass (*Dactylis glomerata*), produced using mass hybridization which rapidly achieved an acceptable, but far from pure, strain—for him, 100 percent purity was pointless. Using mass hybridization between a variety of pure lines or clones to produce *polycrosses*, he took a hit-and-miss approach, with little time for cytology or genetics. Some scientists looked askance but to use the old-fashioned English expression, "the proof of the pudding was in the eating" and many were grateful for Sir George's pragmatism.

SLOWLY AND UNSTEADILY—MENDELIAN PROGRESS IN GERMANY AND FRANCE

In Germany, the development of industry and economic progress was more closely linked to scientific research than anywhere else at the time; this was particularly well illustrated by one of Germany's most successful industries—chemicals. It was not, however, reflected in plant breeding, where the reception to Mendel was mixed, at least until the late 1930s, with considerable disagreement between plant breeders. What is striking are the parallels between these disagreements and some of the contemporary issues of plant breeding politics.

What had become known as the "German method" of continual selection, a refinement of mass selection, was still popular at late as the 1920s. Skeptics of Mendel argued that hybridization was inefficient and slow and that mass selection was better, as farmers could do it themselves; this concern was supported by the fact that by the First World War, landraces were disappearing, so breeders were losing material to work with. Mass selection offered at the very least a simple and cheap way to maintain existing landrace-based varieties, which some felt were higher quality even though they yielded less. Mendelian enthusiasts felt that continued use of this method was a waste of time; as far as they were concerned, landraces contained too many mediocre plants. Many millers and brewers wanted uniformity and easier machine harvesting, both possible with the high yielding and more uniform varieties which Mendelian-guided hybridization made possible. One Erwin Bauer attacked the "stubborn and mindless" farmers who still clung to landraces, though he conceded that they were important to conserve, arguments for what we later began to call "genetic diversity" being recognized even then.[6]

The split was essentially about the differing interests of small farmers and large farmers, as indeed the similar discussions over landraces and hybridization is today. Commercial breeders who established a dominating presence in German plant breeding in the late nineteenth century wanted a single variety of each crop for the whole country—universal varieties (*Universalsorten*). Mendelian skeptics wanted support for local varieties (*Lokalzüchtung*), which they claimed were better for small farmers, as they did not need feeding and were more consistent because their greater genetic variability enabled them to cope with different conditions and seasons. The fact that they could not be machine harvested did not matter, as small farmers did not often have the appropriate equipment.

Commercial breeders in the late nineteenth century were benefiting from expanding agricultural markets, especially that of large farmers in the plains of north and east Germany, where a landscape of vast fields stretches out to distant horizons (climatic conditions are similar over large distances and soils consistent). South Germany and the Rhine Valley, however, were dominated by a hillier and more intimate landscape—a patchwork of different soils and climates with endless variations. It was, and still is, a land of small farms. Commercial breeding was late to start in the south; instead, the early

6. J. Harwood 1997, 190.

twentieth century saw state-supported plant breeding institutions set up in these regions, with an emphasis on improving local landraces, and the teaching of mass selection techniques to farmers. State institutions often worked in collaboration with growers' associations in the testing and bulking up of new or improved varieties. As far as the commercial breeders were concerned, all this was unfair competition. Mendelian supporters of hybridization and commercial breeding all tended to be in universities or institutions in the north or east, opponents in the far west or south; in other words, the attitudes and arguments of the academics were linked to the constituency for whom they worked. It is important to appreciate that opposition to Mendelism in Germany and elsewhere was based on the grounds that it was seen as inappropriate—that it achieved nothing that could not be achieved by other methods.

In some ways, scientific plant breeding developed slowly in Germany—it was not until 1927 that the Research Institute for Plant Improvement was set up. Much German effort went into combing the world for useful introductions for home agriculture, for the most part, new varieties of existing crops; between the wars it was the only European country to do this on any scale.

In France, with the full weight of the Vilmorin company behind it, Mendelian genetics should have gotten off to a good start, but in fact the company, under Henri, and then Philippe de Vilmorin (1872–1917), did not turn to Mendelian methods particularly quickly. Government was also slow off the mark, not establishing an agronomic research institute until 1921. A major separation between agricultural institutions and universities tended to part researchers from practitioners; in addition there was a strong lingering after Lamarck's theories (he was, after all, French) and a general nationalistic distrust of British, American, and German ideas. Instead there was opposition to the predeterminism implied by genetics, the positive side to which was that France avoided much of the ugliness of eugenics.

PLANT BREEDING IN A PACKET—W. ATLEE BURPEE

During the late nineteenth century, the seed house became a distinctive feature of the working lives of arable farmers, professional growers, and amateur gardeners. Sheer longevity has been a notable feature of the industry, with some companies becoming household names. Gartons and other British companies, such as Carters and Suttons, which bred mainly crops and ornamentals for market and private gardening, along with firms elsewhere, like Vilmorin in France and Burpee in America, had in fact developed a pre-

Mendelian system of breeding that was very successful—so successful, in fact, that it must have delayed the take-up of Mendelian principles throughout much of the privately owned sector of the early-twentieth-century plant breeding industry. Workers in these companies had since the 1880s used compound crosses between large numbers of parents as a central feature of their breeding. They collected plants from all over the world, bred them together to "break the type," in contemporary parlance, in order to throw up chance hybrids—similar to the approach used by the Vilmorins in France (see chapter 5). From these, the best could be selected and then be inbred for enough generations to "fix" the characteristics they wanted. These new stable hybrids could then be crossed to bring about the desired combination of characteristics. The whole process was systematic and effective, but was very time consuming. Cheap labor for the drudgery of planting, seed gathering, and sorting must have been an important part of the economics of the whole process. With the First World War, labor costs rose, which must have been a powerful incentive to look to Mendelian principles in order to streamline and rationalize the breeding process. Once they did so, the seed companies were entrenched enough to become the predominant producers of new vegetable varieties.

What made the application of Mendelian principles so attractive in terms of improving efficiency was that they enabled a breeder to calculate what new crosses were needed to "free" a desired characteristic from an undesirable one and therefore fix a character. What appears paradoxical is the amount of time it took the hitherto progressive Vilmorin company to really take this on board. Henri de Vilmorin and his son Philippe helped promote Mendelism in France but did little to advance it scientifically and appeared to have been somewhat slow in making use of it—it was only in 1917 that they released their first wheat bred through Mendelian methods. The company set up a laboratory near Paris, the first one devoted to genetics in France, but this did not lead to major advances.

Those breeders who failed to take cognizance of Mendelism were stuck with the idea of a hybrid as being primarily a source of variation, inherently unstable and therefore needing several generations of pedigree selection to fix desired characteristics. Mendelians saw hybridization as not just a source of new variation but also a way of stabilizing it through planned crosses. Above all, Mendelian principles offered a predictive tool.

One of the most successful of the seed companies has been Burpee. W. Atlee Burpee (1858–1915) was born into a middle-class family in New Brunswick, Canada; his father expected him to become a doctor like himself and his own father. The lad clearly had other ideas; even as a teenager it seemed his heart was set on agricultural improvement, starting with poultry and moving onto other livestock and plants. By the time he was sixteen, he was corresponding with other cattle breeders, including some in England; eminent gentlemen who visited the family home expecting a man of mature years were reportedly astonished at meeting such a youth. A career in raising and selling livestock seemed inevitable, but during the 1880s his fledgling company began to concentrate on seed; partly in response to immigrant farmers writing in for seed varieties they had had back home or simply for seed that stood a decent chance of germinating.

What made Burpee a success was the same factor that also lay behind Vilmorin in France, and Carter and Sutton in Britain—quality and reliability. In an age when so much seed was old or adulterated, he offered free replacement for a year after purchase, which must have reassuring in a land where pioneers in remote places needed to feel secure about the seed they were buying, and where bad seed could even have meant starvation. Burpee made it his business to bring the best of German, Dutch, Scandinavian, and British varieties to the American public, usually the result of long buying trips in the countries concerned. He soon realized though that many European varieties did not perform well in the American climate and so began a lifetime of almost obsessive improvement, much of it based on the recognition of the importance of hybridization. In 1888, Burpee bought Fordhook Farm near Doylestown, Pennsylvania, and began to develop it as a research and development facility. Among his greatest successes were the 'Iceberg' lettuce and the 'Stringless Green Pod Bean,' both released in 1894. The early 1900s saw Burpee produce varieties that were good enough to be in production for decades—among them were 'Burpee's Golden Bantam' sweet corn and 'Burpee's Fordhook' bush lima bean (naming them after the company had the advantage that every single label in the vegetable beds where they grew was a free advertisement, as was every farmer's and gardener's conversation about them). 'Burpee's Golden Bantam' was a success because it germinated in cooler soil than previous sweet corns; it was also the first yellow sweet corn—previous varieties had been white.

When Burpee died in 1915, his business was the world's largest seed company, sending out over a million seed catalogs a year. Burpee remained a staunch conservative in his private life, resisting electric lights, telephones, and cars, but always experimenting in business, instituting an advertising department in Philadelphia and offering cash prizes to the authors of the best advertisements, such as the 1890 "Burpee Seeds Grow." By an intriguing coincidence, one of Burpee's cousins was Luther Burbank—so inevitably they corresponded. Burpee's son David, who took over the firm at the age of twenty-two upon his father's death, made the link closer; after Burbank's death in 1926, he acquired the rights to Burbank's seeds as well as his (rather sketchy) breeding records.

Atlee Burpee had encouraged his sons to take an interest in the seed catalogs through the simple expedient of hiding dollar bills in the pages. David Burpee was apparently a shy and retiring character, but like his father was both shameless and innovatory when it came to business; he consciously borrowed sales techniques from the most celebrated showman of the day, Phineas T. Barnum, dropping leaflets from planes, and also becoming the first seed company to start delivering by air.

The late nineteenth and early twentieth centuries were the high point of the art of the seed catalog; many from this time were richly and creatively illustrated—particularly if they were aimed at the amateur customer. Since this was before the era of the mass-printed photograph, it was all too easy to be creative with illustrations in more ways than one—catalogs frequently featured impossibly large flowers or vegetable plants so overladen with produce that in reality they would have collapsed. As we have seen before (chapter 5), regulation was poor, so outrageous claims were often made, particularly on behalf of varieties bred by the producer of the catalog itself. Reputable companies made a great show of their honesty and plain talking to set them firmly apart from the cowboys; in America, Northrup King joined Burpee in this respect. Company founder Jesse E. Northrup even produced a fifteen-page booklet entitled "Seed Truth" extolling his company's values and wittily satirizing the dishonesty and exaggeration of much of his competition. There are a couple of homely stories to help get the point home, some thumbnail sketches, and a very entertaining page that mocks the overdrawn claims of many of his rivals—an advertisement for "The Wonderful New Bunco Potato."[7]

In Britain, Carter and Sutton were the two most important seedsmen for the horticultural trade. Both their catalogs included vegetables raised by their companies, usually with grandiose names like 'Incomparable Dark Crimson' celery and 'Carter's Champion' cucumber (*Cucumis sativus*). Those not raised by them often mentioned the breeder or the story of how they were picked out by head gardeners. Prizes won at shows and awards given by the RHS were given prominence. Encomiums from royal households and

7. Northrup 1904.

those of other distinguished personages, along with quotations from gardeners to royalty, were also often given; such opinions carried weight in a deeply status-conscious society. Such praise is particularly noteworthy in Carters catalog of 1870, when there was the first serious attempt at promoting a tomato, 'Carter's Greengage.' Now that tomatoes had been bred juicy enough for the salad bowl by both Italian and American breeders, the shiny red fruit would come to occupy more and more pages of seed catalogs with its growing list of varieties.

THE INDEPENDENT SPIRIT LIVES ON—POTATOES

The speculation in tulips in seventeenth-century Netherlands, dubbed tulipomania, has been much written about. But other crops have encouraged manic breeding and overheated investment, followed by the foreclosing of mortgages and bankruptcy—even the humble potato.

The potato is a remarkably variable plant and one that has attracted many a small-scale breeder. Unlike the other important carbohydrate crops, potatoes can be assessed by the most important method of all: on the end of a dining fork within about ten minutes of being dug out of the ground. Scottish growers began to get a good reputation for supplying healthy "seed" potatoes (i.e., tubers for planting) in the late nineteenth century; the prevailing westerly winds ensured that aphids, the main carriers of virus diseases, were kept away from the crops.[8] Eventually, British legislation directed all production of seed potatoes to be restricted to these areas. Until this happened, the only way to stay ahead of virus diseases was to grow new varieties from the true potato seed, which inevitably involved a shuffling of the genes, for the potato cross-pollinates strongly. In other words, disease was driving progress in breeding.

One Scottish grower, Archibald Findlay of Auchtermuchty, Fife, acquired a particularly good reputation—which despite his later vicissitudes, he has kept. He was an enterprising man, relying on a highly intuitive selection of parents for his (pre-Mendelian) crossbreeding program. His first successful variety was 'Up-to-Date,' released in 1892, good enough to allow farmers to make twice as much money per acre out of it than from other varieties. As so often happened in the days before plant breeders' rights, others tended to profit from his varieties *and* mix up the variety names. It was topped by

8. "Seed" potatoes are not seeds at all, but simply tubers planted to start off the new season's crop; all the tubers of a cultivar will be genetically identical.

'Northern Star' in 1900, which rapidly acquired such a good reputation that a speculation began—English farmers were growing on tubers and then selling them for up to £500 a ton, when seed potatoes normally went for no more than £4 a ton. Findlay's next potato was 'Eldorado,' introduced in 1902, which was soon selling for £150 for a single tuber. This accelerated the speculation, particularly when people outside the farming community began buying and selling seed potatoes.

In an attempt to make some more out of his breeding program, Findlay bought a farm in Lincolnshire, the flat and fertile fields of which had by then been established as England's main potato-growing area. However, an established Lincolnshire farmer and potato dealer, Mr. T. Kime, decided—whether through a desire for honesty or to undermine Findlay is not known—that Eldorado was not a new variety but only a previously introduced one, 'Evergood,' with a glossy new name. Findlay tried to sue him in the High Court, but was unsuccessful. The following season, when 'Eldorado' and 'Northern Star' both gave poor crops, the boom collapsed, leaving many bankrupt. Findlay himself lost money and had to sell his Lincolnshire venture and retreat back north of the border. Kime wrote a pamphlet about the potato boom, largely to justify his own role in blowing the whistle on it; in it he claimed that "fathers sacrificed their savings and their children's . . . many mortgaged their life insurances . . . farmers' daughters sold their poultry and ducks, and I will not say sold, but gave something much more valued than poultry or ducks, namely sweet kisses, and all in order to get hold of and grow this 'Eldorado,' this Midas that would bring riches to everyone who could touch it."[9] Findlay had the last laugh though, his 'Majestic,' launched in 1911, became the most widely grown variety in Britain in the twentieth century and played a crucial role during the Second World War, when potatoes helped keep Britain in calories at a time when merchant shipping importing grain was subject to being sent to the bottom by Nazi submarines.

'The Bruce' was another of Findlay's commercially successful varieties, named for the Scottish national hero, although there were accusations that 'The Bruce' was also a renamed older variety—'Magnum Bonum,' and that 'Up-to-Date' was not that original either. Whatever its origins, it was eventually found to have acquired over two hundred different commercial names around the world—a backhanded complement to whoever its original breeder really was. It is difficult to tell now, for then there were no govern-

9. Kime [1917?], 20–21.

ment inspectors to supervise production and distribution, no herbicides to ensure that volunteers did not reemerge from the ground to infiltrate the next year's crop and working methods which relied almost entirely on human labor, much of it poorly trained and poorly paid, with little incentive to care about keeping different stocks separate. All this, and you have to put yourself in the shoes of someone for whom the wind is howling in from the Atlantic, driving freezing rain before it, and has been up since before dawn.

Another Scottish breeder was Donald Mackelvie (1867–1947) of the Isle of Arran, which rises steeply up out of the sea on the west coast. Like many Scots in remote places, he gave up the chance of marrying rather than move to the mainland to find a wife or trying to persuade a woman from the mainland that she would rather live on his rocky, wind-blasted homeland. A breeder of champion Shetland ponies and campaigner for secondary education on the island (so children did not have to cross the water to the mainland), he made his living as a merchant and shopkeeper until 1938, when he sold up to concentrate on potato breeding. He was the first to give all his varieties a common prefix, regarded as vital to branding by many breeders since—'Arran.' His 'Arran Pilot' dominated early potato production for many years, while 'Arran Banner' was a very popular variety in the 1930s. Some are still grown today.

In the days before the registration of varieties and the legal and other mechanisms needed to back it up, there was little way breeders could make money unless they combined their breeding expertise with the opportunism and occasional ruthlessness needed for business. A copy of the *Daily Express* from March 12, 1932, had a feature headlined "Fortune in a Potato Pit, What a Banffshire Farmer Lost," about a James Henry who had bred potatoes and then left for Canada, where he ended up as agricultural editor of the *Ottawa Times*, leaving behind numerous new varieties.[10] One, originally 'The Brae Seedling,' was picked up by a Mr. Kerr, a local businessman, who renamed it 'Kerr's Pink'; it went on to become a great Scottish favorite, a position it still holds.

AN UNPLEASANT DIVERSION—EUGENICS

Eugenics is now almost a swear word. Even without considering its horrific apotheosis in Nazi Germany, its concepts have plenty to make most of us shudder. Yet it was a major intellectual current during the first third

10. Quoted in Dunnett 2000.

of the twentieth century, and by no means a current solely of the political right; many on the social democratic left were enthusiastic, and were sometimes (as in Scandinavia) in a position to put its tenets into practice.

Eugenics originated in the idea of "social Darwinism," and with the origin of genetics and scientific plant and animal breeding, it soon acquired a methodology and plenty of influential supporters. "Degeneracy" has been the worry of a great many cultures. The late nineteenth and early twentieth centuries in particular saw a great upsurge in worries about the "degenerate" or "feeble-minded" elements in their populations, or, to be more accurate, the worrying was being done by the newspaper-reading classes of western Europe and North America. Those involved in genetics and agricultural breeding were, of course, in a strong position to take a stand; and even if they did not, their professional position made it inevitable that they would be pressured to at least have a view. The American Breeders Association established a Eugenics Committee soon after its foundation, with a key figure in Charles Davenport (1866–1944) whose main concern was with restricting the reproduction of the "unfit." Davenport did much to increase membership of the Eugenics Section of the ABA, including directors of various institutions for the insane. The inheritance of "feeble-mindedness" was a frequent theme of articles originating at the Cold Spring Harbor research station where Davenport worked. Like some British researchers, Davenport's eugenics went hand in hand with an interest in biometry. He also wrote about "miscegenation" or "race crossing"—another great obsession of the time being relations between white and black racial groups.

Leading plant geneticist and plant breeder, a pioneer of pure-line breeding and progeny testing in the United States, Willet M. Hays (1890–1985), also climbed onto this unattractive bandwagon; in an address to the 1912 annual session of the National Farmers' Congress on "The Farm, The Home of the Race," he expressed concerns over genetically inefficient people and his belief the countryside as having a "new racial significance."[11] Hays promoted "positive eugenics" as part of the growing Country Life movement in rural America, which for the most part was a very positive and progressive promotion of regeneration through extension services, cooperative agricultural institutions and rural education.

Luther Burbank's widely publicized work also fed into the concerns over the quality of the human population, with one of his admiring biographers,

11. See Troyer and Stoehr (2003) for a biography of Hays, though there is no mention of his interest in eugenics.

W. S. Harwood, making comparisons between run-of-the-mill unimproved plants and "degenerate city slum people" prone to "moral depravity" and sowing "the seeds of anarchy and crime," while the strength and resilience of wild crop relatives is compared to the moral character of country people. The development of improved plant varieties was even described in lurid moral tones, "now and then out of muck of some slum, reeking with moral filth and developing with unwholesome rapidity the seeds of anarchy and crime, a white, pure life springs up."[12]

Edward Murray East (1879–1938), a key figure in the development of hybrid corn (see chapter 10), was deeply interested in eugenics. His influential 1919 book *Inbreeding and Outbreeding: Their Genetic and Sociological Significance*, goes from discussing peas to horses to people with an alarming ease. He shared the general assumptions of the times about African and other non-European peoples, with no consideration given to environmental or other nongenetic factors in the historical developments of their cultures. Above all, he, along with many others, appeared to have been so blinded by the explanatory power of Mendelian genetics in their work with plants and animals that they blithely assumed the same principles could be applied to the complexity of human society. Yet finally, he seems to have had the wisdom to accept that to make a nation out of immigrants, people must mix and intermarry. Over time, his concern seemed less to do with the racial elements of eugenics but more with population growth, arguing that food production simply could not keep up with increasing population. His *Mankind at the Crossroads* (1923) was in a firmly Malthusian mold, arguing that food production simply could not keep up with increasing population; he went on to become involved in the establishment of a number of bodies concerned with eugenics and population growth. Such concerns could be seen as being part of the background to the postwar thinking in America that led to the Green Revolution.[13]

In Britain, the romantic idealizing of country people led Sir George Stapledon to pronounce that they had better genes, which could be used to improve the vigor of the English race. The belief that rural genes were somehow better than urban ones is common to far right and nationalistic ideologies; it was certainly held by the Nazis. These beliefs were very much a part of a groundswell of opinion and thinking about the survival of the British countryside and its farming communities which developed during the

12. See W. S. Harwood 1905, 33.
13. East 1919; Rodgers 1949.

interwar years, associated with what from today's standpoint looks like an odd brew of organizations, people, and belief systems—a fermenting pot of socialism, pacifism, anti-Semitism, Christianity, and neopaganism.

SUMMARY

The introduction of Mendelian methods has a certain parallel with the contemporary discovery of the elements in chemistry; genetics could be used to analyze and then to synthesize new plant varieties. There was also a parallel with the way that the Ford production process made cars, through breaking down a complex task to a mass of simpler ones. Fundamentally it was about a transition from craft to industry and an accompanying professionalization of plant breeding—a task for trained specialists, not just farmers, amateurs, and nurseries. The application of science to plant breeding was slower in more traditional European societies than it was in the United States. There, in contrast to the US government–led approach, there was a variety of approaches to stimulating breeding: private, nongovernmental institutions and government support through universities and research organizations.

The varied acceptance of Mendel highlighted differing relationships between farmers and plant breeders in different places. Not all breeders accepted the analogy with chemistry, arguing that Mendelism was too reductionist—today the word "holistic" would probably have been used to describe their approach. However the fact is that for many years after Mendelian genetics was first introduced to Britain and the other countries of western Europe, a nonscientific, purely "technological," or, "craft," approach to breeding was still yielding very good results, and was supported particularly by farmers in more marginal areas, for which indeed it may have been more appropriate at the time.

Pre-Mendelian techniques also remained popular with seed companies, which were working with large numbers of different species. However because Mendelism offered a predictive tool, it created a short cut to breeding, which eventually became accepted, with rising labor costs probably contributing to its acceptance; much pre-Mendelian breeding work was time consuming and labor intensive. Seed companies became increasingly important as a growing number of consumers were able to afford to buy a wider range of vegetables and other crops from small farms.

EIGHT ❦ LUTHER BURBANK ❦ MIRACLE WORKER OR CHARLATAN?

In downtown Santa Rosa, California, there is a park—the Luther Burbank Home and Gardens. It is more of a garden than a park, with flower borders and lawns around a white painted clapboard house. There are walnuts and fruit trees, too, and a huge clump of cacti. A large greenhouse makes it clear that this was once occupied by someone who took the plant kingdom very seriously. From 1884 to 1906, however, the ground around the house would have looked very different. There would have been rows of vegetables, flowers, fruit bushes, and young trees, and it would have been the scene of considerable industry. For this was the home of one of the greatest names in plant breeding history, a name that appears again and again in the United States: schools and banks are named after him, even a whole city (Burbank, California); and the 'Russet Burbank' potato is the nation's most widely grown variety. For much of the twentieth century, Luther Burbank was regarded as one of the great American heroes.

Most plant breeders are retiring types, happier in the lab or field than facing an audience; the names of only a few become known beyond their professional world—and most of those are in the world of ornamental horticulture. Not so Luther Burbank (1849–1926). He was a larger-than-life character, but like all heroes, his legacy and reputation have come under attack. Separating the myth from the man is not easy, but at least we now have the luxury of distance between his time and ours, and it is easier to make a more balanced assessment. Burbank's life and work is a side branch of our main story; he does not fit into the zeitgeist of turn-of-the-century plant breeding or into the history of plant breeding as a coming together of plant cultivation and genetic science. However, the fact that he so dominated the public image of plant breeding in his time and was regarded as a hero by many for decades afterwards assures him a central place in our story.

THE GENE POOL AS CORNUCOPIA— RISE OF A GENETIC WIZARD

Luther Burbank was born in Lancaster, Massachusetts, in 1849. As a child, the thirteenth in a family of fifteen, he was a voracious reader and apparently had the habit of carrying around a pet cactus. He received only elementary education and ended up working in a market garden, where at the age of twenty-four he made his first selection, of a potato. The story clearly illustrates what was almost certainly his core skill—a great power of observation, an uncanny knack for spotting potential. He had seen a seedpod on a batch of 'Early Rose' potatoes, a variety that rarely produced fertile seed. Twenty-three seeds resulted in a crop of which only two plants were any good. The young Burbank recognized the potential of one of the plants and sold it to a local seedsman for $150.The Russet Burbank potato proved to be a great success, indeed, perhaps the greatest success of his career, which seeing as it was his first, can only be described as truly savage irony. Introduced in 1871, the Burbank came to be better known as the 'Idaho' for the way it came to dominate America's leading potato state. Even today it is still widely grown and has been described as "unsurpassed, a staple of American agriculture." However, it is very blight prone—not a problem in Idaho's dry climate but one that necessitates the use of considerable quantities of fungicide when grown elsewhere. However, since at least one well-known fast-food company insists on it for french fry making, due to its high dry-matter content and oblong shape, it is very widely grown, with obvious environmental consequences in terms of pesticide pollution.[1]

At twenty-one, Burbank went to California, where, as with all good American success stories, he started out practically destitute, reportedly sleeping in a chicken coop while he worked in a nursery north of San Francisco. Eventually he managed to save enough to set up his own small nursery in Santa Rosa. During this time, he began to research plant breeding and was much impressed with the work of Charles Darwin, drawing the conclusion from the naturalist's description of plant fertilization that natural systems could—and should—be harnessed for human benefit.

In 1893, Burbank went into plant breeding full time, raising new fruit and flower varieties and selling them to other nurseries who would raise them in quantities large enough to sell, or to seedsmen. He remained a proponent of Darwin all his life (supporters of Darwin and Mendel tended to disagree at the time) and saw himself as working in a Darwinian mode, creating new species and operating the processes of selection in place of nature. At this time his activities raised a certain amount of religious opposition from a local clergyman, who invited him to his congregation and then denounced

1. Shaw 2007.

him from the pulpit, as he was daring to rival The Almighty in creating new forms of life.

Burbank's nursery in Santa Rosa and the farm in nearby Sebastopol that he acquired in 1885 became famous for the new technique that Burbank pioneered—the large scale trailing of plants, an approach that, apart from anything else, created a very powerful impression on anyone who saw them, especially when applied to ornamental plants; his lily fields were apparently so vast that ranchers a mile away could smell the flowers. Whereas previous plant breeding efforts had been relatively small scale, and selection often involved picking individuals out from existing fields of crops, Burbank's approach was to grow vast numbers of individuals, millions if necessary, simply for the purpose of selecting out the best of whatever characteristic it was he was looking for.

Hugo de Vries, who sailed the Atlantic to visit him in 1904 and again in 1906, was clearly fascinated by his work and declared that "what makes Burbank's work entirely different to other plant breeders is the immense scale." But De Vries also recorded his disappointment that Burbank had no new techniques, no secrets; he was very critical of his destruction of records and of failed plants, making it impossible to see what the ancestry of his hybrids was.[2] Burbank crossed species with great enthusiasm, set out the progeny, selected the best, and discarded the rest. He clearly understood the value of bringing together hitherto unconnected gene pools but appeared to have an approach that was almost random, crossing and re-crossing until something worthwhile came up. His ruthlessness in disposing of material that did not come up to his very high standards was also probably important. Bailey also visited Burbank and came to similar conclusions: that he had no magic, only persistence, and an acute judgment. Bailey was a pragmatist and apparently was only prepared to judge Burbank on his results.

One possible contributing factor in Burbank's success as a breeder was an ability to establish correlations between characteristics visible in young plants and the mature characteristics for which he was breeding them; his ghost writer Henry Williams reported that Burbank could predict the quality of a vine's future fruit from the nature of the tendrils of a young plant and the promise of a plum by the features of its leaves and twigs. Establishing such correlations is an important part of modern plant breeding practice—and invaluable when dealing with plants which may take many years

2. De Vries' comments are quoted in W. L. Howard 1945, 360.

to fruit. Given that Burbank was reported to have made selections of fruit trees at the seedling stage, it is highly likely that he did develop an intuitive sense of correlation, but perhaps he lacked the analytical skills to recognize exactly what it was he was looking at.

The Burbank approach is well illustrated by his work with plums; his innovation was to use Japanese plums, cultivars of *Prunus japonica*, early in his work. He also used the beach plum (*P. maritima*), inedible but hardy and able to thrive in poor soil, so producing varieties that enabled this important crop to be grown in a wider range of climates and soils than had been previously possible. From 1885, when he made his first importation of Japanese plums, he worked off and on with species of *Prunus*: apricots (*P. armeniaca*) and cherries as well as plums, so that by the early years of the twentieth century, he was able to produce hybrids with very complex parentages; De Vries reported that the 'Alhambra' plum was derived from no less than seven distinct natural species. Burbank undoubtedly made a major contribution to the high position that the California fruit industry held for much of first half of the twentieth century as a major earner for the state—by 1945, there were two million Burbank-bred plum trees in Californian orchards. The state not only exported fresh fruit all over the continental United States by rail, it also shipped dried plums (prunes) all over the world. Breeding thin-skinned varieties that would dry easily was one of his objectives; other varieties were bred specifically for canning. There was an almost evangelical fervor in his search for bigger plums, for large fruit had advantages in that the proportion of flesh to stone and skin was higher and picking effort more economical. The resources of the tree were used more wisely—which no doubt appealed to the prevailing work ethic of the time. Burbank was clear and eloquent in seeing food production as being an industrial process—"the fruit grower of today is a manufacturer . . . the manufacturer of pins and nails would not tolerate a machine which failed to produce every season or a rusty product." Industrial production is also profoundly democratic, something which the producers of niche goods often forget; "my purpose," he once declared, "is to make the very best fruits and nuts an everyday food for all, instead of an occasional luxury for the few."[3]

The industrial scale upon which Burbank worked, and the opportunities it created for him to make high quality selections, made a huge impact on his contemporaries. The moral overtones of Burbank's work were dramatically emphasized by the large bonfires he held several times a year in which

3. Quoted in W. S. Harwood 1905, 127.

discarded plants were burnt (these would have been the vast majority of the plants from a trial); the overtones of evangelical religion with the fires of hell ready to receive sinners could not have been far from the minds of many onlookers. Yet Burbank was not a religious man; indeed, he became quite a notorious skeptic in his later years.

There was little in the temperate zone, economic, or ornamental plant world which Burbank did not turn his attention to, although his most important achievements were probably in fruit, of which he introduced over 250 new varieties, including 113 plums and prunes, 10 strawberries, 10 apples, and 11 plumcots, the latter a plum/apricot hybrid.[4] In total, around 800 varieties were introduced by Burbank over a 50-year period, derived from 121 genera. Amongst edibles, his 'Burbank Giant' winter-growing rhubarb (*Rheum officinalis*) was one of the most successful, although his understanding of why it produced so early seems almost laughable now; the pollen parent had been obtained from New Zealand, and it seems as if Burbank believed a widely held myth of the time, that plants from the southern hemisphere had a different biological calendar, which was somehow transmitted to the next generation. Of ornamentals, perhaps his most lasting achievement was the 'Shasta' daisy (*Leucanthemum x superbum*) and the development of large-flowered hippeastrums (often known as amaryllis), the size of the flower being increased from around seven centimeters in the wild species with which he worked (*Hippeastrum aulichum*, *H. vittatum*, *H. reginae*, etc.) to up to thirty centimeters. Whether agricultural or horticultural, he inevitably sold new introductions to other growers or seedsmen to make available to the public, rather than undertaking commercial production himself.

The Shasta daisy is the plant for which Burbank is most closely associated. He crossed wild chrysanthemum stock from North America (it is not clear exactly where from, or even what species), making selections, then crossing with *Leucanthemum maximum* and *L. lacustre* from Europe to produce plants with prolific and large blooms. Finally he used *L. nipponicum* from Japan, whose flowers were small but pure white. Burbank raised over a hundred thousand seedlings over a six-year period; he constantly selected out those that had the biggest flowers (he always carried an ivory ruler with

4. Burbank's plumcots were not particularly successful commercially, although useful for marketing purposes; Floyd Zaiger (1926–), a leading U.S. fruit hybridist has brought modern genetics to bear on plum x apricot hybrids and has had considerably more success—the sweet and juicy "pluot" is now widely available in North America.

him) and those that bloomed the longest. Naming the resulting plant after a local mountain, the Shasta daisy has gone on to become a classic garden plant, its showy free-flowering nature matched by its ability to survive neglect and the competition of weeds and rivals.

Like much of Burbank's selections and hybrids, the exact details of the Shasta daisy's parentage remain obscure, as he did not keep accurate records. Some who worked with him ended up by becoming very critical of his slipshod style and habit of selling plants before they had been fully trialed, while recognizing that he had an amazing ability to discern variety characteristics and make guesses over the parentage of chance or random hybrids. In some instances he has been shown to have launched "new" varieties which had originated elsewhere—the oldest trick in the less scrupulous plant breeder's book; it is difficult to know whether he was at times "economical with the truth" or whether with so many irons in the fire, he occasionally got confused. His ventures into cereals provided plenty of material for his detractors; in 1918 he launched three new wheats, their names redolent of the confidence he now felt in his work: 'Quality,' 'Quantity,' and 'Super.' 'Quality' was eventually described as being identical to an Australian variety called 'Florence,' 'Quantity' never really took off and was also apparently identified as a previously existing variety, and 'Super' turned out to be 'Jones Fife,' a Russian variety which had previously been introduced by USDA in 1893. 'Quality,' however, was a commercial success and dominated wheat production in the north-central states of the United States for several decades.[5]

A REPUTATION IMPALED—ON A SPINELESS CACTUS

The affair of the spineless cactus was perhaps Burbank's most notorious entanglement with publicity. A few Texan ranchers had experienced some success with cacti as a forage crop, which led to speculation that its potential was somewhat greater than it really was. The USDA never took it seriously; the cactus is largely water and needs considerable quantities of fertilizer and irrigation to grow at a rate that makes it economical to feed to cattle. From 1906 onwards, Burbank introduced a total of 34 spineless, or nearly spineless, forage cacti, nearly all of the prickly-pear genus *Opuntia*. His more enthusiastic supporters had a field day; one admiring biographer, W. S. Harwood, quoted Burbank as saying "the population of the globe may be doubled and yet, in the immediate food of the cactus plant itself and

5. These wheats are discussed by W. L. Howard 1945.

in the food animals which may be raised upon it, there would still be enough for all"; Harwood added that "the times of little rain are set at naught, the great flame-hearted sun itself, burning its mighty way across the blistering desert is defied, the whole desert and arable regions of the globe, by the act of one man may become a limitless reservoir of food."[6] The skeptics had plenty of material too and did not have to wait long before feeling vindicated—the cactus revolution never happened.

A lack of accuracy may have been one reason for the mutual distance kept between Burbank and some more scientifically trained workers in the field. He had no formal training and, according to Harwood, he was never able to keep a single graduate in his employ, a fact that the biographer puts down to his intolerance of theory. More likely, it seems as if he found it impossible to delegate. His approach was entirely empirical; he tried crossing species and if they produced offspring, he carried on working with them, but if they failed, he moved onto something else.

A lack of patience with the theories of others extended to a complete distrust of Mendel's work, and despite his early reading of Darwin, he appeared to believe in the inheritance of acquired characteristics. He shared a romantic and antimaterialist viewpoint with many other members of the American elite at the time and was influenced by a number of contemporary writers and thinkers who explored these themes, such as Ralph Waldo Emerson (1803–1882), Henry Thoreau (1817–1862), and the German explorer and naturalist Alexander von Humboldt (1769–1859). Burbank's presentation of plant evolution as a struggle between heredity and environment was also described in a way that was mystical rather than rational, particularly regarding a life force which he believed drove heredity.

One advantage of not understanding or rejecting a materialist conception of science is that it does not hold you back from attempting the seemingly impossible—and occasionally achieving it. Some of his crosses have entered the realm of mythology, with a fog of uncertainty about whether he really made them or not: the apple x blackberry and gooseberry x raspberry certainly belong in this hazy region, as does the 'pomato,' an alleged tomato x potato hybrid. A contemporary joke about him suggested that he crossed a milkweed with an eggplant and got an omelet—Burbank was apparently not amused.[7]

After his death, opinions about Burbank became somewhat divided;

6. W. S. Harwood 1905, 157–58.

7. Quoted in W. L. Howard 1945, 329.

there were those who regarded him, along with Thomas Edison and Henry Ford, as a folk hero who brought the benefits of modern science and technology to the common man, and those, especially in the scientific community, who regarded him as something of an embarrassment. Of Burbank's faults, one of the worst was undoubtedly that he was too easily flattered, and it was the hype and the sales patter of others, notably the nurserymen to whom he sold new varieties, which caused him his worst embarrassments. A good example was the 'Wonderberry,' a cross between two nightshade species, *Solanum nigrum* and *S. africanum*. Both parents were inedible, the offspring scarcely less so. Burbank named it the 'Sunberry,' but nurseryman John Lewis Childs, who had bought the plants to market them renamed it the 'Wonderberry' and launched it upon the world as a tasty new fruit, even publishing a recipe book—*100 Ways of Using the Fruit of the Sunberry or Improved Wonderberry*. It may have been Childs who grossly overplayed the Sunberry's modest hand, but Burbank had an inability to ever admit that any of his vegetable-children was a failure. The bastard nightshade soon came near to ruining Burbank's reputation, with the judges of the 1919 Boston Flower Show declaring it to be worthless and a crusading paper, *The Rural New Yorker*, accusing Burbank of deceiving the public and alleging that the fruit was probably poisonous.

Burbank was not ignored by all contemporary scientists; De Vries, as we have seen, gave him serious attention. And for some six years, from 1905 to 1911, he was funded by the Carnegie Institute to the tune of $10,000 a year. The institute was founded in 1902 and incorporated by an act of Congress in 1904 with a remit to engage in "research . . . discovery and the application of knowledge to the improvement of mankind."[8] George Harrison Shull (1874–1954; see chapter 10), a renowned geneticist and founder of the journal *Genetics*, would work with Burbank for much of this time, in order to try to understand and record how he worked. The relationship was clearly not a happy one, and Shull soon became frustrated at not being able to make much sense of what Burbank was doing. They were two very different men—Shull a hard-headed scientist and Burbank a results-orientated and intuitive entrepreneur—and it is to the great credit of both of them that they managed to work together as long as they did without falling out, or at least without airing their dirty linen in public. The Carnegie Institute kept on promising a report on Burbank's work but then backtracking, saying it needed more time. By 1909, the situation was becoming embarrassing for the Institute, as

8. Quoted in ibid., 433.

the Wonderberry had hit the headlines, as had the spineless cactus. Funding was discontinued in 1911, and a final report was never produced. Apparently, an understanding had been reached between the institute and Burbank that such a report would never be produced during Burbank's lifetime; but after his death the institute was undoubtedly left in a difficult position; they had expended $50,000 with almost nothing to show for it. In the end Shull carried on with his own work, and the institute quietly dropped the whole business. Burbank himself had apparently found working with the institute onerous, complaining of "hampering restrictions and unprofitable conditions," and had been glad to return to "a life of active freedom."[9]

That Burbank's reputation took a dive after his death was not entirely his own fault or that of opportunistic associates. By the mid-1920s, geneticists were beginning to make a complete take over of US plant breeding; science and not artistry was the way forward. Donald Jones, who along with Shull played a major part in the development of hybrid corn (see chapter 10), was apparently particularly dismissive, regarding him only as a producer of novelties.

In a nation hungry for heroes, and heroes of a new kind, not the military ones of old, but leaders in human progress, Luther Burbank was presented to several generations of children through school textbooks as an exemplary figure. A photograph taken in 1915 during the Panama Pacific International Exposition in San Francisco shows him seated with Henry Ford and Thomas Edison: the plant breeder, the industrialist, and the inventor—a triumvirate of the new kind of men who were making America the most dynamic economy and society in the world.

While always ready to receive the notable and famous, such was his fame that ordinary people came in droves, most of them in the summer months when he was fully stretched in doing his breeding work; not surprisingly, he wanted to get on with his work and the less-important visitors were often received with curtness or even downright rudeness. Some of his guests were in a position to spread the word that he was a self-important imposter and fraud, an assessment likely to receive a hearing from those who had bought any of his highly promoted but less-successful products. It should come as no surprise that one study of his work, a couple of decades after his death, was titled "Luther Burbank: A Victim of Hero Worship."[10]

We have seen how a scientific approach to plant breeding, based on Men-

9. Quoted in ibid., 466.
10. This is W. L. Howard 1945.

delian genetics, took a long time to get established in some countries. Even in the United States, which had taken so rapidly at an institutional level to Mendel, out in the fields and orchards, a more critical view of academic institutions was widespread. Many small farmers, influenced by the rural populism of the late–turn-of-the-century period, had a gut reaction resistance to science and government—scientists were seen as most properly working in the field, not doing lab-based research. This love of the "little man," the bold entrepreneur from a humble background, was perhaps part of the explanation for the appeal of Luther Burbank. In the 1930s another independent breeder, the "strawberry king" Albert Etter was offered funding by the state legislature of California. The dean of the College of Agriculture at the University of California objected, wanting him to work as an employee of the university if he received state funding.[11] In the end the dean got his way. The moral of this was that academic institutions wanted control—Burbank had ended up making the venerable Carnegie Institute look foolish, and the men in the mortar boards did not want this happening again.

SUMMARY

Luther Burbank lived and worked at a crucial time in plant breeding history—as it was turning from a craft driven largely by intuitive skill to an applied science. Burbank was an old-fashioned intuitive grower, so it is not surprising that his reputation has faded since his death. Perhaps his greatest legacy was his populist introduction of the very idea of plant breeding to the American general public, the idea that nature is not immutable and that plants can be improved for our appetite and enjoyment. Sometimes a showman is needed to get a new idea across to an audience that does not have the patience or the knowledge to appreciate subtle scientific details. Certainly, plant breeding history would have been much duller without him.

11. Palladino 1994.

NINE ❦ "LET HISTORY JUDGE" ❦ PLANT BREEDING AND POLITICS IN THE USSR

On the fourth of August 1940, Russian botanist Nikolai Vavilov was working in the field, in what was then the Soviet Socialist Republic of Ukraine. It had been a productive day—he had found a new wheat variety. At dusk however, a car drove up. Three men got out; they were from the NKVD, Stalin's political police. Vavilov was arrested. He was tried in July 1941 and convicted of spying (for Great Britain) and of being part of a "Rightist conspiracy" and of sabotage. Sentenced to death, which was commuted, he died in jail in 1942 of malnutrition-induced pneumonia. Like many thousands of others, Vavilov was a victim of a decade of terrifying paranoia; anyone who had foreign contacts was particularly at risk. Vavilov had studied in Great Britain, and being a scientist, he maintained a range of international links—indeed, while he was in jail he was declared a member of the Royal Society of London. Vavilov's death was a tragedy, but it was part of a wider tragedy—and part of the wholesale destruction of Soviet genetics and a vast turning backwards in agriculture.

In 1972, the USSR bought four hundred million bushels of hard red winter wheat from the United States.[1] The irony is that this wheat was descended from winter wheats that originated in Russia and the Ukraine (see chapter 5). A greater irony, perhaps, was that the Soviet Union was unable to feed itself—Communism, for all its rhetoric about scientific progress, had feet of clay. The immediate reason for the failing harvests was the inadequacy of a command economy, but in the background was a massive failure of Soviet science.

DAYS OF HOPE—NIKOLAI VAVILOV

In order to understand the background to the sorry story of plant breeding under Soviet Communism, it is necessary to have some feeling for the atmosphere of the early Soviet Republic. There had not been anything like the Bolshevik Revolution of October 1917 before. A small group who saw themselves as the vanguard of the masses seized power, empowered by an

1. Quisenberry and Reitz 1974.

ideology which, while describing itself as scientific, could more accurately be described as deterministic. Karl Marx's theory of history stressed the inevitable nature of Communism. V. I. Lenin's Bolshevik Party, through its "dictatorship of the proletariat," would carry out this historically determined task, and by unleashing the creative energy of working people would achieve things that had never before been achieved. Under class society, the energy of the mass of the population was wasted; under Communism it would be released and directed by the vanguard party. "If hundreds, thousands, and sometimes millions of individuals participate in a job, then exceptional results are always achieved . . . history forces leading individuals to do exceptional things" said the Soviet head of state Mikhail Kalinin in 1922, addressing agricultural specialists. "We will not only catch up with the agriculture of western Europe, but will also surpass it."[2] There was a feeling that if the party willed something, it would happen. Anyone who expressed doubts about a course of action was all too easily labeled as a traitor.

The October Revolution of 1917 started with so much promise; a new society run for the benefit of working people that also offered progress and opportunity for scientists, intellectuals, and artists, too. One beneficiary of these early days of hope was the botanist Nikolai Vavilov. Educated at the Moscow Academy of Agricultural Sciences, he had had a period abroad in 1913–1914, spent mostly in England studying under Biffen and Bateson at Cambridge (see chapter 7); he also visited Henri de Vilmorin in Paris and the biologist Ernst Häckel in Germany. In 1917, he became professor of genetics at the University of Saratov and soon attracted the interest of Lenin and was appointed director of the All-Union Institute of Applied Botany and New Crops—renamed the N. I. Vavilov Research Institute of Plant Industry (VIR) in 1992.

Vavilov's work, some of the most important ever done on domestic plant origins (see chapter 1), was built on firm foundations—a Bureau of Applied Botany had been established at the Ministry of Agriculture in 1894. Director Robert Regel, appointed in 1905, developed a system for the classification of crops based on a belief that all varieties and landraces of a crop need to be recorded and studied, not just the current commercial strains; he made collections of crops throughout the Russian Empire. As at Svalöf (see chapter 7), the work of the bureau was made easier by technical advances in the machinery for collecting seed, and for cleaning, threshing, grading, and drilling it.

From being primarily interested in pathology and developing strains

2. Quoted in Joravsky 1970, 26.

that were pest and disease resistant, Vavilov's promotion led him to working on a much more ambitious project, the assembling of the "foundation material" for a new program of breeding. With Lenin's support, expeditions were organized, which by 1933 had totaled around 200 in 65 countries, and collecting 150,000 taxa. In order to make sense of this mass of material, it needed to be analyzed, categorized, and classified, a task for which Vavilov's intellectual and organizational abilities were ideally suited. Vavilov's expeditions were not only far more extensive than anything organized before, their conception and rationale was quite different. While the USDA had overseen the introduction of a huge amount of new plant material, this was not done in such a systematic or coordinated way. Vavilov's key conception was the idea of genetic diversity, the sheer scale of this diversity, and the enormity of its implications. He particularly appreciated the need for a systematic approach to the classification of this diversity. His idea, that in order to breed new crops the whole range of genetic material needs to be looked at, was completely new at the time; while Regel's work had probably sown the intellectual seeds, the idea had apparently come to him on a collecting expedition in the Caucasia region while he was still a student.

The Caucasus was an area immensely rich in biodiversity and primitive crop varieties; he often returned there, and it was there that he found the disease-resistant wheat relative *Triticum timopheevii*, which is still being worked on as a source of disease-resistant genes. As well as the massive territory of the USSR, expeditions were sent to all neighboring continents: to North America, South America, and parts of Africa (Ethiopia in particular). The result was that the USSR accumulated the world's finest collections of crop varieties and pioneered the whole process of systematically collecting landraces, a forerunner of the global effort to collect and document landraces and crop relatives during the 1960s.

Expeditions were launched with very little money; Vavilov and his colleagues often had to endure numerous privations for long periods, "he had to sleep in his overcoat for weeks," in the words of American breeder S. C. Harland, who met him in Ethiopia. Harland also described him as having an "extravagant showmanship." He was certainly an original thinker on a big scale, very much a "big picture" man rather than someone who was interested in detail. He was physically big and had "a mind that never slept and a body which in its capacity for enduring physical hardships can seldom have been matched." "Life is short—one must hurry," he used to say.[3]

3. Harland 1945, 622.

Vavilov's expeditions led him to his centerpiece theory, that of geographical centers of origin (see chapter 1). But he was a practical breeder, not just a theoretician, who used his theory of centers of origin in a flexible and pragmatic way to achieve real results. Researchers, he said, should not to be too dogmatic in linking the climate of origin to the likely success of the plants elsewhere, citing Ethiopian wheats thriving in Leningrad. Barley was an early success—by 1939, half the union acreage was of varieties from the All-Union Institute of Plant Breeding.

The All-Union Institute was, from the 1960s on, able to begin to make effective use of the surviving Vavilov collections, since scientists in the USSR were finally able to catch up with the genetics that the rest of the world took for granted. By 1994, over 2,500 cultivars of agricultural crops had been bred using material from the collections, with around 95 percent of new cereal varieties being derived from at least some Vavilov material. With détente between east and west in the 1970s, institute scientists were able to make even more progress, particularly because they were finally able to collaborate with colleagues internationally.

It was only after the fall of Communism that one of Vavilov's greatest secrets was released—the original home of the apple, Alma-Ata (father of the apple) in Kazakhstan in 1929. Notes on the discovery had been lost, but the knowledge had been kept by Aimak Djangaliev, a student of Vavilov's who, after the fall of Communism, invited American plant scientists to visit the area. British DNA research confirmed that the trees in the "fruit forest" of Dzhungariskiy Alatau in Kazakhstan, a small area on the border near China, *Malus sieversii*, were the ancestors of the domestic apple rather than the result of interspecific hybridization among several species, as had always been previously thought. Most intriguingly, the wild populations of *M. sieversii* show considerable variation in the time it takes the fruit to ripen, thought to be an adaptation to ensure that a variety of different animal species, including bears and wild horses, have an opportunity to eat the fruit and so distribute the seed; this genetic variation has been of great value to us, as it has always meant that it is relatively easy to select clones that ripen at different times.[4]

4. One account of the rediscovery of the original apple population, making the somewhat outlandish claim that that it was largely a British discovery (in 1997), is given by Price in an article—"East of Eden" in *The Garden*. There is a fascinating account of wild apples and the historic development of the fruit by Juniper and Mab-

DAYS OF DELUSION—
IVAN MICHURIN AND "SOVIET CREATIVE DARWINISM"

Ivan Vladimirovich Michurin (1855-1935) earned his place as one of the young Soviet Republic's leading plant breeders by his ability to collaborate with the atmosphere of manic energy that the Bolshevik government engendered. Unschooled in scientific method and unattached to any academic establishment, he did not allow inconvenient results to get in the way of his climb to authority. Born in Riazan province, an area of particularly harsh climate, to a family of impoverished noblemen, his education was patchy and he had to work as a railway clerk before he could take over the management of his family's orchards at the age of thirty-five. Not surprisingly, he identified as his most important task the developing of fruit varieties that thrived in the local climate, and began hybridizing, developing a series of novel breeding techniques.

Michurin felt himself to be spurned and insulted by officialdom before the revolution, describing his life as a "pariah" and "outcast," living in poverty and threatened by the clergy who denounced his work with hybridization, claiming that one priest had said to him, "don't turn God's garden into a brothel."[5] Looking for support outside the country, during 1911–1913 he made contact with the USDA, which at the time was undertaking extensive plant-hunting work in Russia and later claimed that he had been offered large sums of money, the directorship of a research station in America, and even his own ship to transport his plant collection to the states.

Michurin's methods can only be described as eccentric, with little scientific underpinning or methodology. His achievements were, perhaps, like those of Luther Burbank, based on good intuition. Central to his methodology was "training," as "the influence of the sum-total of external factors on the structure of the hybrid organism is so great, that in most cases it overrules the action of the characteristics and properties hereditarily transmitted by the parent plants . . . in particular this influence manifests itself in the mother plant at the time when seeds of the future hybrid organism are being formed," he declared[6]—a version of Lamarckism, in other words. His

---

berley (2006), but they have little to say about Djiangaliev. For the story of Djiangaliev, see Browning (1998).

5. Michirun 1949, xvii.

6. Ibid., 189–90.

favorite method was to choose pairs of parent plants, grow them in the conditions for which cultivars were being selected, cross them, and grow the young plants on; the seedlings would "acclimatize" to the local conditions and so be "trained" in such a way as to be able to transmit their new traits, such as the all-important cold hardiness, to any plants grown in turn from their seed. Michurin held that purebred plants produced less adaptable offspring, but that hybrids produced offspring that were more adaptable and easier to "train" to new climatic and other conditions. The more dissimilar the parents, the greater the opportunities for training; thus interspecific hybrids held out greater hope than hybrids of closely related cultivars.

The first fruiting, he believed, produced the best seedlings for acclimatization. The process of development, of adapting to the new environment, continued until characteristics were "fixed" at maturity, but after this, the process of adaptation steadily declined as the plant aged. Young plants were involved in what he described as "gradual form-building," with each gene involved in "a ceaseless struggle for existence . . . with that one conquering which happens to find favorable conditions accidentally existing in the environment, or artificially created by man" while genes "that have been inherited in a lesser degree or such as fail to find the proper basis for development are partly destroyed" or "remain in a latent state."[7] As an example, he describes the development of an apple hybrid, between the 'Niedzwetzkyana' and the 'Antonova,' in which the characteristics of the latter, in growth and fruit pigmentation, gradually spread from one side of the young tree to the other. Another technique in Michurin's armory was the grafting of material from one potential parent onto another before attempting hybridization, which he believed made previously difficult crosses possible, therefore opening up great possibilities. He claimed to have produced apple x pear, apricot x plum, sorbus species x pear, and apple x hawthorn.

Mendel's work was anathema to Michurin; he simply believed that his discoveries were inapplicable to hybridization. He had little time for the work of others, declaring that "it is generally not my custom to sprinkle my work with references to the investigations of others, if only because many of the postulations of authoritative persons are not very sound. . . . [I] consider the propping up of my investigations by references to the works of others a form of needless cowardice in the face of criticism." He had little time for his contemporary Luther Burbank, who he lumped in with "miserable treasure hunters" who used mass plantings and were happy only to discover "isolated

7. Ibid., 164.

chance finds." Instead he described how the "originator must strive . . . not a hundred thousand, but only some tens of seedlings . . . and then by proper training, to improve the largest possible number of them."[8] So, like Burbank, he worked on an industrial scale.

By 1919, Michurin had produced 153 new varieties, including 45 apples, 8 grapes and 9 apricots. After the Revolution, he was able to gather, according to his account, some 800 plant species (probably in fact cultivars) from all over the world with which to work with. Nevertheless, as late as 1931, he did raise 5 cultivars of kiwi fruit, which he thought could have potential as a source of vitamin C; they were introduced to the United States and are now amongst the most widely grown cultivars in the world.[9]

During the early Soviet period, when scientists and other specialists could speak without fear of the firing squad or the Gulag, Michurin was unable to garner much support, although Vavilov was impressed enough by his notes to recommend he be given modest state support. With Kalinin's visit to the nursery in September 1922, his star began to rise, crucially so among the politicians. Most specialists were unimpressed—at the First All-Russia Agriculture Exhibition of 1923, Michurin's contributions hardly got a mention in the papers, and a number of scientists denied that some of his exhibits were authentic. At the end of the exhibition, however, Michurin and his young supporter I. S. Gorshkov, had somehow won the ear of the newspaper *Izvestia* with a dramatic claim that the USDA was offering him support to move to America. Effectively blackmailed by the headline "Kozlov or Washington" in *Izvestia* on October 14, 1923, the Council of People's Commissars decided that the elderly Michurin's nursery was of national importance and deserved full state funding. Perhaps the Soviets felt the need for their own Burbank, for from that point on, he was a hero, one of those who was making possible the great achievements that the party had promised. After his death, a film was even made about him (*Michurin: A Life in Bloom*, 1948) with a score by the leading Soviet composer Shostakovich.[10]

Granted state funding and the grounds of a nearby monastery that had been expropriated, Michurin went on to claim that his methods of breeding needed to be applied across the whole vast country as his own varieties were suitable only for his own climate zone. Possibly this was one way of ex-

8. Ibid., 203, 485

9. Strik and Hummer 2006. Kiwi (*Actinidia chinensis*) is originally from China.

10. "*Michurin*" (movie review), New York Times. http://movies2.nytimes.com/gst/movies/movie.html?v_id=140463 (accessed January 6, 2007).

plaining their poor performance elsewhere. Eventually, a total of fourteen zonal research institutes all over the USSR were modeled on his institute at Kozlov (renamed Michurinsk in 1932), and over a hundred local centers in northern areas. Plant breeding, Michurin believed, should become a mass activity, with horticulturalists, kolkhoz (collective farm) workers, and members of the Komsomol (the Communist Party youth organization) actively involved. One Komsomol expedition to the Altai mountains brought back thirteen onions (probably species of allium), twenty gooseberries, twenty-seven red currants (*Ribes rubrum*), and nine raspberries, an event Michurin described in the bombastic and almost military style of language favored by Stalinism as "a tremendous victory." The selection of plants for decorative purposes was not forgotten in his work either, although little work was in fact carried out in this area.[11]

Michurin died in 1935, honored by the Soviet state, having been personally thanked for his work by Stalin on more than one occasion. Towards the end of his life, as the Stalinist terror began to seize the country, opposition to Michurin was effectively silenced. His own public utterances became more politicized, probably as this old aristocrat began to learn to speak the language that went down well with the party. "Soviet Creative Darwinism" was an expression he developed to describe his work, and given his position as the leading Soviet plant breeder, he was able to expropriate the ideas and the work of others, including early attempts at radiation breeding (see chapter 11). Writing in Pravda in 1934, after extolling Stalin, he declared that "it is no wonder then that I am now working at such problems as the production of a frost-free peach, the possibility of originating new species of plants by applying various kinds of radiant energy—the cosmic ray, the Roentgen and the ultraviolet . . . public attention has been drawn to the necessity of improving the plant. . . . I have no other desire than to pursue side by side with thousands of enthusiasts the cause of the renovation of the Earth."[12]

DAYS OF MADNESS—TROFIM LYSENKO AND "AGROBIOLOGY"

Trofim Lysenko (1898-1976) carried on where Michurin left off—and did incalculably more damage. His name is attached to one of the worst scandals in the history of science, one which inevitably became something of a legend, and like all legends, has begun to serve purposes other than recording a se-

11. Michirun 1949, 482.
12. Ibid., 484–85.

a series of events in an increasingly distant time. One is that he succeeded in demolishing the science of genetics in the Soviet Union because he was close to Stalin—he was not—as Stalin clearly did not understand the science behind the debates between Lysenko and his critics, and for many years made it plain that he was not going to take sides. The other is that Lysenko's attack on genetics was based on an intrinsically Marxist objection to the idea that characteristics can be inherited, with all the implications this has for the perfectibility of the human character and the creation of a new "Socialist man." Marx and Engels had little to say on Darwin or Lamarck, and the attempts by Lysenko and his supporters to add ideological clothing to their belief in the malleability of living material had little serious ideological underpinning, as opposed to the political name-calling at which Marxist-Leninists have always been world masters. The myth that Lysenkoism was an inevitable child of Marxism should be put aside as a product of the cold war.

Lysenko received his education at a school of gardening and earned a degree at the Kiev Agricultural Institute but undertook no scientific research. His career took off in 1930 during the worst period of the collectivization of agriculture with the idea of "vernalization," the chilling of wheat prior to sowing, which improved spring germination and avoided the widespread loss of seedlings that occurred with autumn-sown wheat. The rest of his career consisted of the launching of a series of similar technical fixes, some of which worked tolerably well and gave him sufficient credibility to manipulate the system enough for him and his supporters to gradually take over the life sciences of the USSR. The fact that they did so can only be explained in terms of the struggle between practice and theory. Stalin, and Mao Zedong after him in China, stressed practice—the basic technological knowledge of the farm practitioner, or those who work among them, as against the supposed ivory-tower theories of research-orientated scientists. Science rarely gives the clear-cut answers that dictators like, and research takes too long for people in a hurry. Stalin and the Communist Party preferred "agrobiology," a farm-based system that saw the application of the most basic and often untested intuitive notions to agriculture, and on a huge scale. Plant breeding, along with all other agricultural research, must move onto the farm and involve the masses, the idea being that just as the worker knew his machine better than an engineer, so a farm worker knew more about crops than a botanist. Results supported by peer-reviewed papers were replaced with what can only be described as "proof by testimonial," with the proviso that the more important and powerful the writer of the testimonial, the better the product.

Lysenko entered plant breeding in 1931. The Party Control Commission was determined to speed up the agonizingly slow process of improving the peasants' grain; they declared that certified varieties would replace traditional ones within two years throughout the whole of the USSR and that new varieties must be bred within four to five years, as opposed to the customary ten to twelve. Lysenko reckoned he could shorten the process of breeding by cutting out the years of progeny testing, as he could "predict" the performance of seedlings. The use of greenhouses was included to speed up growth—hardly an effective way to select plants for regional environments.

As Stalinist terror began to take hold, any dispute in any field was rapidly politicized, with the frequent making of unfounded political accusations as a means of settling personal and political scores. Lysenko's followers slowly edged out their critics; in 1934 many research labs were closed, and in the spring of 1937, publication of Stalin's address at the March plenary of the Central Executive Committee was followed by a speech by I. I. Prezent, an associate of Lysenko, who talked of "powers of darkness" opposing the "reconstruction of biological science on the basis of Darwinism raised to the level of Marxism"; he made a xenophobic attack on pure research and described his opponents as "bandits."[13] In 1938, Lysenko became president of the all-important Lenin Academy of Agricultural Sciences. In 1940, he took the presidency of the Institute of Genetics at the Academy of Sciences, a post that had once been Vavilov's. August 1948 saw the decision of the Lenin Academy to support a report by Lysenko: "On the Situation in Biological Science." From that point on, "antiscientific, reactionary" genetics would be suppressed in favor of agrobiology. Critics, fearing accusations of being "wreckers" kept quiet, although it is interesting to note that Yuri Zhdanov, head of the science division of the party Central Committee (the most important body in a Communist state) recorded that "Lysenko has not bred any significant varieties of agricultural plants."[14]

Rather than breeding new plants, Lysenko's approach militated against their production, as indeed did the inefficiency and brutality of the entire Stalinist system. Corn was a case-in-point; realizing the importance of American work with F_1 hybrid corn, party bosses allowed seeds to be imported from the United States by the newly created All-Union Institute of Corn, established at Dnipropetrovs'k in the Ukraine during the early 1930s. Progress, however, was hampered by a sullen workforce of dispossessed

13. Quoted in Medvedev 1969, 46

14. Quoted in Joravsky 1970, 139.

peasants who could not be relied on to completely detassel the crop and therefore emasculate the seed parents to prevent selfing in those rows that needed this done, or, indeed, to even detassel the right ones (see chapter 10). Despite the fact that the workers in the workers' state could not be trusted, some worthwhile hybrids were developed by 1935, but the farm system was in no state to receive them: there was little money to buy seed, and in any case corn was a new and unfamiliar crop for much of the USSR. Hybrid corn needs higher levels of nutrients to perform at its best, so it could be argued that in this respect Soviet agriculture was trying to run before it could walk. The failure of the corn program gave Lysenko and his supporters a field day in their attack on genetics, in particular the distrust he had for anything involving self-pollination. Lysenko's supporters, in language not too far distant from some of today's critics of biotechnology, claimed that genetics in general and hybrid corn in particular were only a swindle by the capitalists to make seed "inaccessible to ordinary farmers" that served only the interests of big seed firms.

Some of Vavilov's pupils tried to keep on working with corn; that they were able to do so was down to their powerful protector (Soviet and Russian politics have always tended to work around patron-client relationships). Nikita Khrushchev was at the time First Secretary of the Communist Party in Ukraine; although he was later a great supporter of Lysenko, Khrushchev was always a maverick, and his belief in corn seemed unshakeable. It was even suggested by leading Soviet dissident historian and biologist Zhores Medvedev that the immediate reason for Vavilov's arrest had been to prevent a meeting between him and Khrushchev to discuss hybrid corn.[15] When Lysenko made a complete triumph in 1948, hybrid corn breeding was stopped altogether, and no amount of political support could save it. After Stalin's death in 1952, Khrushchev was able to "rehabilitate" the reputations of all the faithful Communists who Stalin had had shot or deported to Siberia; he also rehabilitated corn, and research on hybridization began again—in Dnipropetrovs'k, which soon became known as "the corn center of the USSR."

With Lysenko in command, all manner of absurdities were let loose. The Lamarckian and other dubious tendencies of Michurinism had free rein, with a widespread belief that grafting was the best way to transfer characteristics from one plant to another. The belief that environmental influences

15. Medvedev 1969.

on the young plant were paramount even led eager followers to claim to have turned rye into wheat.

Lysenko declared that "Stalin's teaching about gradual, concealed, unnoticeable, quantitative changes leading to rapid, radical, qualitative changes permitted Soviet biologists to discover in plants the realization of such quantitative transitions, the transformation of one species into another."[16] Cabbage-to-turnip and pine-to-fir transformations were attested to, and one collective farm manager even claimed to have turned a chicken into a rabbit—there was nothing that the working class, guided by Comrade Stalin, could not achieve!

In all Communist regimes, music played a role in inculcating appropriate attitudes (as it still does in North Korea); Lysenko had at least one piece of doggerel in his honor, the State Chorus sang a paean of praise to Soviet agriculture that contained the following verses:

Merrily play on, accordion
With my girlfriend let me sing
Of the eternal glory of Academician Lysenko.

He walks the Michurin path
With firm tread;
He protects us from being duped by Mendelist-Morganists.[17]

Soviet plant-breeding textbooks from the 1930s onward became a muddle of basic science and Lysenkoist dogma, with the section on statistics shrinking with each new edition; after all, Lenin had supported the idea that statistics are only important once "basic clarifications" have been made. Self-pollination was seen by Lysenko as a particular problem, so agrobiological workers would stalk the land doing all they could to promote healthy "additional pollination." Since wheat is naturally self-pollinating, it must be forced to breed out, so Lysenko demanded in 1936 at a session of the Lenin All-Union Academy of Agricultural Sciences, that over seventy thousand collective farms involving some eight hundred thousand farmers would engage in a "renewal" of grain varieties that were degenerating through inbreeding. Farm workers were then sent out into the fields to force open

16. Quoted in ibid., 134. The title of this chapter is borrowed from a history of the USSR by Zhores Medvedev's twin brother, Roy, *Let History Judge: The Origins and Consequences of Stalinism* (1971).

17. Quoted in Medvedev 1969, 132.

wheat florets to receive pollen from other plants. The fact that hot dry winds and poor irrigation were responsible for aborted seeds was never considered—key to Lysenko's approach was the "quick fix" rather than any attempt at understanding more fundamental causes.

Stalin's death in 1953 made it easier for geneticists and other critics of Lysenko to start to emerge from almost a generation of secretive work; the lifting of the ban on self-pollination a year later was the first sign of a long slow thaw. However, Lysenko had the ear of the new leadership headed by Nikita Khrushchev, and it was not until a meeting of the Central Committee in 1964, after Khrushchev's overthrow, that Lysenko was finally deprived of his official posts and packed off to run a small experimental farm near Moscow. From then on, geneticists and research stations began to be given the autonomy that is vital for successful work. Lysenko had relatively little effect on the development (or stultification) of genetics elsewhere in the Soviet bloc, although a volume, *The Vegetative Hybridization of Plants*, detailing work on graft hybrids, was translated into German by an East German publisher in 1950 and dedicated by the author to "my teacher, T. D. Lysenko."[18] Chinese genetics was also affected, but with the Sino-Soviet split developing in the mid-to-late 1950s, the Chinese were able to throw off Lysenkoism along with other aspects of Soviet influence (see chapter 11).[19]

The impact of Lysenko on plant breeding and other genetically based technologies was devastating. It has even been suggested that Lysenko's influence so damaged Soviet agriculture that it, more than any other single factor, led to the demise of Communism. Soviet science was clearly able to achieve much—after all, it was all they who got the first satellite and the first astronaut into space; in most areas of scientific endeavor, there *was* relative freedom. Soviet genetics, which had been strong during the 1920s, was the only field to suffer in this way; it lost three decades. Even in the early 1960s Soviet delegates were turning up at international conferences unashamedly discussing Michurin's methods for breeding new crops.[20] Looking back more than half a century later it is difficult to comprehend how Lysenko got as far as he did. The manic impatience of the early Communist period clearly set the tone, and in an atmosphere of terror, with colleagues being sacked

18. Glustschenko 1950.

19. It appears that 1956 was the crucial year in which Chinese geneticists were able to begin to criticise Lysenko's doctrines (see Li 1987).

20. Konstantinova 1960.

or disappearing in the middle of the night, speaking out or expressing any public opinion was fraught with danger. Stalin's hanging back from coming down on one side or the other must have made expressing any opinion positively terrifying. It is possible to speculate on a further reason for Lysenko's long stranglehold—the lack of transparency afforded by a totalitarian political system and the lack of responsiveness shown by a command-driven rather than market-driven economy. Luther Burbank's occasional excesses were stopped in their tracks by journalists and others who were free to condemn them in the press—not so with Lysenko's. Burbank's plant breeding ventures had to find buyers, first of all nurserymen who had to be convinced to invest in propagating and marketing a new plant product, and secondly the public to whom they had to sell. A market economy where businesses and individuals are free to choose soon sorts out the huckster from the worthwhile innovator; in a command economy there is no competition, and decisions on buying and investing are made centrally and high up with very little autonomy for individuals or individual institutions. Additionally, there is the almost feudal way in which Soviet (or more accurately Russian) society worked—through a network of patron-client relationships, which tends to protect individuals, businesses or products on grounds other than their intrinsic worth.

Although "disgraced," Michurin and Lysenko are not completely buried; the possibility exists that some of their work achieved the results it did because of processes in plants that are still not completely understood, and there are those who would prefer we keep an open mind or at least seek to reinterpret some of their results in the light of modern botanical science.[21] There are also still those who maintain that genetic material can be transmitted between different lines through grafting, a belief which continues to hum along in internet gardening chat rooms.[22]

One particular loss from the Lysenko years stands out: the work that was done on perennial wheat, an idea that after having existed on the fringes of agricultural research for years, is now becoming a focus for more widespread discussion among those who farm semiarid regions. The possibilities for reducing soil erosion and promoting more sustainable farming are tantalizing. Work on perennial wheat by A. Derzhavin built on the Soviet interest in making wide crosses between wheat and related grass species. "It's a

21. See Flegr 2002.

22. Liu et al. n.d.. See also King 2001.

shame all this stuff has been lost . . . multiple lifetimes of work is just gone," an American researcher commented in 2004.[23] Derzhavin's wheat was based on crossing with a perennial rye, *Secale montanum*.[24]

Soviet genetics did recover, but suffered another setback with the collapse of the USSR in 1991. The separation of regions once closely linked politically, economically, and scientifically led to the loss of a whole network of scientific exchange of personnel, ideas, research results, and germplasm. Everywhere there was huge reduction in funds, with scientists working for money, best described by one researcher as "symbolic." What was gained, although it took a long time for the benefits of this to be realized, was access to scientists and others on the outside of the old iron curtain. New contacts with international aid agencies and institutions were particularly valuable.[25]

The process was two-way, as is illustrated by the remarkable story of the Sárvári family in Hungary. During the 1950s, Dr. Istvan Sárvári headed a team working on blight- and virus-resistant potatoes; so successful was he that the response of the government was to close down the program. Taking their potato collection with them, Sárvári and his family moved to eastern Hungary, where during the comparatively liberal and economically open 1970s, he began again to engage in breeding. The potatoes were "discovered" by a team of Scottish growers and scientists on a trip to Romania in 1992, where some of the Sárvári potatoes had been included in a trial. Wary of having their hard work "stolen" by outsiders, it took a long time for the Scots and the Danish company Danespo to negotiate a deal. Finally they did, and the potatoes are now being marketed by a trust based in Bangor, Wales, under the name of 'Sárpo' varieties—money earned from sales helps support continuing research by the family in Hungary.

Where did the genes for blight- and virus resistance come from? The original Hungarian team had had access to the Vavilov collections made during the 1920s, which had revealed the incredible diversity of South America's potatoes. The collection had been stored in Leningrad, where they remained, despite the terrible siege of 1941–1944 when Nazi troops tried to starve the city into submission; it is said that research workers starved rather than eat the potato collection.

23. Steury 2004.
24. Vavilov 1951.
25. Aliev et al. 2001.

SEED COLLECTING AT GUNPOINT—NAZI GERMANY AND GENETIC RESOURCES

The theft of art treasures during war is well established—and in the case of much of the art plundered during the Second World War and the Holocaust, is still subject to acrimonious debate. It should come as no surprise that parts of the Vavilov collections, along with other Soviet seed, became war booty during this period. There was a more specific reason behind Nazi interest in Soviet genes, however. This was the *Drang nach Ostern*, the invasion of Poland and European Russia, which were earmarked for settlement with Germans, a project dear to the heart of Heinrich Himmler (1900–1945), commander of the SS. As plant breeder and SS member Heinz Brücher (1915–1991) noted, "the conquest of the East has brought into our possession areas that will be of decisive importance for feeding the German Volk in the future. . . . [T]he climatic conditions of these eastern areas place very special demands on cultivated plants. . . . Falling back on the primitive origins of the cultivated plants that Vavilov collected is all the more important in the current state of plant genetics."[26]

The SS, which was something of a state within a state, set up its own Institute for Plant Genetics in 1943 to work with stolen genetic material, apparently on the direct orders of Himmler. The Nazis had recognized the importance of collecting primitive strains of seed, primarily for breeding work but also as part of their obsessive interest in racial origins. In addition to working with established crops, there was also considerable interest in new crops—such innovative plant selection work was possibly partly due to genuinely creative thinking, but also forced onto the regime by the shortage of raw materials created by the war they had started. Most concentration camps included nursery areas where inmates were used as forced labor to grow crops, including herbs (for the "natural" herbal and homeopathic medicines favored by Himmler) and experimental crops. Auschwitz, for example, specialized in 'Kok-saghyz,' a Russian plant the Nazis believed could be developed as a rubber substitute; dandelion relative *Taraxacum kok-saghyz* also attracted the attention of breeders in Sweden.

In the chaos of the Nazi blitzkrieg, some Soviet research institutions and seed collections were destroyed, which led to the creation of a special commando operation in occupied territory, the *Russland-Sammelkommando*, to seize the important collections that remained within striking distance—to

26. Quoted in Deichmann 1996.

be led by Brücher. His seizure of some of the Vavilov material, as well as new crop varieties grown from them, enabled him to pursue his own goals and elbow his research institute (the Kaiser Wilhelm Institute for Breeding Research) ahead of others in the fight for funding. He did, however, refuse to destroy Soviet research institutes in the Nazi retreat. After the war he went to Sweden and then, like many middle-ranking Nazis, to South America, gradually working his way back into respectability through a series of publications on the origins and diffusion of crops. His crowning achievement was being given a top UNESCO post in 1972, as an expert in biology. He never faced any formal investigation for his crimes.

SUMMARY

Those who live in liberal democratic societies like to think that good science can only be done in societies where there is a certain minimum standard of openness and free discussion. In actual fact, totalitarian states have, and continue to achieve, much good scientific work. The Michurin and Lysenko stories, however, do illustrate a "worst case scenario" about what can happen when science becomes heavily politicized in a society where there is no political or academic freedom. Another way of looking at the Soviet genetics story is to see Michurin and Lysenko as the last bastion of pre-Mendelian thinking about heredity, where ideas long since discarded elsewhere are given free rein in a system where they cannot be effectively challenged.

PART TWO ❦ FLOWERING OF A TECHNOLOGY

TEN ❦ HYBRID! ❦ CORN AND THE BRAVE NEW WORLD OF F_1 HYBRIDIZATION

The red sun of a summer dawn in the American Midwest sees the buses pull up, groups of young people get out and gather for their instructions, then set out for the fields of corn. Walking down the rows, they grab the tops of the plants, pull them over and tug off the pollen-bearing tassels at the top. They do this all day long; at first getting drenched in dew, then scratched from the coarse leaves of the crop and muddy from the occasional flooded part of the field, and, as the sun rises higher, increasingly soaked in sweat. The Midwest in summer is not just hot, but often intensely humid too.

Welcome to detasseling. It's almost a rite of passage for teenagers in some Midwestern states—hard work and a formative first experience of employment. But like any shared experience for young people, it can be made fun, with the intense pleasure of camaraderie, the making of new friends, and a good excuse for a party at the end of the week. Detasseling is organized by a number of companies who sell their services to producers of corn varieties. One of them even organizes an annual poetry competition to celebrate the work:

ODE TO A SUMMER JOB

Sunshine gold on emerald field
Does make my heart grow light;
My jubilation unconcealed
From detasseling day and night.

Whisp'ring breezes in the corn
Intangible as ghosts;
Taunt with cooling air, and warn
Of tassels each row boasts.

Dewdrops bright as molten glass
Drench me head to toe;
As I tread each row, each pass
The mud seems to deeper grow.

I see my cuts and blisters sore
Not as mere marks of trade;

But as the wounds from a worthy war
When the tassels I have stayed.

And though exhausted I may be
I'm comforted to know;
That tassels will be there for me
No matter where my row!
Sarah Tonjes[1]

But what is all this about? To the outsider it seems almost bizarre. Looking back on her youth, one former detassler explains it:

> Corn detasseling is a simple process. In a cornfield, there are four rows of female corn for every row of male plants. The male rows are slightly darker and squatter, making for a pretty sort of geometry when looking at the fields from afar. When the female plants begin to produce ears, their silks gather the pollen from the remaining male rows, a process that produces a hybrid plant with the best characteristics of the male and female plants that made it. Our job was to pull out the tops of the female plants—the tassels—each person taking one row at a time, and dropping them on the ground, where they would fertilize the next season's crop. We used to joke that we were helping the males and females get along better.[2]

Detasseling removes the pollen and the male flowers from the plant which is wanted as the seed parent, preventing it from fertilizing itself and forcing it to accept pollen only from the plants in the alternate rows. This ensures that the seed will all be from a predictable and controlled cross—it will be genuine F_1 hybrid seed (see chapter 4).

The development and use of F_1 hybrid seed was one of the great revolutions in agriculture in the twentieth century. It not only greatly increased yields but also enabled breeders to exercise very tight control over the genetics of the crop. There is also a counter-narrative—that it was unnecessary and benefited business more than the farmer. F_1 hybrid seed will not breed true, so if a farmer wants to grow the same variety again next year, they will have to go back to the seed company and buy it again; farmers are

1. The winner of the 2005 competition run by Not Afraid To Sweat Detasseling, Inc., (based in Lincoln, Nebraska) was Sarah Tonjes; she won $50 to spend in the local mall.

2. E. A. Moore 2005.

thus effectively prevented from saving their own seed. Whatever the arguments, the story of F_1 hybrids has been a dramatic one, involving new models of breeding and business; it all began with corn.

GIFT OF THE GODS—CORN

The achievements of Native Americans in their selection work with corn have been recognized by practically everyone who has worked with the crop. By the time of Columbus, they had already made more changes to corn than humans had to any other plant; all the major types had been developed: flint, flour, pop, dent, and sweet (see chapter 1). Whole tribes and civilizations depended on corn. Those who find it difficult to enter the mindset of those Native American peoples who saw corn as a god should reconsider; the words of James Wilson, US Secretary of Agriculture from 1897 to 1913, convey something of the modern equivalent sense of awe:

> The value of this crop almost surpasses belief. It is $1,615,000,000. This wealth that has grown out of the soil in four months of rain and sunshine, and some drought too, is enough to cancel the interest-bearing debt of the United States and to pay for the Panama Canal and fifty battleships.[3]

Awe expressed by a deeply spiritual Native American and a modern, commercially minded white American might sound very different, but they both express the same fundamental human emotion. The difference lies in the sense of who is in control. The age-old Native American outlook was that the crop was a gift and therefore beyond human control—the gods must be thanked for it. Modern humanity knows that the crop can be changed, guided, and manipulated by us; we may still cherish it, but we do so from a very different position. Such differences in mindset underlie much of the gulf—if not ocean—of misunderstanding between modern breeders and those in traditional communities.

Most of the corn grown by Native Americans in what is today the United States was of the flint type, very similar to what can still be found in Guatemala. Only in the Southwest and Southeast were other types grown. European settlers began to grow corn very early on and as white settlers began to spread out and farm the lands which Native Americans had once cultivated or hunted over, farmers began to experiment with corn in a way which was

3. Quoted in Fitzgerald 1990, 43.

very different to the way that their predecessors had. Rather than keeping strains pure as the native peoples did, they began to experiment with deliberately "mixing" strains. Cotton Mather's observations in early-eighteenth-century Massachusetts have already been noted (see chapter 4).

An early "mixer" of corn was John Lorain (1853–1923), a farmer and storekeeper in Pennsylvania, an area which was on the border between the traditional territories of two Native American corn types, the softer southern 'gourdseed' and the predominant northern flints, the latter favored for their superior taste. Lorain had written to the Philadelphia Agricultural Society in 1812 of his experiences in "forming a judicious mixture with the gourdseed and the flinty corn," forming a new variety which "would yield a third more per acre." Lorain was a patient observer of his crops and recorded his results in a book, posthumously published in 1825 as *Nature and Reason Harmonized in the Practice of Husbandry*. Here he sets out the basic principles of hybridization, noting that different strains do not "mix like wine and water" but are more like "the mixed breeds of animals," where characteristics are not uniformly transmitted between generations.[4] We also know that he met George Washington on several occasions and is known to have discussed corn growing with him; we also know that he was deeply religious and passionately antislavery. Another farmer who recorded their experiences was Joseph Cooper of New Jersey, who in the first volume of the journal of the Philadelphia Agricultural Society, wrote of his having tried a new form of Indian corn and crossing it with his normal variety had produced a strain which yielded well and early.

Gradual improvement of corn by occasional hybridization and trial and error proceeded, but in the 1870s the first sign of a breakthrough occurred, which allowed corn to be improved through leaps and bounds. Science, rather than just the technology of trial and error, was at last brought to bear on the plant; directed by hypotheses, controlled experiments with breeding made for a systematic approach to improvement. The science led to the development of hybrid corn by the 1920s—this was a turning point, not just because of the coming together of science and technology, but also because of the importance of the crop. Corn, as we have seen, is amazingly responsive to its conditions, and given the hot and often wet growing season of much of the central and eastern United States, it could be remarkably productive. While native peoples ate it as a staple, European settlers tended to

4. Quoted in Wallace & Brown 1956, 48–49.

feed it to animals for meat production. Given the fact that it takes several units of vegetable carbohydrate and protein to produce one unit of animal, the need for corn as a feedstuff was considerable.[5]

Hybrid corn did not arrive entirely through science, however, but through a two-pronged attack: without the selection work carried out by farmers in the Midwest, the science would have had relatively low-yielding varieties to breed with, and without the science the farmers could only have taken selection so far. The science was achieved by farsighted and brilliant men, but they were not "corn men." They lacked the interest in the crop to sustain their work on it; for them it was the underlying principles and the methodology which was important. Hybrid corn arrived through two complementary sets of skills: the field-based passion and intuition of the farmer who understands their crop and the abstract intellect of the scientist.

Before launching into the story of hybrid corn, it is worth remembering that many nineteenth-century farmers carried out deliberate crosses between corn varieties. However, early breeding experiments and their methodology remained localized in their impact because the knowledge was never disseminated or systematized by the kind of formal network which modern science has evolved. Even hybridization involving detasseling, once thought to have originated in the 1870s, is now known to have been used by some farmers in the 1830s.

The man generally credited with inventing detasseling is William James Beal (1833–1910), born on a farm in southern Michigan. He got to Harvard, studying under two of the greatest living scientists: the zoologist Louis Agassiz and the botanist Asa Gray, the former an opponent of Darwin, the latter a supporter. These were exciting times: the life sciences were drawn into these two opposing camps—for or against Darwin's theory of evolution through natural selection. There was also the sheer scale of new material to work with, as naturalists and scientists now had the whole globe to explore, with improved communications and imperial expansion bringing all of living creation within an observable distance. Furthermore, teaching methods were changing; Beal, for example, stressed the value of observing original material, not just learning about it from books or the blackboard. He became famous for sending his sometimes bewildered students out into the yard to spend a morning looking at grass, getting them to record everything they could about it.

5. See technical note 24: Feed Conversion Ratio.

Beal worked with corn over a number of years at Michigan Agricultural College, first publishing results in 1876. His great contribution to plant breeding was to grow together the two varieties of corn he wanted to cross, isolated from others, and then to "detassel" every plant of one variety, so removing the pollen-bearing male flowers at the top of the plant—so its female flowers (on the side of the plant) could only be fertilized by the pollen from the other plant. The seed from these flowers would be a first generation hybrid. Beal's work emphasized the input of both parents, arguing that the contemporary selection of seed-corn with reference only to the female was as foolish as taking no account of the quality of the stud in raising animals. He recognized, although did not name, the concept of hybrid vigor and showed how corn could be improved by a program of selective cross-breeding. Darwin was working with corn at the same time, publishing *The Effects of Cross and Self Fertilization in the Vegetable Kingdom* (1877); indeed, the two men had a brief correspondence. Beal may have shown how it was possible to improve corn through controlled hybridization, but the results were nowhere near good enough to justify commercial exploitation.

Meanwhile down on the farm, the pioneers were pushing back the prairie, opening some of the planet's most fertile soils to the plough and the seed-drill. One of them was Robert Reid, who moved to Illinois in the late 1840s, bringing with him a Virginia gourdseed corn he called 'Gordon Hopkins,' named for the man who had given it to him. Germination was patchy from his first sowing, so he filled in the gaps with 'Little Yellow,' an early flint, a local 'Indian' corn, with only eight to ten rows of kernels, which had been widely grown by Native Americans in the Northeast. Needless to say the two crossed (corn is strongly cross-pollinating), and in the years that came after, Reid and his son James refined the resulting hybrid strain. James was an amateur artist, and clearly had an eye for aesthetics that he applied to his family corn. He transformed the crop to an elegant ear with 18 to 24 rows of kernels, uniformly covered, smooth (so as not to roughen the skin when handling) and with a narrow shank (which makes it easier to husk). Native Americans would have understood his sympathy for crop, its aesthetics, his keeping the seed-corn between the mattresses of his bed during the winter, and perhaps explained his improvement of the plant as being assistance from the corn spirit. Not only was the Reid family corn beautiful, it yielded, although James was not selecting primarily for yield. In 1893, 'Reid's Yellow Dent' won a prize at the World's Columbian Exposition in Chicago—and from then on it began to spread.

Reid corn was the starting point for other Midwest farmers, who constantly worked on improving it. One was the Mennonite George Krug, a quiet and retiring—indeed, almost inarticulate and reclusive—Illinois man who had no eye for James Reid's aesthetics. With strains of Reid corn from Nebraska and Iowa, he produced an uneven corn that yielded ten bushels an acre more than the best Reid corn; the rows of kernels may not have been even, but each kernel was plump and each ear weighed in high. Good looks, which meant uniformity, and yield clearly did not always go together. Yet it was looks that many farmers selected for.

Another man who turned out to be very influential was Jake Leaming of Ohio, who spent years picking out the best ears from his fields. He was a careful and intuitive observer of his crops who noticed such things like that long thin ears ripened earliest and plants with two ears grew better in unfavorable conditions. He got yields of over a hundred bushels per acre and in 1878 his corn won the Grand Prix at the Paris World's Fair—the best corn in the world, a status based not on looks or even weight but on a new criterion, one that was set to change the way the world looked at the crops it grew: chemical analysis, or, the proportions of oil, starch, and sugar.

SHOWS, LAB TESTS AND BEAUTY CONTESTS—JUST WHAT MAKES GOOD CORN?

Starting around the turn of the twentieth century, a fashion for corn shows sprang up in the Midwest; farmers and their families would bring in ears to be judged, with prize money to spur the competitive spirit. The competition was at the center of a whole day out for the whole family—the world over, farmers and country people welcome show day as an event to shop, meet friends, and compare notes with colleagues, as well as get hectored by sales representatives. And, like at the British vegetable show, it was looks that counted at the corn show, not such considerations as productivity, practicality, or food value; few corn show judges would consider planting the champion ears to see how well they did as a crop. What got prizes was the evenness of the kernels and a smooth and tapering cob.

A desire for uniformity and good looks had a definite distorting effect on selection, as can be seen from the story of Isaac Hershey, a Pennsylvania Mennonite farmer, who was thoroughly forward-looking when it came to farming, whatever the beliefs of his strict puritan sect may have been regarding rejecting the modern world—many Mennonites, like the old-order

Amish, to this day drive around in horse-drawn buggies. He raised a variety of corns but from 1910 stopped bringing in new stock and concentrated on selecting for earliness, disease resistance, and a tendency for the body of the ear to break cleanly from the shank, determinedly ignoring the fashionable tendency to select for evenness. His 'Lancaster Sure Crop' became famous throughout the northeastern United States, but even Hershey ended up selecting for uniformity of ears, although he knew that selections bearing less uniform ears could yield ten or fifteen bushels an acre more.

At one corn show in Iowa in 1903, Professor P. G. Holden, noted as a great corn evangelist, was invited to judge by Henry C. Wallace (1866–1924), one of the staff at Iowa State College of Agricultural and Mechanical Arts (now Iowa State University) who was also editor of *Wallace's Farmer*, an influential family-owned paper. Wallace's son Henry A., still in high school at the time, was a keen corn grower too and no doubt hoping to carry off a prize. He took advantage of his unique access to a leading corn judge to ask him some questions. Holden suggested that the best way to grow a prize cob for the following year would be to grow from the winner. Wallace senior suggested to the boy that he plant out some samples on some spare land next spring, so in 1904 young Henry planted seed from fifty ears, half from high and half from low ranking show entries. To the astonishment of everyone—especially Holden—the tailenders yielded higher, the winners all less than the average of the losers.

Any pretense that the value of the corn show was anything much more than a beauty contest or an excuse for a social occasion would be brushed aside with the work of Professor H. D. Hughes at Iowa State University, who from 1914 onward undertook to trial and measure the yield of a variety of different strains of Reid's Yellow Dent. Hughes, and just about everybody else, was astonished to find that there was a difference of twenty bushels to the acre between the highest and the lowest, and that winning at a corn show said nothing about a strain's productivity.

Donald F. Jones (1890–1963), who went on to play a crucial role in corn improvement, was in no doubt about the lack of science behind the corn shows, writing that:

> The choosing of seedcorn [*sic*] still proceeds in the primitive way of selection based solely on appearance. The choice can be made on only one side of the family, for no matter how excellent an ear of corn may be, there is no way of judging the qualities of the plants which furnished the pollen to fertilize the seeds on that ear. . . . By propagating elaborate score-

cards for judging seed corn, agronomists have thrown a cloak of pseudoscience over an antiquated system that is anything but scientific.[6]

In 1920, Henry A. Wallace, now associate editor of *Wallace's Farmer*, took the idea of the corn show a stage farther and suggested that the Iowa State Department of Agronomy collect samples from corn show winners and losers and trial them for yield; the suggestion was taken up and the Iowa Corn Yield Test was born. The department divided the state into districts and ran trials in each district, awarding prizes to winners, who were announced at the annual Iowa Corn and Small Grain Growers Association banquet. In later years this test was useful for selecting varieties for developing into inbreds for use in hybrids and for proving hybrids against open-pollinated varieties. At this time, however, before the days when practically the whole of the Midwest was planted up with a limited number of varieties, all bred by corporate R&D departments, what every farmer grew was potentially genetically different from what their neighbors did. In seeking to find the best, the Iowa Corn Yield Test was selecting not just farmers who were skilled at growing good and bountiful crops but farmers who had intuitive skill at selecting good corn, and refining that selection every year. It is also known that some of the winning farmers were men who had been conducting their own crossings.

Meanwhile, in next-door Illinois, another series of tests was now well established at the Illinois Agricultural Experimental Station. Originally a chemist, Cyril G. Hopkins (1866–1919) became involved with corn after the college dean, Eugene Davenport, suggested that he might look at the low protein and oil content of corn. Corn, as was then the case, was high in carbohydrate, but for it to be a good feedstuff it needed more protein and more oil. As part of his research, Hopkins developed the "ear to row technique," which involved planting rows from the kernels of single ears. He also conducted detasseling experiments so that he could assess the different contribution made by male or female parents. From Louis de Vilmorin's work with sugar beet he also took the idea of the progeny test.

Starting in 1896 with 'Burr's White,' he analyzed 163 ears for oil and protein, and then began the trial with those that had the highest protein and highest oil (24 ears each) and the lowest (12). The process was then repeated every year and has been ever since, apart from four years during the Second World War. Hopkins found that improvements in the chemical quali-

6. Quoted in Kloppenburg 1988, 298.

ties were maintained from one generation to another, although there was little improvements in yields. Conversely, other work showed that increased yields did not accompany improvements in chemistry. A recent evaluation suggests that the lowest levels have been reached for both oil and protein, but that the highest levels have not yet reached.[7]

The Illinois test has been described as "a turning point in scientific plant breeding,"[8] and the fact that it has continued for so long makes it an experiment unmatched in the history of plant breeding. It has shown that the corn under test has made astonishing changes, with continuing gains being made from selection despite a narrowing of the genetic base and consequent inbreeding.

Hopkins had a reputation for being old-fashioned, puritanical, and difficult if you disagreed with him; some of those who he took on had their own ideas, so clashes were inevitable, especially since Hopkins had no real interest in plants—"I am not a plant breeder, horticulturalist, or farmer, I am a chemist" he is once reported to have declared.[9] One of the young men he took on to help him was Edward Murray East (see chapter 7), who had also trained as a chemist but had taken courses in botany at the University of Illinois from Dr. Charles Hottes, recently back from Europe and full of enthusiasm for Mendel. Another youngster who went on to do great things was Harold H. Love. East and Love shared a poky little office, united in their relaxed manners as against those of Hopkins.

In 1904, East noticed that all the high protein corn could be traced back to a single ear and that there were similarly narrow pedigrees for many other strains. East and Love also found that protein content and yield were inversely related. In order to investigate these observations farther, East persuaded Hopkins to start an experiment with inbreeding, but it would only last a year before Hopkins terminated it, much to East's disappointment; apparently Hopkins had recently become much more interested in soils, and saw no point in continuing with corn.

East, stymied at Illinois, not surprisingly leapt at a new opportunity; Dr. E. H. Jenkins of the Connecticut Research Station at New Haven invited him to join their staff and continue with his work on self-pollinating corn (i.e., inbreeding). Jenkins and East, however, needed to get hold of some of the kernels of the inbred corn they had been working with at Illinois. East was

7. Dudley and Lambert 2004.

8. Goldman 2004, 61.

9. Quoted in ibid., 66.

worried that Hopkins might not accede to their request, so he contacted Love, who without telling anyone, removed some kernels from each bag, labeled them and sent them off. East kept this a secret for forty-one years.

East used a modified version of Hopkins's ear-to-row cross-fertilization system in which two unrelated strains of corn were crossed, after one of the parents had been detasseled. What he got were F_1 hybrid crosses, but unlike the F_1 hybrids developed later, which were based on inbred strains grown solely for the purpose of breeding, the parental material used here were standard open-pollinated varieties available to everyone. East found that his "varietal crosses" unleashed a hybrid vigor that brought an average increase of ten bushels an acre and improved quality. From that point on, there was no going back for East—he would unquestioningly promote the concept of hybridization for the rest of his life. East was not by any means the first to be aware of, or to use, hybridizing to improve corn, but in publishing his results and promoting hybrids as the only way forward, he had made a decisive break with selection as a means of improving the crop. However, East's varietal crosses were not a great commercial success—the yield increase was not great enough to justify the work for farmers involved in producing the seed—they are best thought of as a halfway house on the road to true F_1 hybrids.

TOO FANTASTIC A CONCEPT? THE BIRTH OF HYBRID CORN

The New Haven station was in the next state to another experimental farm—the Carnegie Institute's Cold Spring Harbor research station on Long Island. There, George Shull (see chapter 8), from a poor country background in Ohio, had started work with corn in 1905, his first task being to clear a virtually derelict plot. He may not have been the first to realize the fact and the value of hybrid vigor in corn, but he was the first to outline a systematic program for its use in a plant breeding program. He discovered, with only limited data, that through inbreeding the normally outbreeding corn plant, it was possible to sort out the heterogeneous plant into a number of breeding lines each of which had remarkable internal consistency—pure lines.[10] The inbreds, however, were puny plants; Darwin, writing back in 1876, had noted the weakness of corn when self-pollinated but that this could be restored by crossing the inbreds.[11] Shull noted that not only did

10. See technical note 25: Inbreds.
11. Darwin 1875.

crossed inbreds produce notably vigorous progeny but that this crossing process gave the breeder a great deal of control over the results. He also took a further step and made some three-way and some double crosses. By 1908, Shull was ready with some results, reading them in January of that year to the American Breeders Association meeting in Washington, DC. Although he saw the potential, his own interest was primarily a theoretical, rather than a practical, one, as can be appreciated from the paper's title, "The Composition of a Field of Corn." What he was really interested in was the fact that a typical field of corn was an enormous assemblage of complex hybrids. He told his audience that "the question is whether we are going to use our new found knowledge of heredity and grow corn hybrids of our own choosing, or are we going to go along as the Indians have done for thousands of years, depending upon the chance hybrid combinations the winds happen to present to us?"[12] What Shull had thrown open was the possibility of sifting through a field of corn, sorting it into pure lines, and then recombining the results according to specific criteria.

In 1908, East met Shull and realized that Shull's work and conclusions were similar to his own, but that Shull had made an additional key discovery—that inbreeding was a way to select for particular characteristics. East, however, did not consider it to be of commercial importance, because the weak inbreds simply did not produce much seed. Despite a variety of disagreements, the two men realized that they had both reached the same conclusions and made a gentleman's agreement in that they would never let their personalities get in the way of their research or its discussion.

Next year, at another meeting of the American Breeders Association in Columbia, Missouri, Shull read a second paper, "A Pure Line Method in Corn Breeding," elaborating twelve laws of hybrid vigor—which, to some extent reflected ideas he had discussed with East. After three generations of inbreeding, corn will yield only half as much as the first generation, but it will be very uniform; such a "pure line" can then be crossed with another to produce a high-yielding and predictable F_1 hybrid. Although Shull's publications were received with great interest, there were simply not enough staff at state experimental stations with the right frame of mind to make a success of F_1 hybrid breeding, and there were many practical details that needed to be worked out. Shull did not have the commitment either; he was first and foremost a scientist rather than a practically orientated breeder and saw the new technique as being of primarily theoretical interest.

12. Quoted in Crabb 1947, 57.

Shull grew his last crop of hybrid corn in 1913 and then moved to Germany; he became a founder of the journal *Genetics* and one of the first people to use the word "gene." He was also certainly the inventor of the word *heterosis* to describe hybrid vigor. He never returned to active plant breeding. Hybrid corn had to wait for others to make it a realistic crop.

East, meanwhile, had gone to Harvard but still had links with the New Haven station. Here he was to influence, if not directly teach, a whole generation of plant breeders and geneticists—one of whom was Donald F. Jones (1890–1963), who had come from the Kansas State Agricultural College. At Kansas State, Jones had been taught by staff who had heard Shull deliver the second of his three papers to the American Breeders Association and had reported back to the students; most had thought Shull's ideas too fantastic, but they had struck a chord with Jones. East encouraged him to try working with his own inbreds, but he was frustrated at how little seed the puny plants produced—so he had an idea: why not produce two single crosses, and then cross them? In other words, he would not rely directly on the inbreds for the quantities of seed needed to make the process commercially viable. But it was not at all clear if a double cross would be a throwback to the inbreds or if it would combine the qualities of all four inbred varieties. So in the spring of 1917, he made two crosses with all that was at his disposal: two between old Leaming hybrids that had been sent to East by Love from Illinois and two between Burr hybrids developed by colleague Herbert K. Hayes at New Haven; he then crossed the progeny. Planted in the spring of 1918, the harvest yielded 116 bushels per acre, 20 percent more than the best Leaming open-pollinated; the crop stood well and the ears were remarkably uniform—and these were the first crosses Jones had ever made! He tested the new hybrid for five successive seasons, and the average of 20 percent better yield than open-pollinated varieties was maintained.[13] But, there was an element of luck—as Jones later learned that only occasional combinations of single crosses will blend well into a double cross. Whether this was beginner's luck or incredibly good intuition we shall never know.

The importance of the double cross, was, as Jones realized, not just that this was a way of making incredibly productive, predictable, and uniform corn plants, but also a way in which one variety could combine the best qualities of four. In 1918 Jones wrote his doctoral thesis, "The Effects of Inbreeding and Crossbreeding Upon Development," which explained hybrid vigor in Mendelian terms and in terms of chromosomes. In 1920, the first

13. See technical note 26: Open-Pollinated.

New England farmers were growing the new double-cross hybrid corn, which was commercially launched as 'Burr-Leaming' the next year.

It seemed, however, as if the potential of F_1 hybrid corn was recognized only by Jones, Hayes, and East. At the Minnesota State Experimental Station, where he now worked, Hayes was prevented by his superiors from trying the new crops, and in Indiana, that the experimental station would not have anything to do with them either. Part of the problem was that the New England–bred hybrids did not perform well in the Midwest, and it was an uphill struggle to convince Midwesterners to start their own breeding. In an attempt to get people interested, Jones started writing for the popular farming press, which at least kept the issue in the public eye. F_1 hybrid corn was, until well into the 1920s, largely developed outside the system of state research; many agronomists regarded it as unrealistic—the whole concept was somehow almost too fantastic for them to grasp. Many of the surviving corn shows were closed to them, while in Indiana, it was not until the 1930s that the Department of Agronomy would have hybrid corn in their test plots at the Purdue Agricultural Experimental Station.

In November 1920, the ebb and flow of electoral politics threw up an opportunity—Warren Harding was elected president. He appointed Henry C. Wallace to the post of Secretary of Agriculture. Wallace was interested in a clean sweep of the department and in being more proactive in corn breeding than the previous Principal Agronomist in Charge of Corn Investigations, C. P. Hartley. Hartley's work on improving open-pollinated lines was wound down, and serious work on inbreds started, with Jones being asked by the Office of Cereal Investigations to tour the Corn Belt and give the work a boost.

The slow takeoff of F_1 hybrid technology is perhaps a reflection of the variety of different ideas around at the time about how to improve corn and other crops. The basic concept of the hybrid was, in fact, in current parlance and regarded as innovative enough to have been used to describe and market corn by the Funk Brothers Seed Company as early as 1916. Around 1900, Funk Farms had been developing corn "families" through a refined form of mass selection—strains of corn, of consistent characteristics, continuously and rigorously selected. Around thirty strains were being propagated on widely separate farms; the separation ensured that they remained true. Eugene Funk, the company head (the shortening of his name to "Gene" would not have had the humorous associations it does now) had heard Shull speak in 1908, which had inspired him to start crossing between the "families." He found increased vigor—a bit like East's varietal crosses. Funk then decided he would try to bring together in one plant the best qualities of his best families, so he made a simple cross between two, and the next year crossed the progeny with another one of his best families. This combination of three outstanding strains was then sold as "hybrid corn" in 1916—though not a true F_1 hybrid, it was the first attempt to use the hybrid concept in marketing.

Novel ideas attract the young and forward-looking, ready to bet their futures on a new idea. One of those who were won over by the F_1 hybrid corn idea was James Holbert from Indiana, who had met Hayes at the Minnesota Experimental Station in August 1915. Hayes had inspired him in less than an hour of discussion, apparently telling him that "from the time I stood there in Connecticut in 1910 and 1911 and saw those East single crosses—the most beautiful corn I have ever seen, every ear so large and looking like the one right next to it—I was sure that somehow, some way and some day a practical use for it would be found, I never doubted from that time on that hybrid corn would come."[14] Looking around for a good place to work on corn, he asked his professor if he could approach Gene Funk. His reputation at college had been brilliant (he had researched fungal diseases of oats), so when he arrived at Funks', it was not to stay in the staff quarters with the hired hands, but at the Funk family home. Funk and Holbert stayed up many nights talking; he particularly remembered Funk as saying that "our corn today is good when everything is favorable, but if corn disease strikes, or if we have some dry weather, or if it rains too much, then the corn we have now just goes to pieces . . . we must have better corn."[15]

It wasn't realized at the time just what an uphill task hybridizing was, in particular at just how many hybrids were in fact completely useless. In addition less than 1 percent of open-pollinated material ever produced inbreds worth working on; it was also believed at the time that potential inbreds had to be selfed (i.e. self-pollinated) for four or five years before they could be tested, a process which later was speeded up considerably. The result of all these factors was an enormous expenditure of time for little outcome—not surprisingly the 1920s were almost entirely given over to work on inbreds. The scale of the project was later likened by Henry A. Wallace to the Manhattan Project, which worked on the atom bomb. Those who were working with developing promising inbred strains from a host of open-pollinated varieties came to the conclusion that the breeders had done an amazingly good job, as nearly all their good material came from varieties bred over the

14. Crabb 1947, 104.
15. Ibid., 106.

previous few decades: 'Kansas Sunflower,' 'Lancaster Sure Crop,' corns from Reid, Funk, and Leaming. Landrace materials from Mexico, South America, and Asia were tried, but no outstanding lines were found.[16]

Like many other workers with corn, Holbert (now with Funk Brothers in Illinois) worked on a dawn to dusk schedule examining corn plants through stifling summer heat, inspecting every plant in fields up to sixty acres; he also went through thousands of ears in processing plants. Finding very few plants that matched the standards he had set, he came to the conclusion that it was almost impossible to find examples of high-yielding, disease-free plants in the varieties represented by the old show corns; in fact, the healthiest and most productive plants tended to have narrower and longer ears. He ended up with five thousand ears, from which he germinated the kernels, plowing and sowing the crop himself, discarding three out of five as not up to scratch.

So anxious was Holbert to get started that he selected a few plants for selfing simply on the basis of their mid-season appearance—in this he was vindicated, as among these was 'Inbred A,' still widely used twenty years later. Finally, only twenty plants, 1 percent of the original sowing, were deemed good enough; applying germination tests, he narrowed these down to twelve—all were from derived from 'Funk's Yellow Dent.' The two following years were very difficult ones, with storms one year, drought the next, but the upshot of this was that his selections faced severe tests—most remained standing.

Funk recognized that the higher yielding open-pollinated varieties were particularly disease prone, so Holbert's turning up some inbreds that were healthy while all around them were collapsing with fungus was a particularly encouraging development. Aerial photographs that were published at the time showing rows of resistant corn amidst rows of pest or disease devastated ones made a big impact. Holbert also discovered that insect resistance was genetic, albeit through a complex combination of genes. Ensuring that resistant inbreds would have their resistance carried into their hybrids became a particular focus. Vitally for future production, he also found that plants varied in how well they would grow when the weather turned cold, which it can do suddenly and dramatically in the Midwest, with icy blasts howling down from the Arctic a matter of days after a summer of sultry heat.

Increasing numbers came to see what Holbert was up to; he always had time for his visitors, often entertaining different groups for breakfast, lunch,

16. Crabb 1947; Troyer 1999.

and dinner. He would press samples onto the farmers, which was a way of getting his varieties trialed on a wide range of different sites and soils; most were only too keen to cooperate with him for no reward save that of improving the crop.

In 1925, Funk Farms released its first double-cross hybrid onto the market. Two years later Funk set up the first department for hybrid seed production. Over the next few years, the message began to spread, with farming and country magazines spreading the word with boosterish headlines like "The Day of Super Crops Has Come."[17]

NEW CROPS, NEW BUSINESS

Of all the people involved with launching hybrid corn onto the world, one name stands out—that of the Iowan Henry A. Wallace (1888–1965), who as a boy had asked that searching question of the corn show judge. As a boy, Wallace junior had hiked the marshes and fields around Ames, Iowa, with George Washington Carver (1864–1943), the remarkable African American botanist and agricultural scientist who was then studying at Iowa State University; together they collected samples of plants, fungi, and plant diseases (see chapter 11). Carver also apparently gave young Henry his first lessons in how to breed plants.[18]

Wallace had one of the most dramatic and interesting political careers of anyone in twentieth-century America; writer Studs Terkel went so far as to say that "there are three great Americans of the twentieth century, two are household names, Franklin D. Roosevelt and Martin Luther King. The third should be Henry A. Wallace."[19] Hybrid corn was only one part of an extraordinary career. Wallace is said to have had the respect and love for corn that the Indians had; like his father, he became Secretary of Agriculture (under Franklin D. Roosevelt), and in his address to the Corn Research Institute at Iowa State College, he summed up his feelings towards the crop:

> Of all the annual crops, corn is the most efficient in transforming sun energy, soil fertility and man labor into a maximum of food suitable for animals and human beings. It is to be regretted that so few of the millions

17. This headed an article by a leading farming writer of the time, Sidney Gates, in the March 1929 issue of *Country Gentleman*. This magazine was apparently one of the most widely read of the time.

18. Four Iowans Who Fed The World 2003.

19. Quoted in Iowa Public Television 2006, 1.

whose prosperity rests on the corn plant should have so little knowledge or appreciation of it. Even those who work most with corn display little of the genuine reverence for it which characterized the majority of the corn growing Indians up until this century.[20]

In 1924, Wallace entered the Iowa Corn Yield Test with a strange-looking corn, which he had called 'Copper Cross,' after its color, bred from an old Leaming inbred line sent to him from the Connecticut Experimental Station by Donald Jones, and an open-pollinated strain known as 'Bloody Butcher,' which had originated in Asia. It won a gold medal. Jones wrote to Wallace to congratulate him and jokingly suggested that since so much of the parentage had been developed at his research station maybe the medal belonged to him. Wallace did not see the joke and posted him the medal. Deeply embarrassed, it took Jones a week to compose a two-page letter to accompany the medal back to Iowa. Later that year, fifteen bushels were advertised for sale to farmers who wanted to try it.

'Copper Cross' went into commercial production through the efforts of Wallace's friend George Kurtzweil. That next summer, Kurtzweil bred his first crop, a process that had to ensure that only the pollen of one parent, 'Bloody Butcher,' could get to the female flowers of the Leaming inbred, so the latter had to be detasseled—all done by his sister Ruth. Later in life, when F_1 hybrid corn had become big business, and detasseling crews involved hundreds of people on one farm working dawn to dusk, Ruth loved to astonish people with her story of how she single-handedly detasseled all the hybrid seed corn fields of Iowa.

During the 1920s, Wallace devoted much time to thinking through how hybrid corn seed could be produced commercially. There was simply no model. Two sets of crosses had to be produced and then combined into a third. The seed of each cross had to be produced from fields sufficiently separate from other corn crops to ensure that absolutely no "foreign" pollen could blow in, and which were planted with each parent in alternate rows. Next, detasseling personnel had to be organized and trained—they had to be trusted to detassel the right plants! Detasseling had to start before any of the seed parent's pollen could land on its silks, and once it started, the job had to be done in double-quick time. Pollen that escaped the detasselers would mean seed corn that was not a hybrid but a first-generation inbred, and therefore, not the superior item for which farmers were being asked

20. Quoted in Crabb 1947, 142–43.

to pay premium prices. There was little margin for error. After considering various options, Wallace realized that only a special organization dedicated to corn breeding, with crews especially trained and equipped, could handle the task. So, in 1926, Wallace and some colleagues launched the first ever company formed for the exclusive purpose of developing new strains of hybrid corn, their production and distribution—the Hi-Bred Corn Company, later the Pioneer Hi-Bred Corn Company.

The rest of Henry Wallace's career had relatively little to do directly with hybrid corn, but it is an extraordinary one and deserves to be far better known. A highly successful Secretary for Agriculture under Roosevelt during the New Deal administration, he was US vice president from 1941 to 1945. Losing to Harry Truman in a bitterly (and possibly dishonestly) contested Democratic Party election for the vice presidential candidate, he became Secretary for Commerce. Wallace, however, became increasingly left-wing in his politics at a time when the atmosphere in the United States was moving to the Right; disagreeing with Truman over the latter's increasingly (and mutually) belligerent policy to the Soviet Union, he was fired. In the 1948 presidential election he stood as an independent on a platform of ending racial segregation, equal rights for women, and many other progressive proposals. Like many of America's best and brightest, he ended up being harassed by Senator McCarthy's House Committee on Un-American Activities. Returning to farming, he became an incredibly successful breeder of poultry—at one stage over half the egg-laying chickens in the world were descended from birds bred by him. Throughout his life he maintained an interest in the mystical and was for a time associated with the Russian artist and spiritual leader Nicholas Roerich (1874–1947), who, like many "spiritual leaders," had a distinctly shady side; Wallace is known to have funded Roerich on expeditions to central Asia for both seed collecting and meeting with mysterious spiritual adepts. Given how Truman only became US president on the death of Roosevelt, it might be said that Wallace came within a hair's breadth of the US presidency—if he had, the history of the postwar world would have been very different.

During the 1920s, Wallace was faced with not only setting up a new company to develop and maintain a completely new system of breeding but also the task of developing quality inbred lines of corn, for it is the quality of these upon which the whole system of F_1 hybridization and the exploitation of heterosis depends. He cannily encouraged young farmers to take his inbred lines and cross them into single cross or three-way cross hybrids and submit the results to the Iowa yield tests, with him paying the entrance fees.

This way he not only got some of the laborious work done by others but also built up a network of interested and sympathetic farmers. The irony is that in doing so he was building on the intuitive knowledge of farmers about what strains grew well on their land, and in particular their ability to "read" corn, to link the appearance of one year's crop with its potential to be a good parent for another year's, and yet the end result was a "de-skilling" of farmers as breeding was taken over by the professionals.

Hybrid corn was not taken up immediately; its advantages were not always immediately apparent—some double-cross F_1s were very productive, but others, such as 'Copper Cross,' were not. Instead their advantages lay in the uniformity of F_1s, and the predictability and level of control which breeders could achieve by combining different inbreds, each of which had particular traits which could contribute to the final product.[21] The Depression, starting in the late 1920s, made it a very bad time to innovate. There were also practical problems; for example, the seeds of inbred corn did not fit most planting machinery.[22]

Some breeders had a real struggle keeping going during the depression years; Lester Pfister of Illinois was one who endured real hardship—and the opprobrium of neighbors who thought he had gone mad. He loved mechanics and devised various machines for dealing with seed, including the Jitterbug Grading Machine, which sorted out kernels by size; he also developed a detasseling machine. In the end Pfister was saved by publicity, when *Life* magazine rather melodramatically wrote up his story. He went on to become a major force with the Pfister Hybrid Corn Company, though his success was not based on his own initial work with corn—for he had discarded his own hybrids when he came across the open-pollinated work of George Krug. Krug had always ignored the corn shows and concentrated simply on heavy solid ears. Pfister enjoyed one of the ultimate pleasures of the rags to riches entrepreneur—to be a major shareholder of a bank that had once refused him a loan.

Early commercial producers of hybrid corn needed the strength of their convictions to carry them through the worst years of the slump, for they had few resources to plough into marketing or advertising. Fortunately, farmers who had done well with hybrids were usually happy to promote them to other farmers. Farmer salesmen were employed by the new companies, men who knew the land and the language; Russell Rasmusen of the De Kalb

21. See technical note 27: Combining Ability.

22. R. Murphy 2006.

Company went one better, encouraging farmers who bought company seed to grow it in prominent places so the whole neighborhood should know about it.

As the New Deal rescued the U.S. economy and much of its social fabric, Wallace was at the helm in Washington, DC, so not only did hybrid corn have political support, but also there were increasingly economic reasons to invest in it. There was a need too to boost productivity—one option was to take in more farmland from forest and other natural habitat. But this has high costs—in clearing land, fertilizer, labor, relocating and housing that labor, and so forth. It made more sense to increase productivity through higher-yielding plants and get more value out of land already in production.

By 1934, only 0.4 percent of the total U.S. corn acreage was down to hybrids; by 1944, 56 percent was, and 90 percent was in the Midwest's Corn Belt. In the early years, yields were generally 15–20 percent higher across all areas. The fact that farmers had to buy fresh seed every year must have militated against its faster take-up during the Depression, but the higher yields and the greater resilience in the face of disease and stress shown by most F_1s helped them; during droughts of 1934 and 1936, hybrids did particularly well, dramatically emphasizing their superiority.

What slowed the acceptance of hybrid corn was the fact, noted before, that hybrids, which grow tall and have nice fat ears in one place, have a tendency to grow wan and yield badly in another. Funk Brothers in 1929 sent around a thousand bushels to growers in thirty-five states and Canada and Cuba—virtually all grew poorly. Reacting quickly, the company diversified its breeding program and ten years later was selling more than thirty of its own regional hybrids and a similar number developed by others, notably agricultural colleges.

The adoption of hybrid corn is a classic example of what economists have identified as the "S"-shaped curve on a graph—where in the beginning only a few, especially innovative or risk-taking users try the new product, then large numbers do, and finally a few laggards (often the isolated or the poor) catch up. As a general rule, those regions with the biggest corn acreages had not only invested in developing hybrids earlier, but farmers took them up more quickly. Hence the tendency, seen so often seen with innovations, that initially the effect was to increase regional disparities, before evening them out. The Green Revolution would see a very similar effect in many regions.[23]

23. Griliches 1960.

It has even been suggested that the diffusion of hybrid corn was so dramatic that it helped stimulate a whole area of sociological research—"diffusion-adoption theory."[24]

THE SCIENTIST AS MYSTIC? THE BARBARA MCCLINTOCK LEGEND

Barbara McClintock (1902–1992), one of the premier women scientists of the twentieth century, enters our story not because she was actively involved in plant breeding but because her experimental and theoretical work played a key part in driving forward the genetics which increasingly lay behind plant breeding. The big conceptual break she made was to show how flexible and dynamic genes were, in particular, how they could move around on the chromosome and how they could work in relationship to each other. In many ways she was ahead of her time, and some of what she proposed could not be proved with the technology of the period in which she was most actively working, but her work helped to create the intellectual climate in which the research necessary for today's biotechnology could take place. She, and many others, worked with corn, not just because of its economic importance, but because the corn plant has easily observable traits and relatively large chromosomes, which makes it a useful experimental tool.

Such a combination of characteristics has made the modern corn plant the result of an almost perfect marriage of the worlds of commerce and pure research. The relationship, like all strong relationships, is mutual and synergistic. If corn were not so easy to research, it would not have been possible to breed it so inventively and creatively; if it were less important commercially, research funds might have gone to other crops. Perhaps we should look at this relationship in a broader historical context and see it as a continuation of the corn plant's earlier history—the genetic acrobatics of its origins and development that made it such a useful plant to Native Americans, who had the ability to recognize and develop the new forms which were thrown up by the unconscious breeding they carried out.

McClintock's focus was on the cytogenetics of corn, the discipline that brought together the study of genetics and of what actually went on inside cells. Her first major discovery was of an explanation at a cellular level of how genetic material can cross over from one chromosome to another so that genes are shuffled, a key mechanism for the generation of variation dur-

24. Kloppenburg 1988.

ing reproduction. In the spring of 1936, McClintock was invited to become an assistant professor at the University of Missouri at Columbia, where she worked on X-rays and corn. The use of X-rays was popular among geneticists, as they could be used to hasten mutation rates which could then be used in mapping chromosomes. Sometimes X-rays resulted in actual breakage of chromosomes; on occasion, irradiation led to a whole cycle of chromosome breakage and fusion. This and other work led her to challenge the conventional wisdom that the gene was essentially unchanging, and suggest a new idea—that genes could be *transposed*—move around on the chromosome. During the 1940s and 1950s, she showed how particular genes could be linked to particular characteristics of the phenotype, such leaf shape or kernel color. She went on to postulate that some genes were involved in controlling the suppression or expression of other genes—again a radical idea for the time.

Many of McClintock's findings were greeted with coolness from colleagues, leading her to stop publishing; a later generation of feminist historians argued that she was sidelined and not taken seriously because she was a woman. To younger people in the 1980s, she achieved an almost mythic status: a prophet in the wilderness, a mystic who admonished coworkers to "listen to the plant" and who had a uniquely feminine, "holistic" approach to the gene. Such thinking may owe much to the contemporary ideological concerns of the time in which it was made, but she undoubtedly had "a special talent to recognize the underlying order and provide an explanation for the most perplexing observations," in the words of one staff member of the Cold Spring Harbor Institute, where she did much of her work.[25] A less politically charged explanation of her apparent sidelining in the 1950s is that much of her work was simply too radical and ahead of its time for less imaginative colleagues to comprehend—the problem faced by many great scientists who make conceptual leaps.

During the 1960s, however, her work was seen as having been highly prescient, as a younger generation of geneticists grappled with the problems of how genes are controlled. Her theory of transposition was finally shown to be a fact in a wide variety of life forms, opening the door to the development of technologies for the manipulation of genetic material. Recognition for McClintock finally came, albeit late in her career. In May 1971, President Richard Nixon awarded McClintock the National Medal of Science, confessing in his address to her and the other scientists who won awards "I have

25. Quoted in Peterson 1992.

read [explanations of your scientific work] and I want you to know that I do not understand them." In 1983, at the age of 81, she became the first woman to receive an unshared Nobel Prize for Medicine for her work on "mobile genetic elements" (i.e., genetic transposition).[26]

"CRUNCH!" THE IRRESISTIBLE SAVOR OF SUMMER—SWEET CORN

Late summer is when the first fresh sweet corn appears in the shops. Boiled or barbecued, dabbed with butter, sprinkled with a little salt and maybe pepper, it is one of the greatest and most elemental gastronomic delights to those with a full set of their own teeth—and a deep frustration to those without. There is something atavistic about sinking your teeth into something so big and that you cannot hold with a knife and fork. The combination of sweet and savory with its soft but also crunchy texture, to say nothing of the succulent appearance of a plump yellow cob, makes for an eating experience that demands the full concentration of *all* the senses. No one can talk or be social when eating sweet corn—the cob is all we can think about.

Sweet corn is the result of a rare mutation where sugars replace starch. Native Americans, at least the Iroquois, had sweet corn landraces, and it was first recorded as being grown by European settlers in 1779. It soon became popular, at least in corn-growing states and was gradually improved to become more succulent and sweeter. W. Atlee Burpee's introduction of 'Golden Bantam' in 1902, a variety with extra long ears and a particularly succulent texture, was a notable landmark, although the Shakers had produced a widely praised variety in 1886—'Shaker's Early Sweet.'[27] But there was a fatal flaw—the sugars turned to starch rapidly after harvest: store or transport sweet corn and it lost its flavor—fast!

The breakthrough that led to modern "supersweet" corns was made by assistant professor of botany at the University of Illinois, John Laughnan (1919–1994), in the 1950s. Investigating a gene that caused kernels to be shrunken, he found that they also had far less starch and four to ten times as much sugar as "normal" kernels. Attempting to raise interest among breeders was an uphill struggle, so initially he had to work on his own, beginning a program to turn several well-established sweet corn varieties into "supersweets" by incorporating the gene he had discovered—*sh2*. Eventually, he did per-

26. Quoted in ibid.
27. Sommer 1966.

suade a company to back his idea, Illinois Foundation Seeds, Inc. (IFSI), and in 1961 released a supersweet version of Golden Cross Bantam called 'Illini Chief.'

For the next two decades, IFSI and the Crookham Company in Idaho were the only producers of superweet varieties, for which they were slowly but surely finding a market. Meanwhile, another University of Illinois man, Professor A. M. Rhodes, had found a "sugary enhancer" (se) gene in an inbred corn line of Bolivian ancestry. The se trait made for cobs that were not as sweet as *sh2* plants but a lot more tender—the result of the presence of a high level of polysaccharides rather than simple sugars. From the 1980s on, supersweet corn took off, with many more companies investing in research that combined the traits of high sugar levels (and associated reduction in the rate of post-harvest conversion of sugar to starch) and succulence. Supersweet corn is now something that can be enjoyed by everyone from their local supermarket, not just country dwellers who buy from the farm shop or gardeners who grow their own.[28]

POWER POLITICS, PESTICIDES, AND OBESITY—THE F_1 CORN STORY CONTINUES

Did F_1 hybrid corn win the cold war—the struggle between the Soviet-dominated Eastern block and the US-led Free World between 1945 and 1990? We have seen how failures in plant breeding undermined Soviet agriculture (see chapter 8) and hence Communism's ability to feed itself. The mirror image is that the success of plant breeding, in F_1 corn in particular, enabled the United States and its allies to not only feed themselves but also to export food and technology to the rest of the world. The implications of F_1 hybrid corn for geopolitics has perhaps been underestimated. Agricultural writer Sidney Cates wrote during the Second World War that "hybrid corn has done more to put us out farther in front, releasing our strengths for tasks other than food getting, than has any other nation in modern times. Corn is the strength of the nation. And the hybrid gives heretofore undreamed-of strength to corn."[29] Surpluses could be exported—good for the United States, both politically and for the balance of payments. Throughout the 1950s and 1960s, opportunities to export grain as food aid were rarely lost. Of course, part of this was a simple humanitarian impulse, but it also en-

28. Pataky 2003.
29. Quoted in Crabb 1947, 292.

abled the United States to project itself as a force for good in the world. The ability to export grain to those on the both sides of the iron curtain impressed non-Communist populations and humiliated the governments of Communist-led countries. The importance of corn was recognized in the postwar Marshall plan for the rebuilding of Europe, part of which involved the export of hybrid corn and the development of new varieties for Europe, in particular, that of varieties which would germinate at lower soil temperatures. This breeding program began to bring results to northern Europe during the 1960s when silage corn became popular—the whole crop is harvested and fed to stock rather than just the cobs.

The story of corn since the development of F_1 hybrids has been one of continued progress, but with a downside too. At the time of this writing, more than 80 percent of the corn grown in North America is one of many single-cross hybrids; the other 20 percent consisting of double, three-way, and other crosses. Single-cross hybrids began to take over from double crosses in the 1960s—they are the result of the great deal of effort put into developing good inbred lines and make considerable economic sense, as if they can produce enough seed to make double crossing unnecessary, then a great deal of time can be avoided.

Detasseling was to some extent replaced by the discovery of cytoplasmic male sterility (CMS), first found in corn in 1933.[30] The discovery of a corn line with no pollen was a boon—introduce the gene for the control of this quality into the lines a breeder wants to work with, and seed production becomes much easier. Donald Jones was one of those who worked on CMS, and in 1956, he applied to register a restorer gene and became the first person to be granted a patent on a genetic technique. Jones wanted a portion of the royalties to go to the research bodies which had done the work, rather than benefit personally, so patent money was sent to a body he set up—called Research Corporation (RC). Jones's patent broke the "rule" whereby public work was freely available. In doing so he faced opposition from both scientists and industry, and was denounced at the 1956 annual meeting of the American Society of Agronomy, along with colleague Paul Mangelsdorf. After seven years of litigation, the patent was granted. By the late 1960s, a limited number of CMS and restorer genes were included in nearly all U.S. production, and it seemed as if detasseling might become a thing of the past outside the experimental plot. As we shall discover later, this nearly led to disaster (see chapter 14).

30. See technical note 28: Cytoplasmic Male Sterility.

The F_1 breeding process enabled breeders to concentrate on developing strains of the crop which were resistant to particular stresses, and then be sure that these would be consistently expressed in farmers' fields—hybrid corn enabled the Corn Belt to move north into areas where it had previously not been a viable crop, such as Wisconsin and Minnesota. It also made mechanical picking a possibility—as the crop was so uniform in size with the ear positioned in the same place on each plant, that machinery could harvest it easily. In 1935, only 15 percent of Iowa corn was machine harvested; by 1945, 75 percent was. Only larger farms could invest in machinery, and given that the price of corn has tended to fall over time while the cost of inputs has stayed the same or increased, this has added to the many pressures that have hastened the trend to larger farms.

Among the impacts of F_1 hybrid corn first noted was its effect on the soil; its larger root system took so much moisture and nutrients out of the soil that it limited what could be grown afterwards—there was often not enough moisture for winter wheat for example. However, the other side of the coin was that the greater productivity of F_1 hybrid corn meant that land less suitable for the crop could be retired or given over to other crops. The fact that it needs high fertility meant that farmers looked after the soil better, as they had an incentive to practice good soil management. Writing in the 1940s, A. L. Lang, a soil expert at the University of Illinois, was of the opinion that "the coming of hybrid corn has been the greatest stimulus ever given our soil-conservation and soil-building programs in the Corn Belt."[31]

After the Second World War, nitrate-based chemical fertilizers became more widely available, partly because capacity had been built up for nitrate-based explosives during wartime, so lack of nutrients became less of a problem. The tendency then became to grow the crop at higher and higher densities, as one breeder recalls, "the density of corn is a major factor in yield, early in my career you could have 28,000 plants per acre, above this and you'd get barren stocks, now you can have 40,000 plants per acre with no barren stocks."[32] Higher densities made attacks of fungal diseases, which spread in still, humid air, more severe, so there was a need for more fungicide. Increasing yields of corn made it cheaper and cheaper to use as animal feed, which worked well for the U.S. food industry; there is a fundamental rule of agricultural economics that good harvests depress prices. "Adding value" to a product is how the industry makes money—feed corn to pigs, then process

31. Quoted in Crabb 1947, 294.
32. R. Murphy 2006.

the pigs not into cuts of meat but sausages and burgers, and value is added all the way along the food production chain.

There has been a downside to all this productivity: intensive pesticide use during the 1950s and 1960s is now accepted as having caused not only environmental damage (mostly to bird populations) but harm to human health, too. Obesity, resulting largely from overconsumption of meat-based "junk food" is now an epidemic, spreading through the industrialized and industrializing world. The "vertical integration" of corporations during the 1980s onward, which increasingly brought together plant breeders, seed producers, suppliers of agrochemicals, and food producers, may make business sense, but to critics of agribusiness it appear as deeply sinister, as if the whole human food chain is now under the control of only a small number of immensely powerful players. The Slow Food movement, community-supported agriculture, and farmers' markets are the most visible manifestations of the constructive reaction against the apparent corporate takeover of the food industry. For the most part those involved in these niche sectors use the same products of the plant breeder as the big boys, but there is a constant throb of criticism that once plant breeding is so thoroughly integrated into large corporations it does not reflect the needs of smaller growers and businesses, those which have to survive in atypical conditions: agricultural, cultural, social, economic, or commercial, or the needs and expectations of niche markets. Even in the US Corn Belt, there is a growing feeling that the land has lost simply too many farms and too many farming families—without them, small rural towns are dying on their feet. Having been pulled into the corporate world, partly through its own success, plant breeding now finds itself in a situation where its role and continued development is politically contested.

The extraordinary flexibility of corn is now set to grow farther, with the rapid expansion of the biofuel industry. The industry promises new outlets for farmers' corn, and greater profits for seed producers (and no doubt breeders) but arguably will distort the world market, as the staple diet of the world's poor rises in price as it competes with the needs of the rich world's desire for fuel. Corn will remain an intensely political crop for a long time to come.

A CON-TRICK WITH CORN? A CRITICAL PERSPECTIVE ON HYBRID CORN

By the time Mendelian genetics became established, religious scruples concerning hybridization had almost entirely died away, so hybrid corn attracted very little opposition. Those involved in breeding may have had disagreements about how appropriate Mendelian genetics may have been for crop improvement, but there appears to have been almost no opposition or concerns expressed about the wisdom or the ethics of the new methods. The USDA plant breeder, G. N. Collins, who in 1910 declared Shull's work to be "dangerous . . . and doing violence to the nature of the plant" appears almost exceptional.[33]

Hybrid corn has clearly delivered. But there has also been a long-running rumble of a contrary opinion, not about the morality of hybridization per se, but on relying on it as a commercial undertaking.

East and Jones apparently believed that mass selection of open-pollinated varieties would ultimately produce lines as good or better than hybrids.[34] East, however, had another agenda. In *Inbreeding and Outbreeding*, published in 1919, he wrote the following about F_1 hybrids:

> It is not a method that will interest most farmers, but it is something that may easily be taken up by seedsmen; in fact, it is the first time in agricultural history that a seedsman is enabled to gain the full benefit from a desirable origination of his own or something that he has purchased. The man who originates devices to open our boxes of shoe polish or to autograph our camera negatives, is able to patent his product and gain the full reward of his inventiveness. The man who originates a new plant, which may be of incalculable benefit to the whole country, gets nothing—not even fame—for his pains, as the plants can be propagated by anyone. There is correspondingly less incentive for the production of improved types. The utilization of first generation hybrids enables the originator to keep the parental types and give out only the crossed seeds, which are less valuable for continued propagation.[35]

East and Jones believed that commercial breeders would not have sufficient incentive to improve plants until they could prevent farmers from using their own crops to propagate the next generation, so in effect what the hybrids were doing was conferring the equivalent of a patent right on F_1 varieties of seed. It has been suggested that they knew that, in theory, hybrids were not the only, or even the best route, to the improvement of corn, but without the commercial incentive provided by hybrids, they believed that corn

33. Quoted in Kloppenburg 1988, 98.
34. Paul and Kimmelman 1988.
35. East 1919, 224.

could not be improved. They also thought it fundamentally unjust that the creators of new plants and animals should fail to profit by their inventions.

There are two arguments here: one is over whether hybrids were or were not necessary to improve corn, the other over whether or not forcing farmers to buy new seed every year is morally right or not. Opinions on the latter, though, tend to color opinions on the former. Norman Simmonds (1922–2002), one of the most widely respected and knowledgeable plant breeding experts of the latter half of the twentieth century, was of the opinion that improvement of populations through open-pollination could have produced the same results as hybridization if the same effort had been made and that hybridization only seemed a more productive route because other methods were not giving good results at the time.[36]

It has been suggested that corn development in the early twentieth century had been stagnant, the beauty contest mentality favored by the corn shows, along with a poor knowledge of statistics, inadequately designed test plots, and the sheer difficulty of working with such a promiscuous crop all slowed down progress. Alternatives to hybridization were put forward; in 1919, Hayes and a coworker, Garber, published a paper in which they outlined a method which they claimed gave results as good: a process of selfing plants, selecting lines, intercrossing superior lines in bulk, followed by recurrent selection, with the outcome a superior selection that could be propagated indefinitely by crossing within itself. It offered another way of utilizing inbreeding and heterosis but with seed that could be saved from one generation to another. Many farmers used only selection to continually improve their stock until the 1930s, some research workers continued to undertake basic variety crossing; both techniques showed considerable potential. It is possible that F_1 hybrids achieved the success they did when they did because of commercial pressures—specifically the desire of seed companies to make themselves indispensable in the marketplace and to ensure that they continued to make good profits. This was the beginning of a great divide in thinking about plant breeding—between farm-based selection and breeding based on non-F_1 varieties on the one hand and company research plot and laboratory-based hybrid breeding on the other; the conflict continues today and is behind a great deal of the contemporary disagreements in plant breeding. The former approach can be pursued with a profit motive or not, the latter lends itself to a strongly profit-driven system.

Donald Duvick, a veteran corn breeder with Pioneer Hi-Bred and a staff

36. Simmonds 1979.

member of Iowa State University, argued that hybridization achieved amazingly quick results and that success tends to feed on itself. He also pointed out that farmers happily paid for hybrid corn—*every year*—no one was making them buy it any more than anyone is making farmers buy GM crop seed today. In addition, he interpreted the rush to hybridize in psychological terms:

> The capacity to produce repeatedly a handsome hybrid in exactly the same form and to make diversity unfold in front of your eyes as you developed hundreds of recognizably distinct inbred lines out of ragged open-pollinated varieties produced a kind of disease—a continuing and nearly uncontrollable impulse to breed, test, and release new corn hybrids.[37]

Perhaps the final word on the subject should be left with H. V. Harlan, who in 1957 wrote about what he considered to be the downside of the discovery and exploitation of Mendelian principles—an opinion which many today, in the stampede to invest in molecular-level biotechnology, should perhaps have framed on their laboratory walls.

> The field of plant breeding actually suffered in a way from the greater knowledge we had acquired. Mendel's work was quickly accepted as an enormous advantage in plant science. It was a definite, tangible thing that seemed to take plant breeding from the arts and place it as a science overnight. It captured the imagination of all workers, and genetics at once became a field offering prestige that both soothed and satisfied.
>
> A genetic paper gave new dignity to the author. We boys began to get our hair cut and our shoes shined. The effect on plant breeding was calamitous. Good varieties were still produced, but explorations in the field of practical plant breeding were wholly neglected.
>
> A few of us eventually realized that there would come a day when the world would recognize the difference between a good geneticist and a poor one, so we went back to thinking about plant breeding. We have undoubtedly lost the resources of many good minds from this field for a time, but they will be back.[38]

"Traditional" methods of corn breeding were also hamstrung by a shortage of influential supporters. During the early 1920s, the USDA's Bureau of Plant Introductions appeared to commit itself to work on hybrids. A key

37. Duvick 1996, 541–42.
38. Harlan 1992.

person here was Frederick D. Richey, a USDA breeder of key importance who is known to have corresponded extensively with Jones and Henry A. Wallace—the three together could be seen as representing all major U.S. corn interests: farmers, seed producers, geneticists, and experimental station workers. There was clearly a synergy of interests: scientific, technological, political, and commercial.

Initially there was a widespread assumption that many farmers would produce their own hybrid corn, *not* buy it from specialist seed producers; during the mid-1920s, Iowa and Wisconsin experimental stations ran short courses for farmers to enable them to do so. The University of Wisconsin developed drying and grading equipment for farmers, and by 1939, the state had 436 farmers engaged in seed production for small-scale local sales. The corporate production of F_1 hybrid corn was not then set in stone by historical necessity, but given the growing scale and complexity of the relationship between seed production and plant breeding, it is difficult to see how farm-produced F_1 hybrid seed could have continued for long—especially if the seed industry was determined to gain control over the whole process.

Historians and geneticists can argue endlessly over whether, if the dice of history had landed differently, the road of corn development would have been any different. The second argument, the ethical one, is highly subjective and is inevitably colored by a wide variety of a priori assumptions often based in particular ideologies; it also leads directly onto a discussion of the ownership of and control over plant varieties. There is a fundamental ideological, perhaps temperamental, split between those who see market forces as being the best driver to progress in plant breeding, and therefore the patent-effect enforced by F_1 hybrids as being desirable, and those who have a gut reaction against bringing the profit motive into food production.

There is another aspect to the politics of F_1 hybrid corn production—its production requires genetic knowledge and a highly organized system of involving farmers in its production. It is therefore not a good means of breeding for poor countries. Despite this, many have promoted it as *the* way forward; in 1962 for example, the FAO of the United Nations declared it to be the most important method of breeding.[39] Many who work in development would beg to differ. The story of corn in southern Africa is a case in point. Corn became a staple in much of the region during the twentieth century, displacing local crops as well as an important crop for the white farmers who dominated the economies of the British colonies. In 1960 in one colony,

39. Pistorius and Van Wijk 1999.

Southern Rhodesia (now Zimbabwe), scientists at the Salisbury Agricultural Research Station produced a hybrid of quite exceptional quality, which they named 'SR-52.' This was a single cross, unusual at a time when commercial single crosses had yet to make their full impact in the United States. SR-52's level of productivity was extraordinary, giving a yield of nearly 50 percent more than the previously most popular variety on white-owned farms. Yet it needed constant moisture, had a long growing season, did not function well when grown alongside other crops, *and* if the seed was saved for sowing next year, the F_2 generation's yield was as low as 50 percent less. Clearly, this was not a crop for the black African peasants. Just like the colony in which it grew, the corn was deeply racist.

SR-52 served its white masters well. In 1965, the white minority community of Southern Rhodesia, dominated by farmers, made a unilateral declaration of independence (UDI), in order to stop the British government handing over power to the black majority. For the next fifteen years, the minority regime weathered a regime of sanctions imposed by an outraged international community. One of the factors which enabled it to withstand sanctions was the productivity of its corn.

Corn breeding in Rhodesia was aimed at the white commercial farmers, not, needless to say, black smallholders. Nevertheless, they were trying the new varieties too, and by the late 1960s had spotted that one of the newer hybrids, 'R201,' was early-maturing and flourished on drier and sandier soils. The colonization of Rhodesia in the late nineteenth century had resulted in white settlers bagging all the good land in high rainfall areas, with the black inhabitants pushed onto marginal land. R201 may not have been bred for them, but it gave the smallholders a very good new plant. Without any government or corporate help or promotion, black smallholders were investing in hybrid seeds, and reaping the benefits.

In next-door Zambia (formerly northern Rhodesia), however, production of SR-52 never really took off, as farmers responsible for seed production muddled the seed, and the young nation's starter stock became genetically corrupted. Zambian breeders at the Mazabuka Corn Research Institute during the 1980s did, however, manage to replace it with double and triple cross hybrids, which were cheaper and easier to produce than the single-cross SR-52, and which when grown on to an F_2 generation, did not suffer a huge drop in yield—so allowing farmers to buy new seed only every two or three years. The lesson seemed to be that *if* hybrid strategies are adopted for developing nations, then they need to stress robustness and resilience, and be forgiving of occasional lapses in organization.

SUMMARY

The production of F_1 hybrid corn is the most obvious outcome of the revolution in thinking which Mendelian genetics introduced to plant breeding. While it has been disputed as to whether it really was the best route to follow for improving the crop, the fact is that corn breeding was in a depressed state in the early twentieth century, and the development of F_1 hybrids and production technology galvanized some of the greatest minds in genetics and most dynamic rural entrepreneurs of the time. F_1 hybrids allow for a very tight degree of control over the crop, as well as tending toward the higher productivity typical of hybrids (heterosis) and a very high level of consistency.

Corn is naturally outbreeding, so forcing it to inbreed for the purposes of F_1 production is somewhat counterintuitive. The development of the process of F_1 production took many decades and a variety of stages: detasseling to force one variety to cross with another, an understanding of the type of characteristics which are desired in the crop (particularly yield and composition over good looks), the recognition of the value of producing inbreds as a way of controlling the outcome of crosses, and finally the double-cross method whereby the genes of four inbreds are used to produce a commercially viable seed crop. Later in the twentieth century, given the greater knowledge of corn genetics, it was possible to introduce single crosses.

Despite being introduced in the United States during a period of great economic uncertainty, F_1 hybrids went on to become a great commercial success, arguably revolutionizing U.S. agriculture. Corn breeding throughout the rest of the world followed suit—although tempered by recognition that the technique is not always the most appropriate method of crop improvement for poorer countries. F_1 hybrid corn has enabled an enormous increase in productivity in farming systems the world over.

ELEVEN ❦ CORNUCOPIA ❦ GENETICS OPENS UP THE HORN OF PLENTY

"My first encounter with plant breeding was as a student at Cambridge University," recounted a distinguished retired plant breeder. "We were shown around the greenhouses, where they were working with tomatoes . . . all the features of the prospective new varieties were explained . . . being the sort of young chap I was, I asked an awkward question—'do they still taste like tomatoes?' . . . The answer I got was to the effect that 'the customers don't buy them for their taste, it's what they look like that's important.'"[1]

That fruit and vegetables should be bred for their cosmetic appearance might seem a luxury. In fact, it seems to be a distinct human trait, perhaps a failing, that we fall for appearance so often—consider the history of the American corn shows or the way that size and perfection are the benchmarks of quality in British produce shows. All those giant vegetables don't actually taste of very much! Historically, farmers, gardeners, and breeders have first of all aimed for reliability—the central characteristic of landraces. Yield became more and more important from the eighteenth century onward, but with it an emphasis on "quality," a large part of which was not just eating quality or being fit for a particular purpose but also fashion-show good looks. During the nineteenth and twentieth centuries, the growing geographic separation between consumers and producers meant that a whole new set of traits to breed for became important: the ability to survive transport and storage. New industries, such as canning and jam making, also required produce with distinctive traits too.

Continuing the trends established in the nineteenth century, the twentieth century saw the range of traits that breeders select for gradually expand. Breeding had to become more and more specialized, there needed to be different tomatoes for canning, picking fresh, drying, picking, selling "on the vine," growing in greenhouses, and growing in Alaska, as well as for consumers who wanted big "beef" tomatoes, little "cherry" tomatoes, very tasty sweet tomatoes, plum tomatoes, or those who want them grown organically. This specialization is within the context of agricultural systems which are industrial in scale, a scale which creates its own set of demands, for

1. Leakey 2007.

plants which fit machines, that survive conveyor belts and roll neatly onto supermarket shelves.

With more and more traits to breed for, breeders needed to be increasingly inventive in seeking out traits, and combining them. And, towering above all else, is the ominous growth in human population, and our tendency to consume more and more as we grow richer and richer. There is also the ever-present threat of pests and diseases. What gave breeders the ability to produce so many and so varied a range of new crop varieties has been genetics. The twentieth century saw immense changes in most aspects of human life, but few fields define the century as neatly as genetics and plant breeding: 1900 was the year of the "discovery" of Mendel and the last decade of the century saw the first, albeit unsteady, steps of commercial plant varieties with recombinant DNA (GM crops). The discovery of heterosis and its commercial development in the first thirty years of the century was a harbinger for an almost exponential rate of farther development. Advances in genetics, mathematics, and a variety of support technologies enabled a wide range of plant breeding strategies to be developed. From now on, we can only cover the techniques involved in broad outline. To summarize, these have been:

- the exploitation of F_1 hybridization and heterosis in an increasing range of crops
- techniques to increase the levels of genetic variation available to breeders
- techniques to cross progressively less closely-related species with each other
- the increasing level of control over both plant chemistry and physical structure

The bottom line is that during the twentieth century, food production kept up with population growth. Most analyses show the roles of various factors: fertilizers, irrigation, pesticides, farmer education, and so forth, and tend to put the share of the increase in yield due to plant breeding at around half. It is this yield increase that is the biggest single factor between us and starvation. In many cases, per-acre yields doubled between the 1930s and the 1980s.[2] After breeding, the next largest factor has been the use of synthetic fertilizers.

While the term "agricultural revolution" is frequently assigned to the era

2. One of many who gives a figure around the 50 percent mark is Lupton (1985).

during the eighteenth century when European agriculture underwent major changes, it might be argued that if the term were used to define more strictly increases in yields, the period from the 1930s onward might better deserve it. Though the eighteenth century saw many structural and qualitative changes, it only saw a modest increase in yields. It was the 1930s, however, which saw a definite and quite remarkable change in gradient for the line on the graph, for cereal crops especially.[3]

CROSSING THE BOUNDARIES—THE HYBRID STORY CONTINUES

F_1 hybrids, for which growers have to buy fresh seed every year, were keenly sought by commercial seed producers, for this and their many other advantages. Whether a particular crop would oblige by being amenable to the hybrid seed production process was another matter. Consequently the progress of F_1 production has followed a path between two extremes: one, that of species whose reproductive processes make inbreeding and controlled hybridization easy—but which may not be intrinsically such useful plants—and the other, species where F_1 hybrids may be very desirable but where they are difficult to achieve. In some cases, F_1 hybrid production may transform a crop and greatly increase its agricultural and economic importance. Sorghum was one such crop. During the 1940s and 1950s, the inbred/hybrid method was applied very successfully to this cereal, once a staple only for poor farming communities in Africa, but now, thanks to hybrid breeding using cytoplasmic male sterility (CMS), a popular new feed for cattle in the United States, usefully drought tolerant. The crop went through an enormous transformation, from being 2.5–3 meters tall with a very high proportion of foliage to grain to being 1–1.5 meters tall with a far better grain yield.[4]

Onions were the first crop after corn to be improved through F_1 hybridization, when H. A. Jones at the University of California found one with male sterility; a commercial hybrid appeared in 1952.[5] Jones was a pioneer in attempting to exploit heterosis in a variety of crops, moving on to success with asparagus (*Asparagus officinalis*), strawberries, and squashes. Tomatoes were greatly improved from the 1950s on, with varieties that set fruit in less

3. Grigg 1984.
4. W. L. Brown 1990.
5. McCollum 1976.

than ideal conditions a great favorite with amateur gardeners and small producers in higher latitudes.[6] Tomatoes are a good example of a crop where the time-consuming business of hand pollination proved financially worthwhile; in most cases, hand emasculation or pollination is simply not commercially viable, hence the dependence on CMS or other large-scale ways of preventing production of pollen. In the 1960s, dramatic progress was made with sunflowers (*Helianthus annus*) after the discovery of one with CMS—the plant then became an important crop for oil and feed; in Europe, it displaced efforts to breed hardier soybeans.[7]

F_1 hybrid wheat, however, has been an elusive goal. There is evidence from Australia that perseverance will bring not just higher yields but a general all-round ability to flourish across a range of environments and specific advantages, such as early cropping—a bonus in drought-prone areas. Breeding F_1 wheat has proved slow and costly, with many companies dipping their toes in the water and getting out again. Breeders working with hybrid wheat have attempted to make use of chemical hybridizing agents (CHA), a class of compounds that can be used to cause either male or female sterility; they were first developed in the 1950s. Used in hybridization, they do have several advantages: compared to CMS, a restorer line (to reinstate fertility) is not needed, the necessity for extensive searching for male-sterile lines is removed, and a large number of breeding lines can be developed. The compounds have proved difficult to use in practice for a wide range of reasons, not least of which that many have proved toxic to plants, not just gametes.[8]

More successful has been hybrid rice. One man's name dominates the story: Yuan Longping. Yuan was born to a poor urban family in 1931, but Communism assured him and many others of an education. To be a geneticist in 1950s China was not easy, however, because of the survival of Lysenkoite ideas (see chapter 9). Yuan's disillusionment with Lysenko and Soviet "genetics" began after the failure of the graft hybrids he had made at university. Having learned English and Russian, he was able to explore further and discover Mendelian genetics, which he taught to his students despite the considerable personal risk. No sooner than Chinese genetics had made the break from the stranglehold of Lysenkoism, the economy was convulsed by Mao Zedong's "Great Leap Forward" of 1958–1962, intended to thrust China into the modern age. Modeled on Stalin's attempts to achieve rapid

6. Rick 1976.
7. Leakey 2007.
8. Edwards 2001.

industrialization in the USSR, the results were similar—chaos and mass starvation. It was during these difficult years that Yuan became interested in breeding hybrid rice; in 1960 he had discovered a rare natural rice hybrid, which got him thinking about the possibilities of using heterosis.

In 1964, Yuan was able to establish a hybrid rice research project in Hunan Province. A CMS plant was founded in 1970, apparently by a peasant woman in Shanxi Province, but in the absence of a restorer gene, commercial production of seed was impossible. This is where the all-powerful Chinese state came up with what was needed—a nationwide program to find rice plants with this quality—a search for a needle in a haystack if ever there was one. Over twenty research institutes in several provinces were mobilized, with thousands of researchers taking part. In 1973, the first rice with a restorer gene was found. A hybrid with distinct heterosis was developed in 1974, followed by regional production tests conducted simultaneously in hundreds of counties in 1975. In 1976 it was released—'Nan-You No.2.' This early work was on *indica* rices, but *japonica* varieties more suitable for northern China followed soon after. By 1987, about 30 percent of the rice area of China was planted with F_1 hybrid varieties. Yields were around 15 percent higher than conventional varieties, but there were problems, and there was a decline in acreage during the late 1990s.

Producing F_1 hybrid rice has been a task of enormous complexity, a task that perhaps could only have been undertaken in China, where not only are millions dependent upon rice but where there are also large numbers of people who can be drafted in to work on the many aspects of the task; in 1977, no less than thirty thousand people were sent to Hainan Island to work on pollinating the parent crop under Yuan's supervision. There were plenty of practical teething problems in the early years, and the seed was ten times the price of normal rice (although its seeding rate was much lower), and its later maturity reduced the ability of farmers to grow more than one crop a year. Political pressure was used to persuade farmers to grow it, but after the death of Mao in 1976 and the introduction of incentives for farmers by premier Deng Xiaoping (in office 1980–1997), production soared.[9]

The main problems with F_1 hybrid rice, at least in these early days, were poor flavor and cooking qualities. Subsequent research has indeed targeted quality. Certainly in terms of yield, hybrid rice has been described as "one bright point in a world of plateauing yields," feeding as many as a

9. Hsu 1982; Lin 1994; Li and Xin 2000.

hundred million extra people in China.[10] The amount of land planted with it will probably not increase though until consumers feel it to be a better product.

Outside of China, hybrid rice has not been successful to date and has sometimes proved controversial. The International Rice Research Institute (IRRI) and the FAO made a big push for hybrid rice research in "The Year of Rice" in 2004 in an effort to give a boost to a flagging program. Varieties developed for China have not proved successful elsewhere and have provoked resentment among other Asian countries. Even in China there is opposition; one researcher from a minority ethnic community in Yunnan has complained that "the paddy field seems to have got addicted to heroin, the more rice output you want from it, the more chemicals you have to give it."[11] As with corn, but with perhaps more justification—now that we know far more about genetics than we did in the early days of F_1 corn, there is the feeling that there are alternative routes to increasing rice yield, and that, at least outside China, hybrids are being promoted first and foremost by the seed companies who can afford to invest in the research. For now, the verdict is uncertain.

CASTING THE NET EVER WIDER—HYBRIDIZING BETWEEN SPECIES

Hybridization between species was one of the fundamental early processes involved in the domestication of crops. It has continued to be one of the most important ways of increasing the level of variation for plant breeders, in both agriculture and horticulture. One clear trend in plant breeding history is that as time has gone on, it has become possible to cross species ever wider apart in their relationship to each other.

Plant species can be separated by geographic, ecological, or physiological factors—barriers that the endless human search for novelty or progress have progressively overcome. Species or other taxonomic groups may also be separated genetically. Generally speaking, species within a genus will tend to hybridize (some botanists have even suggested that a genus is partly defined by the ability to cross within it), but ability to cross beyond the genus is rare. There are exceptions, particularly of closely related species that do

10. Quoted in Normile 1999, 313.

11. Quoted in GRAIN 2005, 13. This report provides a good overview of hybrid rice's problems, albeit from a partisan standpoint.

not cross. "Incompatibility" is the name given to the physiological phenomenon that prevents plants with normal pollen and ovules from setting viable seed. Overcoming self-incompatibility allows breeders to self-pollinate in order to obtain inbreds; overcoming cross-incompatibility allows crosses to be made between related species.

Much of twentieth-century plant breeding has been concerned with getting progressively less and less well-related relatives to cross, moving up to crosses between apparently close genera. Methods of overcoming cross-incompatibility by forcing reluctant stigmas to accept alien pollen have become increasingly bizarre (or desperate): passing electric currents across stigmas during pollination to get brussels sprouts and savoy cabbage to cross, "mutilating" (to use the research scientist's words) cabbage stigmas with a wire brush, heat treating of lily (*lilium* species) styles prior to pollination.[12]

As laboratory-based biotechnology began to make advances, techniques to achieve wide crosses became more sophisticated still—embryo rescue has been particularly useful. Some species will cross, in that ovules can be fertilized by pollen, but seeds do not develop. Embryo rescue involves extracting the immature hybrid plant embryos from the mother plant ovaries before they can abort and growing them on in nutrient solutions. Some species or hybrids act as "bridges"—accepting pollen from related species that will not cross with each other; desired genes can then be transferred from A to C via B. In conjunction with embryo rescue, this can then make the seemingly impossible possible.

Wide crosses have been particularly important in wheat improvement throughout the twentieth century, ever since an agronomist at the South Dakota Experimental Station in 1929 succeeded in crossing Yaroslav emmer wheat—tough and immune to stem rust, with common wheat, despite the former having twenty-eight chromosomes and the latter forty-two.[13] The door was now open to increasingly complex crosses involving related grass genera: *Aegilops*, *Elymyus*, and *Secale*, crosses useful for bringing in genes for disease and stress resistance. Advances in cytology and the ability to identify single genes during the 1950s allowed geneticists and breeders to focus more clearly on this area. Linked with chromosome manipulation, it became increasingly possible to create intergeneric hybrids, which while not being intrinsically worthwhile, as were weak corn inbreds, but are immensely use-

12. Roggen et al. 1972; Roggen and Van Dijk 1972; Hopper, Ascher, and Peloquin 1967.

13. McFadden 1930.

ful for their ability to transfer genes into other cultivars.[14] Without these advances, the problems created by rust in particular would have had a major impact on world wheat production and the use of chemical fungicides would also have been incomparably greater. Such hybrids are about introducing one particular trait, *not* about creating genuinely interspecific hybrids that share a wide range of traits.

An intergeneric example that does involve sharing a wide range of traits is "triticale" (*x Triticosecale*) a wheat-rye hybrid, the two genera being *Triticum* and *Secale*. Natural wheat-rye hybrids have been noticed a number of times; Rimpau was the first to publish a description of such a plant in 1888 and obtain a few seeds. Intergeneric hybrids, being at the outer limits of what it is naturally possible to cross are nearly always sterile, so Rimpau's seeds were almost the proverbial hen's teeth. Researchers at northerly latitudes have always regarded the potential of a wheat-rye hybrid as considerable for forage or feed grain but with possibly some potential as a source of bread flour too. Canadians, Soviets, Swedes, Americans—all have sought the holy grail, and emboldened by the discovery of polyploidy in the 1930s, which allows an odd number of chromosomes to be doubled (see below), the hybrid has been attempted on many different occasions—indeed, every time a new technique in laboratory-based breeding work is launched, someone in a white coat somewhere has another go at triticale. German workers had produced some worthwhile cultivars by 1950, the Soviets by the late 1960s, in both cases after decades of work. The Canadians had usable varieties by the late 1960s too, and even launched some experimental breakfast cereals, with a distinct "nutty" taste.[15] During the 1990s, triticale research could draw on an impressively wide range of laboratory techniques, allowing many different wheats and ryes to be crossed, so building up a wide genetic diversity for further work.[16] If it not already there, triticale is definitely coming to a health food shop near you.

Soviet breeders were particularly interested in wide crosses, a function perhaps of the range of material, which Vavilov had collected, and the confidence Communism placed in science. Such research became rarer as Lysenkoism took hold, but some interspecific and intergeneric crossing programs did survive the Lysenko years, with major efforts being made with *Aegilops* and *Agropyron* in the postwar period. Complex hybridizations, often involv-

14. Lupton 1985.

15. Hunter and Leake 1933; Larter 1976; Merker 1986.

16. Holden et al. 1993.

ing the use of bridging species, were needed to get useful genes, primarily for disease resistance and high protein content, into useable bread wheats. Several were commercially viable by the 1980s—just as the USSR was about to bow out of history.[17]

Possibly one of the most beneficial applications of wide crosses has been in developing rice for Africa. African rices are nowhere near as productive as Asian ones, so researchers at the West Africa Rice Development Association in the 1990s tried to marry African *Oryza glaberrima* and Asian *O. sativa*. Embryo rescue had to be used to overcome genetic barriers, but by 2004, the head of the research team, Monty Jones of Sierra Leone, could claim the World Food Prize (see chapter 12). The *NERICA* rices not only yield well but also have good protein levels, and thrive in difficult upland environments.[18]

Interspecific hybrids, not surprisingly, have been especially important for those crops where there are a large number of related wild species—potatoes and tomatoes are good examples, with the seeking of genes which give resistance to pathogens a particular priority. Potato cyst eelworm (a nematode) is one of the worst pests of the crop, building up on land when potatoes are grown continuously or in short rotation. Between 1941 and 1952, Conrad Ellenby at the University of Newcastle upon Tyne in northern England tested more than 1,200 accessions from the Commonwealth Potato Collection, which comprised collections originating from more than sixty wild and cultivated potato species. At a time of war, rationing, and shortages, he had no funds to rent land but found an elegant solution—to grow his test plants on allotments (community gardens) rented to amateur gardeners. Potatoes had been grown annually for the best part of the last fifty years, so the presence of eelworm was assured. At around the same time, Dutch researchers showed that resistance was genetically controlled.[19]

The history of interspecific hybrids has not always been happy—that of the hybrid grape has been particularly problematic. The "true" grape is *Vitis vinifera*, a species with a great deal of natural variation which produced thousands of cultivars in the Old World—much of the interest of wine is down to the individuality conferred on it by the variety from which it is made. *Vitis vinifera* does not perform very well in North America, but given that there are very similar *Vitis* species native to much of the continent, accidental hy-

17. Merezhko 2001.
18. Manners 2001.
19. Holden et al. 1993.

bridization soon occurred (see chapter 1), followed in the late nineteenth and early twentieth centuries by deliberate hybridization. French breeders, desperate to limit the spread of the phylloxera pest, also tried hybridization. In the United States, hybridization produced vines that grew better in American conditions, and in some cases showed resistance to a number of other pests and diseases, which encouraged European growers to plant them.

There was a problem, however—wine made from hybrid grapes did not taste like true *vinifera* wine. All matters of taste, of course, are highly subjective, and many New World drinkers appreciated the "wild strawberry" aroma of varieties with *Vitis labrusca* parentage, while the "herbaceous nose with flavors reminiscent of black currants" of *V. riparia* crosses had its aficionados too.[20] The French, ever protective of the product that was central to their sense of culture and nationality soon turned against it, and hybrids were banned from quality wines beginning in 1927. Although hybrid grapes were used extensively in French table wines, they too were eventually excluded. American growers were more enthusiastic; some, such as Philip and Jocelyn Wagner of Maryland, going to great lengths to collect obscure hybrids in France to take back to the United States. Extensive plantings of hybrid grapes were made in both America and Canada, although in the last two decades of the twentieth century these have often been replaced by *vinifera* varieties.[21]

Within the European Union, the French antipathy to hybrid grapes makes life difficult for British wine growers, whose very existence is regarded by many in the French wine trade as being akin to that of flying pigs. Hybrids such as 'Seyval Blanc,' bred originally by Frenchman Bertille Seyve in the 1930s, do well in Britain, particularly as they tend to be resistant to fungal diseases in a wet climate. However, the European Union does not allow wines made from hybrid grapes to be given the coveted "quality" status, only the inferior "table wine."[22]

While the French may have scorned and eventually eliminated hybrid grapes, they were hugely indebted to a Texan, T. V. Munson (1843–1913), who helped breed rootstocks which were resistant to phylloxera—traditional varieties could then be grafted onto them. Hybrid grapes may have been unacceptable, but hybrid roots were received like manna from heaven—

20. "Hybrid Grapes," Encyclovine, http://encyclowine.org/index.php/Hybrid (accessed March 15, 2007).

21. G. A. Cahoon 1996.

22. Boze Down Vineyard 2006.

phylloxera had devastated the French vineyards. Munson's patient classification of American grape species and varieties and his exporting tons of rootstocks to France did much to help rescue the industry. A grateful nation even awarded him the Legion d'Honneur, the nation's highest award.[23]

IN FULL FLOWER—PLANT BREEDING AS AN APPLIED SCIENCE

Once Mendelian genetics was accepted and F_1 breeding in corn showed the exciting possibilities offered by its understanding and use, scientific plant breeding took off. In doing so it developed a wide range of systems, from pure line selection with maximum inbreeding to mass selection with minimum inbreeding. Much of the progress made in plant breeding during the twentieth century has been made not just through the application of genetics per se, but through the use of statistics. Statistics enabled the differences between plants to be analyzed, the characters of progeny to be predicted, and the differences that resulted from genes and from the environment to be clarified.

One immediate use for advances in statistics grew out of the realization that much inheritance was governed in a way that appeared to contradict Mendel—through multiple genes. Nilsson-Ehle in 1909 and East in 1916 (see chapter 6) both published papers arguing the case for a "multiple factor hypothesis": that characters that vary quantitatively are governed by several genes. With the clarification that a simple plant trait, such as yield, can be governed by several genes, the yawning gap between the Mendelians and the biometricians suddenly seemed to close, allowing for synthesis of biometrics with not only a Mendelian genetics but also with the Darwinian theory of evolution. So, in the late 1910s, the stage was set for the application of the study of minute graduated differences to the products of a Mendelian-based breeding program. Quantitative genetics, the fruit of détente between the Mendelians and the biometricians, was born.[24]

A key figure in the marriage of plant breeding and statistics was the myopic and brilliant Ronald Fisher (1890–1962). He was not a plant breeder, but his contributions to genetics and evolutionary biology were immense and his statistical methodologies acted as a foundation for much work in plant breeding. That Fisher had such poor eyesight was in fact part of the background to his brilliance—as a child he had received mathematics tu-

23. McEachern 2003.
24. See technical note 29: Multiple-Factor Hypothesis and Polygenes.

toring, but finding it so difficult to see what was written down in poor light, he developed the ability to imagine mathematical problems in a geometric form and developed a very powerful sense of intuition. A scholarship to Cambridge in 1909 brought him to town at just the time for the debates between and among Mendelians, Darwinists, social Darwinists, eugenicists, and biometricians.[25] It was his ability to see that biometrics could be applied to Mendelian patterns of inheritance that was instrumental in enabling the biometry-Mendelian feud to be healed; in 1918, he published a paper, "The Correlation between Relatives on the Supposition of Mendelian Inheritance," which laid the foundations for biometrical genetics. The following year he began work at the Rothamsted Experimental Station, Britain's leading agricultural research farm, where for seventy-three years data had been collected on a number of plots undergoing different kinds of feeding—"raking over the muckheap" as Fisher apparently described it.[26]

At Rothamsted, Fisher gained a reputation as being number obsessive—in the days before computers the ability to solve complex problems mentally counted a great deal. Even more important than his work on variation in plants were the design of experiments and the analysis of their results; his *Statistical Methods for Research Workers*, published in 1925, was a vitally important book for all the life sciences, plant breeders in particular.

One of the fundamental realizations made by pre-Mendelian breeders was appreciating the difference to a crop made by the plant itself and the environment it was growing in. After Johannsen had made the distinction between genotype and phenotype, growers and breeders had to face the next problem—how to separate out these two distinct contributions. The carrots in this plot are so much bigger and juicier than the ones in that plot—but why? Is it because they are from different sources of seed, or is it because one plot is slightly downhill of the next, and they are getting more moisture? One way around this problem is to grow lots of plots and sow the seed from different sources in plots chosen at random—so evening out any possible environmental contribution to size and juiciness.

Again, statistics comes to the rescue—the elucidation of the differing contributions of environment and genes needs statistical analysis. What is more, different varieties may respond differently to the same environment, for example, variety A may be better than variety B on fertile soils, equal on average soils, and worse on infertile ones. Such relationships are known

25. Fisher in fact became a particularly enthusiastic eugenicist.

26. Quoted in Yates and Mather 1963, 93.

as genotype-environment interactions (GxE). By the 1940s, plant breeding had incorporated a range of statistical techniques to analyze these kinds of problem.

Statistics as a science has continued to advance and contribute to plant breeding—as has the science of research plot design. Advances in computers have also made a great impact. The computer revolutionized the application of statistics, allowing far more complex number-crunching and therefore making it useful for breeders to collect more and more data on plants, secure in the knowledge that they can be not only stored securely but made sense of—if not now then on another occasion. It also makes the art of prediction easier—for this is where plant breeding is still an art, in trying to assess just what crosses to make in order to move towards a desired outcome. Computers can run endless theoretical breeding programs and compare outcomes; while they cannot predict the future, they can at least point towards likely scenarios and the probability of achieving each one.[27] Increasingly advanced equipment for measuring vital plant statistics or reducing the labor involved in doing so also made data collection easier.[28] Breeding gradually became less of an art and more of a mathematically driven science.

COMPOUNDING THE COMPOUNDS—BREEDING FOR PLANT CHEMISTRY

Many of the fields of northern Europe are now bright yellow for several weeks every summer.[29] The crop is canola, or, rapeseed (*Brassica napus*), a mustard relative with acid-yellow flowers in early summer and an oil-rich seed. It is a relatively new crop, and attitudes to its impact on the landscape are mixed: some appreciate the vivid color it gives to the patchwork of traditionally shaped fields, others resent its brightness and peppery aroma. That canola is so much a part of the landscape is a result of breeding for edible oil; it is widely used in the food industry. Originally canola, however, was quite unsuitable for human consumption; it contained ingredients that affect the functioning of the thyroid gland. With their reduction through breeding in the 1960s and 1970s, it became possible to make food-quality oil from this easy-to-grow crop, an attractive proposition for people seeking lower-fat alternatives to butter and animal-derived fats.

27. Ferguson and Garretsen 1968.
28. Goldman 2000.
29. Williams 1985.

Taming the oils in canola is an example of one of the age-old goals of plant breeding—the elimination of chemicals in the plant that are toxic or reduce digestibility. Pulses, such as beans and lentils, are another example. They are relatively high-protein foods, widely eaten in many countries. A problem is that they tend to contain variety of compounds which that inhibit human digestion, with the well-known results that cause them to be the butt of endless jokes. British agriculture expert Colin Leakey was first asked to look at digestion problems in Uganda in the early 1970s, with particular regard to infant feeding. This led him on to a lifelong search for the non-flatulent bean. Discovering that tannins in the coat are a good indicator of a high content of digestion-inhibiting compounds, he has focused on breeding white beans (hybrids of *Phaseolus vulgaris*), measuring their effects on digestion—with what effects on the Leakey family life, we can only guess at. His best result so far has been 'Prim,' a primrose-yellow bean with excellent flavor and eating characteristics.[30]

Plants are chemical factories—photosynthesis turns CO_2 and water into sugars through using light as an energy source. These can then be converted into the physical structure of leaf, stem and root vital for plants. They can also be converted into *secondary metabolites*, compounds not necessary for plant survival but vital for certain other functions, for example, the deterrence of insects or fungal diseases. Some of these compounds can be beneficial to human health, some toxic, some inedible but useful for nonfood purposes. The analysis of plant chemistry and the use of breeding to manipulate the levels of plant chemicals has been an important part of twentieth-century applied plant science.

In using plants, humanity makes use of a huge range of plant-derived compounds. Historically these have tended to be used crudely—herbal medicine makes use of secondary metabolites, but not usually extracted from the plant. As technology has advanced, uses are discovered for more and more plant chemicals, which can now be extracted to whatever degree of purity is required. To continue with the example of herbal medicine, many medicinal herbs now grown are cultivars or hybrids bred to achieve high levels of certain desired compounds, for use either as raw material for the extraction of key compounds in conventional medicine, or for lower levels of processing for use in a variety of herbal medical traditions, Eastern and Western.

The chemistry of food crops, in fact, had been recognized as important since the pioneering work carried out in Illinois by Hopkins in the late nine-

30. Leakey 2007. See also http://www.colinleakey.com.

teenth century (see chapter 10). A major aspect of twentieth-century plant breeding has been to maximize the nutritional elements and minimize the harmful in both food and fodder crops. Corn is widely used as cattle feed, yet it is deficient in some of the amino acids vital for animal nutrition: lysine and tryptophan—the same reason corn is a poor source of protein in a human diet. In the early 1960s, Paul Mangelsdorf identified the single recessive gene that enabled him to develop higher lysine strains.[31] This proved useful in both making the crop more valuable as a cattle feed and in improving the protein intake of populations, particularly in Africa, where people depended on corn as a staple. Centro Internacional de Mejoramiento de Mais y Trigo (CIMMYT, International Corn and Wheat Improvement Centre) continued with the work, reflecting a major concern of the international agricultural agencies that new varieties of staple foods for developing countries need higher nutritional content than traditional ones. Greater availability of minor nutrients can also be bred into plants, such as zinc, for example (zinc deficiency can cause a variety of health problems, particularly for women and children, and is common on many alkaline soils). A particular focus in Turkey during the 1990s was on breeding wheats with high zinc concentrations to try to overcome this.[32]

A key figure in a technological exploitation of plants beyond foodstuffs was George Washington Carver (1864–1943), one of the outstanding American plant scientists of the twentieth century. He was born into slavery but managed to get an education in the bitterly divided post–Civil War South. Denied college admission because he was black, he traveled to Iowa, where he was eventually accepted at Simpson College then transferred to Iowa State University. Carver did some breeding work, but his main interest and importance lay in the emerging field of chemurgy—the development of chemical-based products from agricultural raw materials. A simple example might be the growing of potatoes for the manufacture of starch, or for distilling into industrial alcohol. During the twentieth century, chemurgy spawned a huge range of products which can be made from crops—this has inevitably added to the work of plant breeders, who needed to produce varieties with particular chemical characteristics.

Carver left Iowa in 1896 to lead the agriculture department at Tuskegee University in Alabama. Here he threw himself into the improvement of the lives of poor farmers throughout the South, researching and introducing

31. Mangelsdorf 1974.
32. Braun et al. 2001.

new methods of cultivation and, vitally for the rural economy, developing new products that could be made from crops, thus enabling farmers to diversify. The industrial research laboratory he and his assistants ran is credited with developing 300 applications for peanuts (*Arachis hypogaea*) and 118 for sweet potatoes, as well as applications for many other crops, including cowpeas (*Vigna unguiculata*), pecans, and soybeans (*Glycine max*). The products included dyes, polishes, cooking oils, varnishes, glues, synthetic rubber, talcum powder, and shaving cream. He virtually never sought patents, for he believed that plant products were the gifts of God and it was wrong to earn money from them.[33]

THE GENETICS OF THE ROULETTE WHEEL—MUTATION BREEDING

Anyone today selling a food called 'Nuclear Rice' would face severe marketing problems. But in the 1950s and early 1960s, nuclear energy was part of the technological revolution that, it was widely hoped, would transform the world. Despite its role in the immensely destructive power of the atomic bomb, nuclear energy was generally seen as good, to the extent that it could be used to sell products as modern and efficient; the author once had a set of kitchen scales called the "Nuclear Scales" from this period—the logo was the name surrounded by a stylized explosion. It was the Hungarians who came up with the high-yielding 'Nucleoryza'—during the 1960s, it covered 80 percent of the Hungarian rice-growing area.[34] It was the result of radiation breeding, one of the most extraordinary chapters in the whole history of plant breeding. If it had never been invented and someone suggested it today, the outcry would be enormous; popular fears of radiation and chemical residues would combine with all the arguments raised against GM. Unlike GM, the change induced in the genetic material is random and unpredictable, although the technique has greatly improved its level of precision over the years. The fact that almost no one outside the world of the plant sciences knows about it illustrates just how little popular (or pressure group) interest there was in plant breeding until relatively recently. A basic knowledge of the story of induced mutation can do much to put the alarmist fears raised by some in the GM debate into a very different perspective.

Mutations, or sports, have long been recognized as a source of sudden

33. Four Iowans Who Fed The World 2003.

34. Gottschalk and Wolff 1983.

and spontaneous variation in plants of all kinds. Breeders were often frustrated—Mendelian genetics enabled them to work with the level of variation they already had, and what they had could perhaps be increased by new introductions or crossing with more distant relatives. But breeders have always on the lookout for that extra level of variation, that magic trait that could enable them to transform the plant and raise a new variety that would take the world by storm. Mutations offer a source of novelty—but natural mutations are rare. Any method had to be seized which offered the possibility of improving on the rather ungenerous hand that nature had given.

Ancient Chinese texts record mutant cereal crops as early as 300 BCE.[35] Hugo de Vries (see chapter 6) coined the term in 1901 with the publication of his first researches on mutations, which he believed played a major part in evolution. De Vries also predicted that mutations might be generated artificially. The discovery that chromosomes can be changed by X-rays, and that these changes are permanent, was made in 1927 by Hermann J. Muller (1890–1967), an associate of Thomas H. Morgan (1866–1945), the American geneticist who had, earlier in the century, shown that it was chromosomes which were the carriers of heredity. The obvious next step was to speed up the natural slow rate of mutation—using X-rays and other forms of radiation which increases the natural frequency of mutation. The first results were obtained with the geneticist's favorite subject, the fruit fly, but corn and barley were soon tried.[36]

Initially, X-rays and gamma rays were used to bombard seeds in order to stimulate mutations. Later, cutting material or other plant parts that could be used for vegetative propagation were used. During the 1940s, it was discovered that a variety of chemical reagents could be used to induce mutations; compounds with highly reactive alkyl groups for example which react with DNA.[37] The most useful has been ethylmethane sulfonate (EMS)—a compound related to mustard oils.

Mutation breeding's high point was probably the early 1990s—before GM technology began to threaten its hard-earned position as a viable source of variation. A survey conducted in 1990 showed that a total of around a thousand commercially viable crop varieties had originated through induced mutation: of the 998 whose origin could be definitely traced, gamma rays were overwhelmingly the source of most (68 percent of seed-propagated

35. Van Harten 1998.
36. Hagberg and Åkerberg 1961.
37. Chahal and Gosal 2002.

crops, 44 percent of vegetatively), X-rays next (12 percent and 49 percent), while chemical mutagens accounted for only a small minority (15 percent and 3 percent).

Swedish researchers were among the early enthusiasts, with Nilsson-Ehle and Åke Gustafsson (1908–1988) beginning to work with a variety of crops in the 1930s. Gustafsson became known as the "father of mutation breeding"; he was a gifted speaker, which put him in good stead in debates with skeptical colleagues.[38] Barley was a particular interest and focus of his researches; he was also instrumental in establishing a gamma field at the Bålsgard Fruit Breeding Station—from which came many useful cultivars.[39] Intriguingly, he was also noted as an essayist and poet.

The International Atomic Energy Agency, established in Vienna in 1957, has played a major role in promoting radiation breeding. Mutation breeding's first commercial success was a tobacco—'Chlorina'—achieved with X-rays in 1934 by a Dutch worker at a research station in the then-Dutch colony of Java. By 1936, it and plants bred from it were grown on 10 percent of the total tobacco area—its success being its light leaf color, useful for use as a wrapper for cigars. Barley has responded particularly well, the first commercial variety was released in the Soviet-satellite German Democratic Republic in 1955: 'Jutta,' which was also the result of bombardment with X-rays. By 1981, a total of sixty-one barley varieties had been released; many were early maturing and had increased yield, shorter and stiffer straw, increased drought resistance, or higher protein content. Perhaps one of the greatest individual successes has been the durum wheat 'Creso,' which covered about one-third of the total acreage for the crop in Italy in the early 1990s. Those who might be worried about "irradiated genes" will by now be choking on their spaghetti. Accountants were certainly not choking—it has been calculated that during ten years of production, 'Creso' generated $1.8 billion in additional value for Italian farmers—the total costs of the mutation breeding program on durum at the Casaccia center near Rome over a fifteen-year period came to only $3.5million.[40]

The problem with mutations is that the vast majority of them are deleterious—most likely to be dysfunctional dwarfs, to have distorted leaves or imperfect floral parts, than to be the super high yielding, drought resistant, lustily growing plants of a breeder's dreams. When seeds are irradi-

38. Van Harten 1998.
39. Leakey 2007.
40. Van Harten 1998.

ated, nearly all the possibly useful traits turn out to be recessive, and the vast majority of what appear to be mutants in the M_1 generation (the first generation to be treated) are not part of the stable genetic makeup of the plant. Stable mutants are only revealed in the M_2 generation (i.e., after the treated plant is selfed or crossed), which necessitates raising large numbers of plants. In one case, in a peanut-breeding program in North Carolina in 1959, around one million M_2 plants were grown to produce one commercially useful variety.[41]

Mutants themselves may be of little value. What *is* potentially valuable is their genes—the main value of mutants is in crossbreeding program, where their new characteristics can be incorporated into active breeding gene pools. The interactions between mutant genes and the background of the rest of the genotype cannot be predicted, so mutants need to be crossed with a wide range of other varieties in order to discern positive and negative interactions—which can be very time and labor intensive, of course. One particularly useful trait is that of male sterility, which can then be made use of in F_1 breeding programs, such as capsicum in Bulgaria and rice in Japan, both used during the 1970s. Leaf "distortion" has had its value too—nearly all the commercially important "leafless" peas originated following induced mutations.[42]

Where mutation breeding has come into its own is the field of breeding clonal crops, where achieving change through sexual reproduction is either very slow or difficult. Good mutations are effectively "sports" derived from proven varieties and so do not need any further breeding, as is inevitably the case with seed-grown crops; they can simply be vegetatively propagated (see chapter 5). Combined with meristem culture and other laboratory techniques that cause young plants to mass-multiply, mutation breeding undeniably has the capacity to be commercially very useful for such crops. Induced mutations have been very useful for producing single character breaks in existing varieties without affecting the rest of the genetic makeup or situations where there are desirable genes linked very closely to undesirable ones.[43] The Dutch cut flower and pot plant industry, for example, has made extensive use of the technique to produce color breaks or new petal shapes in flowers, particularly chrysanthemums. In some cases these have been produced by generating mutations from already mutant derived vari-

41. Ibid.
42. Leakey 2007.
43. Chahal and Gosal 2002.

eties.[44] Chrysanthemums (*Chrysanthemum x morifolium*) are the extreme case in economy of effort—two hundred to five hundred cuttings are needed to produce a worthwhile new cultivar, which may take less than two years to come into production. These, however, are not genetic mutants, but bud sports.

Radiation breeding has considerable support in Japan, something of an irony given the country's history as the only recipient of nuclear weaponry. Among other facilities, the country's Institute for Radiation Breeding maintains a radiation field, comprising a circular field of 100 meters in radius with an 88.8-terabecquerel Cobalt-60 source at the center and surrounded by a shielding dyke 8 meters high. Globally, around twenty such fields were constructed during the 1950s and 1960s, but few now remain. Successes include new, and widely distributed, varieties of Nashi or Asian pear (*Pyrus pyrifolia*) with enhanced disease resistance.[45] Fruit, which has often proved hard and slow to breed, has been a particular target of induced mutation.

"Don't try this at home" is what TV popular science presenters are fond of saying, as they show us yet another spectacular experiment. Radiation breeding is certainly something that we would think twice about before attempting on the workshop bench. But it was not always so—the 1950s, as we have seen, was an era when the industrialized world was in love with nuclear power and radiation technology. Known as "Frankenstein of the flowers" for his work with radiation, John James of Ohio was an inventor; having worked in the development of color photography and television production, he also clearly had a literary streak, as he wrote for television and film. He apparently became interested in radiation and using it to breed new plant varieties when he was in the hospital for wounds received while a marine in World War II. In his book, *Create New Plants and Flowers—Indoors and Out*, published in 1964, he described how he scraped the luminous paint off every watch or clock "I could lay my hands on"—this was at a time when luminous watches contained appreciable quantities of radium. He tried out the radium on rose buds and a variety of garden seeds, producing a rose which he called 'Soeur Therese,' which even he admitted was deformed and disease prone. Legally obtaining lower-level radioactive materials was not particularly difficult in the 1950s, so he was able to further his researches

44. Gottschalk and Wolff 1983.

45. Institute for Rice Breeding Web site, http://www.irb.affrc.go.jp/index-E.html (accessed January 16, 2007); McGloughlin 2001.

with Cobalt 60 and Iodine 131. He did, however, recommend not working with food plants and limiting the amount of time spent with radioactive material. Among his other breeding successes, he claimed to have produced a plum-sized cherry and a rose with a smell that puts off Japanese beetles.[46]

Opinions on the use of radiation and chemicals to induce mutations for breeding programs remain divided—indeed, the topic seems to have a remarkable ability to polarize opinions. Many practitioners have been openly critical, pointing out the enormous effort required to produce a small number of real breakthroughs. Norman Simmonds in particular was very critical, describing induced mutation breeding as a "useless, even baneful activity" that "generated mountains of mostly disreputable literature, some trivial ornamental mutants and nothing of any practical consequence."[47]

Induced mutation has its supporters though. According to one plant breeder, "mutation breeding has achieved so much."[48] Its advantages, he argues, lie in the fact that the technique allows the breeder to knock out or change single genes and leave the rest of the genome untouched. He points out that "we now know so much more about chromosomes than we did in the early days, and that different mutagens [i.e., different types of radiation or different chemicals] have very different effects." Inducing mutation, in other words, is no longer a random scattergun business.

Intriguingly, induced mutation never attracted worries over health or the possible effects of "random mutations." The fact that it did not, compared to the furor which erupted over food irradiation in the 1980s and GM in the 1990s, indicates how times have changed (see chapter 13).One possible reason for the lack of public concern over mutation breeding is that it has always remained in the public sector; the institutions undertaking it are all publicly funded, results are published, and all the work done is clearly for the public good. There are no corporate owned gamma fields, producing new crops to mesh with company-produced agrochemicals, no private secrets, no shareholders clamoring for higher dividends. The fact that mutation breeding is out in the public domain, that it is relatively "low-tech," and that sifting through the results can be quite labor intensive, make it eminently suitable for developing countries. Given the tight grip that multinational corporations seem to be intent on gaining over the genomes of crop

46. James 1964.
47. Simmonds 1991, 7.
48. Leakey 2007.

plants which will inevitably restrict the ability of others to exploit GM technology, mutation breeding may be with us for a long while yet.

ONE SET OF CHROMOSOMES GOOD, TWO SETS BETTER—POLYPLOIDY

A related field to that of induced mutation, and which began to evolve at a similar time is that of induced polyploidy. A team led by Morgan in 1915 had established the physical basis of heredity, locating it in genes linked on linear chromosomes. But it was not only changes in and among genes, which could affect evolution. Whole chromosomes could change in number too. Cyril D. Darlington (1903–1981) initially worked under Bateson at the John Innes Institute as an unpaid technician, having had a rather indifferent career as a student of agriculture. His first paper was on tetraploidy of the sour cherry (*Prunus cerasifera*); he went on to show with *Recent Advances in Cytology*, published in 1931, that changes at the chromosonal level, such as polyploidy, could also effect major evolutionary changes.

Polyploidy we have already seen (see chapter 1) as a source of new species. With Darlington's discovery, breeders tried to induce polyploidy by shocking plants with heat or by propagating from cells from the callus tissue formed from damaged shoots. In 1937, it was discovered that colchicine, a compound derived from *Colchicum autumnale* (a small bulbous plant, often incorrectly called the autumn crocus), inhibits the function of the spindle fibers that separate the chromosomes during mitosis; cell division fails to take place, so the new cell has twice the number of chromosomes.[49] Instant polyploidy, or, "colchiploidy," seemed to promise a whole new step forward in breeding, but as noted in one modern textbook, it "has not lived up to the expectation of the breeder."[50] Even at the time it was first touted as the solution to many breeders' problems, it was described as a "fad."[51] From 1934 there was a veritable explosion of research papers on colchicine, but by the 1950s, interest had largely died out. John James, needless to say, promoted the use of colchicine at home by amateur plant breeders in the 1950s and 1960s, and in this he was not alone.[52] Its main uses have been in the opportunity it offers in "fixing" heterosis in the hybrids of many species, and in intro-

49. Hagberg and Åkerberg 1961.
50. Quoted in Chahal and Gosal 2002, 428.
51. Wellensike 1939.
52. James 1964.

gressing genes from wild species into cultivated ones.[53] It has been of great use in breeding triticale, for example, which inherits different numbers of chromosomes from its parents. In doing so the technique has perhaps found its obvious home, as a valued laboratory technique—not a source of novelty in itself.

THE JUGGERNAUT OF PROGRESS ROLLS ON—THE GASTRONOMIC AND SOCIAL IMPACTS OF PLANT BREEDING

I was once visiting a friend in Wisconsin, touring his orchard, when he offered me a 'Red Delicious' apple fresh from the tree. Knowing how tasteless this variety normally is, I bit into it rather dutifully but was astonished at how rich and aromatic it turned out to be. The irony is that when fresh, this is a wonderful apple. The fact that those bought in the shops are so disappointing cannot be blamed on either apple or breeder. What has gone wrong?

During the twentieth century, agriculture became agroindustry. The scale of everything increased: farms got bigger; supply chains longer, companies larger. The tendency of humans to live in cities, far from the sources of their food, has forced the industry to transport food over longer distances and to store it for longer. Wealthier populations demand more, and enabled by cheap transport costs and the desire of supermarket owners to offer a cornucopia of produce, it became increasingly possible for food to be offered that was either impossible to obtain in the country in which it was sold (such as bananas in Europe) or that was out of season. A whole generation has now grown up who expect tomatoes in the depths of winter and leeks at the height of summer. They also expect that what is in front of them in the shops to be perfect. Our grandparents may well have been happy just to get *any* potatoes; we expect ours uniformly graded, scrubbed clean, polished and all too often, neatly presented in a Styrofoam box. Oh—and cheap, as well.

Fruit and vegetable crops now have to *designed* for consumers and retailers who expect nothing less than perfection. Over time, however, a reaction has set in, in particular with customers increasingly complaining about lack of flavor. Partly this may be put down to a romantic misremembering of the past, of childhood memories where every tomato was redder and every cherry sweeter. But anyone who grows their own vegetables will know that with some crops, there is a real difference. Many of the reasons for this are

53. See technical note 30: Introgression and Backcrossing.

to do with growing conditions or the ways produce is stored, or simply the fact that it has been stored at all. Some however is down to variety. Breeding has concentrated on the ability of food to be picked at one end of Europe, chilled, put in a truck, driven and ferried a couple of thousand miles, mechanically fed into boxes, and still look scrumptious.

Flavor has not been a priority in breeding. To be fair to plant breeders, there are problems in breeding for flavor. One is that flavor is generally created by sugars and other compounds that attract pests. Another is that flavor is appreciated very subjectively—one person's delicious apple is "too tart by half" for someone else. It is commercially more sensible to breed for blandness than it is for flavors that some consumers may find too strong or in other ways disagreeable—breeding for the lowest common denominator, in other words.

Looks matter because customers see fruit before they taste it, and good looks are very seductive. There are major regional preferences, however. Consumers in southern Europe like striped apples, but not in the north, where they like them green with red shoulders.[54] Americans, however, like them big, red and looking like something in a child's picture book. It is no surprise that U.S. supermarket shelves groan with 'Red Delicious,' which has the ability to be transported and sit in storage, and on the shelf looking good for months after its insides have turned to tasteless pap. It was originally plain 'Delicious,' but needed to distinguish itself from the 'Golden Delicious' (see chapter 5); it was not originally red, either, but yellow with red markings. The red apples of today are descended from sports thrown up by the original variety over the years, or are sports of sports. Not only did the wholesale trade and the supermarkets love its indestructibility, but the growers in Washington State did too. But eventually, the customers stopped coming and started to buy tastier apples, with the result that its percentage of the harvest in Washington fell from 75 percent in the 1980s to 37 percent in 2003.[55]

There is a more serious problem—reduction in flavor may be accompanied by a reduction in nutritional content. This is a hotly contested area, with radical critics of agribusiness and the organic food lobby claiming that "industrial" food not only tastes bland but is also low in nutrients putting the food industry on the defensive. It is interesting to note that attempts by

54. Eyssen 1994.

55. A. Higgins, "Why the Red Delicious No Longer Is, Decades of Makeovers Alter Apple to Its Core," Washington Post, August 5, 2005.

the USDA to mandate levels of nutritional value in 1970 were resisted by the industry.[56] By the 1990s, the issue had crept up the agenda, and the industry began to accept that there was a problem.[57] The turn of the century has certainly begun to see improvements, as supermarkets turn to stocking a wider range of crop varieties for more-discerning customers, with a premium price on those with a superior flavor. The days of the tasteless 'Red Delicious' are probably numbered.

With a huge growth in interest in health, particularly in the public awareness of micronutrients such as vitamins and antioxidants, industrial scale agriculture is under great pressure to produce more nutritious produce. There is no reason to doubt that it will rise to the challenge. However another byproduct of industrialization seems much more intractable—that of unemployed farm labor. The case of the California tomato industry provides a dramatic example.

Small tomato farms were once common in California's central valley—much of their success depending on cheap immigrant labor from Mexico. Worries that this might come to an end during the late 1930s prompted an interest in mechanical harvesting. In one man's mind at least, that of G. C. "Jack" Hanna, an extension worker at the University of California at Davis, one of the world's leading agricultural research centers. His colleagues begged to differ and regarded his search for a tomato that could be harvested by machine as something of a personal eccentricity. No one had actually designed such a machine or even proposed building a prototype.

Traditionally, the pace of tomato picking was dictated by the fruit, which set over many weeks. Machine picking would require a tomato plant that set its fruit all at once, was firm enough to take insensitive mechanical handling, and had a "high rate of abscission" (i.e., the ease with which it is possible to pull the ripe fruit away from the stem). Despite misgivings, the university gave Hanna its support, and let him work with engineer Coby Lorentzon on developing a harvester. Hanna did find his tomato, in New York State, and used it in a breeding program—machine and tomato evolving together.

Lack of real interest in what seemed like a fantasy project led to foot dragging until the mid-1950s, when labor unions began to put increasing pressure on the U.S. government to restrict the flow of Mexican migrant workers. Employees had to pay higher wages to local workers, which rekindled interest in the idea of machine harvesting. By 1959 the prototype was ready

56. Kloppenburg 1988.
57. Morris and Sands 2006.

at a cost of $25,000—then a big sum. When further restrictions were made on migration in 1964, the machine took off.

While in 1964 only a tiny proportion of tomatoes were mechanically harvested, in 1968 virtually all were. What the designers of the machine and the tomato had not realized was the cavalcade of social and economic change they had unleashed. The machines were expensive, so only large farms could use them, so the number of farms rapidly shrank, with devastating consequences for small family businesses; grower concentration increased 90 percent between 1962 and 1976. Another effect was "vertical integration" in tomato growing, the process whereby one company controls the entire process of production, from growing to harvesting to processing to selling. This was an early example of what has become a much more common process in agriculture. Far fewer workers were employed—thirty-two thousand fewer, according to one estimate—so both Mexican Americans and white Americans suffered. An organization representing farm workers even tried taking the University of California to court on the grounds that designing machines that put people out of work was an inappropriate use of taxpayer's money. They lost. There were changes in who did the work, too; women tended to take over from men, as the machine work involved much less heavy lifting. However, the machine did prevent the development of what could have been a major trend, the moving of tomato production to Mexico. The cost may have been high, but the tomato was saved for California.[58]

The "hard tomatoes, hard times" story has been repeated many times. Crops often need to be bred specifically for machine harvesting, which usually takes longer than designing the machines. But once both are up and running, the human costs are repeated. There have been costs to consumers too, as the breeding of machine-harvestable crops tended to restrict consumer choice. Sometimes, however, machines made possible what was not possible before and therefore created a need for labor, as in the case of winter wheat and barley in northern Europe during the last quarter of the twentieth century, as more powerful and larger machinery made working in heavy soils in winter easier. Breeding and production of winter-sown barley in particular took off—in the early 1970s, only 10 percent of the British barley crop was winter grown; by 1992, it was 75 percent.[59]

Mechanization, however, does generally reduce the need for labor. Yet by doing so it improves efficiency, which means more money elsewhere in the

58. Friedland and Barton 1976; De Baedemaeker 1994; Winner 1986.

59. Silvey 1994.

economy, and therefore more work for others, elsewhere. Such a trade-off is part of the dilemma of technical and commercial progress. Balancing the costs and benefits of such changes is not part of the remit of the plant breeders' work, but there is no excuse for being unaware. Plant breeding does not happen on an island, rather, it sets off complex chains of cause and effect. There has always been a tendency for those in science to work in isolation, and remain aloof from the social and other impacts of their work; it is rare to come across such a forthright recognition of these issues as that made by Boysie E. Day, a University of California plant physiologist in an address to the 1977 annual meeting of the American Society of Agronomy, as he discussed the role of agronomists in changing the livelihoods of farm workers, "no meeting of social reformers . . . modern-day Jacobins or anarchists will cause as much change in the social structure of the country as the ASA meeting of crop and soil scientists."[60]

CEREAL MAKEOVER—REDESIGNING WHEAT

The harvest has been a favorite subject of artists down the ages. Historic illustrations, such as *The Harvesters* (1565), by Pieter Brueghel the Elder, are a fascinating insight into the days before combine harvesters, when men and women cut the wheat by hand with sickles. Looking closer, we also notice that the wheat is up to the shoulders of the harvesters. People were smaller in the past—but the wheat was also taller. Traditional wheat varieties grow to around 120–140 centimeters; modern ones can be as short as 60 centimeters. Some feel of what it used to be like walking through a field of wheat can be gained from walking through rye, which has not yet experienced the "dwarfing revolution."

Over the twentieth century, the shape of many plants has often been hugely changed: peas have lost leaves and gained tendrils, former bush beans have been made to climb and climbing ones made to squat, gangly "indeterminate" tomatoes have produced varieties that are compact enough to fit into a window box, or even a hanging basket. Plant "architecture" has been changed to suit a whole variety of purposes: mechanization, disease control, the need to fit more plants into a limited space. The changes in wheat have been among the most important and dramatic.[61]

An early American agricultural development worker in Japan, Horace

60. Reprinted in B. E. Day 1978.

61. See technical note 31: Partition and Harvest Index.

Capron (1804–1885), noticed productive wheats as short as 50 centimeters. Short wheats depend on dwarfing genes, which limit how much stem there is between the leaves, so producing shorter, stiffer straw, which has the advantage of being less likely to lodge and is very responsive to increased fertilizer (see chapter 3). In 1874, the U.S. Commissioner for Agriculture in Japan reported officially on high-yielding, semidwarf wheat. The first Westerners to try to make use of Japanese wheats were the Italians, who imported 'Akakomugi' seed in 1911, using it to start a wheat program; the varieties they bred were still useful for wheat breeding by CIMMYT in the 1960s.[62]

The Italian breeding program did not seem to attract much attention. In the United States, there was a widespread belief that tall wheat was needed to produce good yields. Wheat breeding had somewhat stalled, and from 1919 to 1944 production was dominated by a few varieties, notably 'Turkey' and 'Marquis.' The last burst of good work had been done in the Pacific Northwest around the turn of the century, with William J. Spillman (1863–1931) concentrating on breeding for rust resistance and stiff straw. Spillman was another one of the pantheon of people, who "almost reached the same conclusions as Mendel," but in his case was held back by the enormous complexity of wheat genetics—derived as it is from several species. Others in the region concentrated on drought-resistance, using Australian varieties, but smut became a major problem. Resistance was bred into many new varieties, but the fungus soon adapted. 'Elgin,' bred in Oregon in the 1930s was one of the first wheats noted as being particularly responsive to fertilizers, but it was notoriously smut prone. A combine harvesting a field could be seen to be followed by a black cloud as if on fire.[63]

Japan's advances in wheat only really began to make a global impact after World War II. Under the American occupation, agricultural advisor Samuel Salmon (1885–1975) noticed Japanese farmers growing good-yielding short, stiff-strawed wheat. Salmon had been born in a sod house in South Dakota and had grown up surrounded by wheat and worked with it all his life. He brought sixteen varieties back from Japan and made them available to researchers in western states—these were the 'Norin' varieties, Norin being an acronym composed of the first letter of each word in the romanized title of five Japanese experimental stations. Of these, one stood out: 'Norin 10,' a variety partly derived from Turkey wheats that had originally been imported from America (and had before that been Russian).

62. Perkins 1997; Gale and Youssefian 1985.

63. J. Shepherd 1980, 57.

The University of Washington's Pullman was one of several research stations which received the Norin wheats. Key was Orville A. Vogel (1907–1991), a Nebraska farm boy who ended up in Washington State as a researcher in 1930. He produced a stream of new varieties that were used throughout the breadbasket of the Northwest. Driven partly by a deep hatred of the tedium involved in much plant breeding, as well as being a lifelong lover of tinkering with machinery, Vogel designed and built more than a dozen specialized machines, which greatly sped up breeding work. In 1932, he developed a combination one-row or three-row seeder, replaced in 1959 with an improved semiautomatic eight-row seeder—enabling three men to seed five thousand eight-foot plots per hour. His best-known machine was the Vogel Nursery Plot Thresher, a self-propelled combine able to self-clean inside of ten seconds. It came to be used around the world by cereal, pulse, and oilseed breeders.

The university was the logical place for the Norin varieties to be made the most use of—as Vogel's team had been working for more than a decade on combining higher yields, good nitrogen response, and resistance to lodging. In 1949, Vogel and his team released 'Brevor' and 'Elmar,' two relatively short-strawed wheats that gave 5–12 percent higher yields than the previous dominant variety for the region, the smut-prone 'Elgin.' Vogel's crossing Norin 10 with Brevor brought the Japanese short-straw character into the wheat-breeding mainstream—this was the material that Norman Borlaug was to use with such great effect in Mexico in the 1950s. Vogel and the team at Washington spent the 1950s perfecting Norin 10 progeny for the Northwest; in 1961 he released the first semidwarf winter wheat for commercial production—'Gaines.'[64] Yields were up by as much as 20 percent, from plants that were 65 centimeters in height compared to Brevor's 100 centimeters; the sheer efficiency of the new wheat ensured that from now on, commercial wheat breeding began to work towards dwarfing. The new genes made rapid progress around the globe.

THE GRASS KEEPS ON GETTING GREENER— A SHORT JOURNEY ALONG THE BACK ROADS OF BREEDING

In the early 1950s, England's once-mighty soccer team was thrashed in several key matches. Foreign players, often those from South America, played the "beautiful game" with much more élan; the reason was that they had

64. Perkins 1997; Dalrymple 1988; Lupton 2005; Kral 2005.

learnt to play on dried earth pitches, where there was a better foothold than on turf. Turf is always slightly slippery, and with too much top growth it slows down the players and the spectators get a less exciting game as a result.

It was time for plant breeding to rescue England's national game. Better sports turf was simply the next logical step on from breeding better grasses for grazing. Turfgrass breeders have had to come up with varieties that produce plentiful top growth even at low cutting heights, hardwearing strains that produce shoots rather than leafy growth. Staff at the Welsh Plant Breeding Station (now the Institute for Grassland and Environmental Research) were already world leaders in the improvement of fodder grass, oats, and clovers, moving on to forage *brassicas* in the 1950s. Grasses had to be tested, so a machine was developed to simulate studded boots. By the 1960s, sports pitches were holding the players better and in 1966, England won the World Cup 4–2 against West Germany, wresting the title from the Brazilians, who had won the previous two championships. In the United States, grass breeders facing similar problems went hunting for grasses where "natural" selection had been at work—footpaths. One of the first was from New York's Central Park—it was marketed as 'Manhattan.'[65]

Plant breeding as an applied science had established itself in the industrialized world. After its initial success at improving the most important crops and adapting them for a more industrialized agricultural future, the skills and techniques used in plant breeding soon spread, like a successful gene, throughout the world of horticulture and agriculture. Nothing illustrates how far it penetrated than the production of recreational drugs. Estimates of the size of the global scale of the illicit drugs trade vary, but they tend to agree that it is very big.

During the 1990s, people who had been using cannabis (hybrids of *Cannabis sativa*) since they were hippies thirty years earlier began to complain that the stuff they were getting from their dealers was so strong it was practically knocking them out. Called "skunk," the new drug was super strong, sometimes with effects closer to hallucinogens like LSD than the mellow drug they were used to. It was not simply that they were getting older; the drug was certainly getting stronger—so much so that worries began to be voiced that it was having harmful effects on some users, such as the inducement of psychosis (the marijuana of old had been regarded as a pretty safe drug with few side effects).

65. G. Harvey 2001.

The culprits, or geniuses, depending on point of view (or perhaps state of intoxication), were mostly growers on the American West Coast, and later on in the Netherlands, a country with a reputation for tolerance of cannabis. They had discovered hybridization, crossing robust and hardy *Cannabis indica* clones with the warmer-climate and more potent *C. sativa.* During the 1980s, a range of hybrids appeared on the West Coast: 'Northern Lights,' 'Skunk no.1,' and 'Big Bud,' which were the basis for further work, much of it done by American exiles working in the Netherlands. The plants were noted for their efficient harvest index—dwarfs were highly valued for the ease of growing them under lights and for their incredible productivity with levels of the main active ingredient THC (as high as 20 percent, compared to less than 5 percent for "ordinary" marijuana). Cannabis breeders seem like any other plant breeders, making hundreds of experimental crossings to get one new variety and breeding for a range of traits: hardiness, early cropping, high yielding (no pun intended), and a range of qualities, including mellow, cerebral, physical, and trippy.[66]

Unlike other applied sciences, plant breeding has always maintained an amateur fringe. Most people have a garden, many are passionate about plants, and so the raw materials for breeding work are close to hand. Anyone can do plant breeding; indeed, anyone who saves their own seed from garden vegetables or flowers is doing just what our Neolithic ancestors did, practicing unconscious selection. Amateur plant breeders continue to play an important role in ornamental horticulture. They are not entirely absent from edible crop breeding either.

Books on do-it-yourself plant breeding have been around since the 1920s, with their heyday being the 1950s, the era when the Western world was most in love with science and technology. More seemed to appear in the United States than anywhere else, possibly a reflection of the confident "can do" attitude of Americans, the Baconian idea that the nature is there to be changed for our benefit, combined with a dose of populism—anyone can do it! Such is the message of John Beaty, author of *Plant Breeding for Everyone* (1954), who is described on the title page as being "a former associate of Luther Burbank." The tone throughout is hearty, with plenty of stories of amateur gardeners finding new forms of delphiniums, gladioli, and other flowers and selling them for large sums to nurserymen. He reminds readers of the four-figure sums earned by those who found many of the most successful American apple varieties. Morality is not missing either: "you are benefit-

66. Pollan 2001.

ing thousands—probably millions of people, some still unborn . . . a new variety may continue to serve mankind for years," he proclaims. Beaty offers a self-help guide to encourage amateurs to recognize new varieties in the wild, sports from garden trees, or exceptional, superior, or novel plants from batches of seedlings. He lists superior and inferior characteristics, teaches the basics of hybridization, emphasizes record keeping, instructs on home trialing, and then gives guidance on how to select a nursery or seedsman to approach to sell the variety.[67]

We have almost certainly not seen the last of the amateur breeder, even in the world of major commercial crops. Plant breeding is a bit like astronomy, a field where amateurs can still make important discoveries. There is as much genetic variation among plants as there are stars in the sky, and the more you look, the more you see. And it will never cease to be an art as well as a science.

SUMMARY

The success of F_1 hybrids in corn stimulated a search for the ways to apply the technology to many other crops—with varying degrees of success. As the twentieth century advanced, a great many other techniques evolved, founded on a growing knowledge of genetics. Among the most important of these have attempted to get species divided by various barriers of incompatibility to cross. Attempts at increasing the source of variation with which breeders can work has also been a major focus, with "induced mutation" through the use of radiation and chemical mutagens of considerable interest.

Plant breeding has had to work to produce plants for a steadily expanding range of purposes, often connected with the increasingly industrialized scale of agriculture and the need for bulk handling, transportation, and storage of fruit, vegetables, and other plant products. Breeding for chemical composition has also become increasingly important, either for industrial purposes or for increased nutritional value, or to reduce the levels of toxins in plants. Changes in plant architecture have meant major productivity gains in some species, which allied to increased use of fertilizers, and improved crop protection, have resulted in major increases in yield through the century. It has been calculated that plant breeding has accounted for

67. Beaty 1954, 4.

roughly half the increases—but breeding itself has been transformed by the use of statistics, improved research methodologies, and computers.

Finally, the industrialization of agriculture, while undoubtedly of great benefit, has had negative social consequences for many working in the industry, as humans have been displaced by machines and small farms are unable to invest in new technologies. A period of consolidation of small farms into larger ones has been one of the outcomes, as has vertical integration of the food and farming industry, with large multinational corporations beginning to dominate food production from top to bottom.

TWELVE ❦ GREEN REVOLUTION ❦ CAN PLANT BREEDING FEED THE WORLD?

The Punjab, divided between India and Pakistan, lies flat and lush; for much of the year its landscape of luxuriant green croplands makes a striking contrast to the dry and eroded hills of the neighboring regions. This is the breadbasket for the populations of northern India and Pakistan, who have historically eaten wheat flour–based flatbreads like *chappati*, *puri*, and *naan*. For the Western visitor, however, it looks very different than what we expect of major agricultural regions—despite the flatness and the big sky, it feels intimate. It is a landscape of innumerable villages and small farms, each one divided between a multiplicity of fields; a variety of crops is always visible, and there are a surprising number of trees. For those used to the vast fields and mathematically scattered farms of the American Midwest or the expanses of cropland with almost no human habitation in sight for mile after mile, as with the plains of northern France or eastern Germany, it feels almost crowded. Harvest time sees roads jammed with tractors, trucks and bullock carts. It should come as no surprise that harvest is celebrated—with an exuberant dance called *bhangra*, the music for which has inspired much of the liveliest of the modern Indian pop music scene.

That the wheat fields of the Punjab have remained a reliable breadbasket is one result of what has become known as the Green Revolution. The term was first used by William Gaud, the director of the United States Agency for International Development, in March 1968. It is a very telling one, for what America feared at the time was *Red* revolution. Gaud's term alerts us to the fact that the origins of the Green Revolution are inextricably linked to cold war politics.

The Green Revolution was the first time in history that an agricultural policy centered on plant breeding became a major part of a political strategy. While the expenditures involved were almost trivial compared to those spent on armaments, espionage, or the outright buying of political support, the effects have perhaps been far greater. Compared to the main historical events of the period—the Suez crisis, the Cuban Missile Crisis, the Vietnam War—the Green Revolution received remarkably little publicity. Its very quietness and lack of obvious drama has meant that historians are all too likely to overlook it. Yet consider what could have happened if there had been no Green Revolution, if food production had not kept up with the rapidly burgeoning population of the world during the latter half of the twentieth century—the historians may well have had far more to write about—famine on an unimaginable scale, with catastrophic global political consequences.

Some basic facts about the Green Revolution speak for themselves:

- Wheat production in Pakistan increased by 60 percent from 1967 to 1969.[1]
- By 1969 the Philippines started to export rice for the first time in a century.[2]
- India, during the 1967–1968 season, achieved a wheat harvest of 16.6 million metric tons, one-third higher than the previous best year, despite a drought. The following season exceeded this. In 1969–1970, the harvest was again up, to 20 million metric tons. [3]
- India achieved self sufficiency in cereals around 1974, a situation that had been widely regarded as inconceivable fifteen years earlier.[4]
- Indonesia's rice yields nearly doubled from 1965 to 1989, with production growing annually during this period at around 4.5 percent.[5]
- By 1982–1983, over half the wheat and rice area in developing countries was planted with HYVs.[6]

In the words of one commentator, "history records no increase in food production that was remotely comparable in scale, speed, spread and duration."[7] Despite these obvious successes, the Green Revolution has been accompanied by a constant throb of criticism. First and foremost, we are entitled to ask: even if malnutrition has been reduced and serious famines are now a thing of the past, why is there still so much malnutrition in the developing world? Eight hundred million people were defined as suffering from chronic hunger in 2006. More specifically, a whole range of problems has been attributed to the Green Revolution, covering agricultural, economic,

1. Sen 1974.
2. Ibid.
3. Frankel 1971.
4. Hesser 2006.
5. Sudaryanto and Kasyro 1994.
6. Lipton 1989.
7. Lipton 1989, 1.

social, and environmental issues. Those who raise awareness of these problems have done so from a wide variety of standpoints. On the one hand are those in the business of development who are used to being self-critical and only too aware that solving one big problem often results in the creation of a multiplicity of other lesser problems. On the other hand are those who have taken against the Green Revolution for reasons that can only be described as ideological: from a standpoint rooted in opposition to capitalism, to science, to the whole idea of supposedly Western ideas of development. In many cases, these critics have gone on to become the core of the opposition to biotechnology.

The aim here is to look primarily at the Green Revolution from the standpoint of plant breeding and at how institutional and policy initiatives promoted plant breeding as the central plank of development projects centered on agriculture. However, given that the Green Revolution has become a contested area of history, we shall also look at its results—which means taking into consideration a wider range of the economic and social impacts of plant breeding than we have tended to elsewhere. To many outside the world of agriculture, the reputation of plant breeding as a discipline stands or falls on the Green Revolution, so it is important that its many and varied impacts are clearly understood.

Understanding and assessing the successes and failures of the Green Revolution requires an attitude of objectivity—but it first and foremost requires an understanding of what the Green Revolution actually was. The popular conception is that it was about new high-yield varieties of grains—usually dubbed the HYVs. It was much more than this. One study, based on the eight countries responsible for 85 percent of the Asian rice crop concluded that the dramatic rise in yields was divided more or less equally between increased fertilizer use, irrigation and HYVs.[8] The Green Revolution was not just about magic new varieties, but about the whole package of agricultural modernity: HYVs, irrigation, fertilizer, pesticides, extension services, *and* the improved availability of credit so that farmers could fund improvements. In some cases, dramatic improvements were made *without* new varieties. Rice production in India doubled between 1950 and 1970; this was largely through improved irrigation, as HYV rice varieties did not make much of an appearance in the country until after 1970.[9] Indeed, the Green Revolution has been described as being "fundamentally a fertility revolution

8. Discussed in Conway 1997.

9. Hutchinson 1974a.

. . . with the part of the plant breeder in it . . . to supply new varieties capable of exploiting the higher fertility levels that can now be established."[10] HYV varieties are not intrinsically high yielding, but instead, they are capable of responding to environments that make available high levels of nutrition and moisture with higher grain yields. Older or non-HYV varieties can be fed and watered well, but may not respond with higher yields—they may not respond at all or only grow tall and lush and then topple over.

So, the Green Revolution was not just about plant breeding; it was about an all-round modernization of agriculture. Indeed, it makes sense to see it as part of the whole process of modernization and development. The Green Revolution was also about farmers having access to sources of finance to invest in new seeds, fertilizers and water pumps, and very often new markets in which to sell their products. It was not just about technical innovations but the whole process of the change from local, community-based, and often self-sufficient economies to cash-based markets linked to the global economy. Such huge changes almost inevitably involve major social ruptures, so it is not surprising that opinions about the Green Revolution reflect heavily the opinions about globalization and the politics of development.

The Green Revolution needs to be seen in context as one part of a series of momentous changes affecting economies and societies—nearly always with ramifications for politics and for the wider environment. These changes have on balance led to greater opportunities for people to improve their standard of living, their education, and their health. But there have been losers as well as winners—and it is the responsibility of anyone dealing with these issues to keep this foremost in their mind.

THE BALEFUL EYE—THOMAS MALTHUS AND FOOD POLICY

Any discussion of food and population sooner or later evokes the name of Thomas Malthus (1766–1834), an English economist and pioneer demographer. Malthus made a series of predictions concerning population growth that have haunted us ever since. Chief among these was his belief that while food supplies only increase arithmetically but population growth increases exponentially—the inevitable result would be that that the latter would outpace the former. Malthus was concerned that the crunch would come in the mid-nineteenth century—it clearly did not! Extremely influential on economics, history, and evolutionary science, Malthus's work has been

10. Ibid., 104.

fuel for both sober analysis and rash sensationalism—among the latter was Paul Ehrlich's 1968 book *The Population Bomb*, which predicted hundreds of millions of deaths by starvation by the mid-1970s. Ehrlich also claimed that India would never be able to feed itself.

The Green Revolution began with postwar U.S. foreign policy, although the concerns with population growth had begun earlier, growing out of the eugenics movement (see chapter 7). With the end of the Second World War in 1945, the United States was left as the only participant left standing—facing a world of great uncertainty. Foreign policy concerns rapidly became dominated by the fear of Communism: former ally Stalin was imposing his will on much of Eastern Europe, and in China Mao Zedong's Red Army was clearly on its way to victory. Hostile Communist-led governments and guerrilla movements led by Marxists began to appear elsewhere. It must have seemed a terrifying world; no sooner had two expansionist totalitarian regimes (Nazi Germany and Imperial Japan) been defeated than liberty and free enterprise seemed threatened yet again, and on more fronts. Those concerned with national security in the west were faced with a world where the dislocations of war and increasing populations in many poor nations were placing great pressure on food stocks.

During the 1950s, national security (i.e., facing down and suppressing the supposed Communist threat) dominated any discussion of foreign or development policy. Everything became subsumed to it. It was assumed that Communist politicians and guerrilla leaders arose from situations of poverty, and that poverty is most obviously a shortage of food. Ensure enough food, and the threat of Marxist-Leninist soapbox orators influencing the poor would be greatly weakened—a simple enough equation. So, increasing the supply of food in poor countries became a national security issue, not just a humanitarian one. It was in this political climate that what could be termed the "escalation" of the Green Revolution during the 1950s and 1960s began.

The development of F_1 hybrids in corn and the exploitation of heterosis led to an understanding of just what plant breeding was capable of. Not surprisingly, the standing of plant breeding as a discipline began to attract more notice among policy makers. Here might be the technology that would enable poor countries to feed themselves, resist the blandishments of subversion, *and* build their economies—and so become trading partners for the United States. It would be wrong, however, to see the origins of U.S. involvement with the start of the Green Revolution entirely in cynical foreign policy terms. The humanitarian impulse was also very strong and was particularly linked to food policy. With the opening up of the West, America had become the world's leading agricultural nation; the successes of American farmers and scientists in feeding not only their own nation, but their trading partners, and increasingly, countries in crisis, contributed to an almost evangelical sense of mission—America could feed the world! And had a duty to! Cynics might point to the "white man's burden" of the European imperial powers, the attitude that the powerful had a duty to arrogate the right to organize the lives of the less powerful, but there is also something very moving about the conviction that a nation that can produce so much, has a duty to help others achieve the same.

If one person can be said to have provided the germ of the idea behind the collaborative engagement in crop improvement that became the Green Revolution, it was Henry A. Wallace. Shortly after being elected vice president, he went to Mexico for the presidential inauguration of Manuel Ávila Camacho (in office 1940–1946). Wallace drove himself around the country and kept stopping at farms along the way to talk to people and to look at corn (he was a good Spanish speaker). The Mexicans loved his interest in their country, and by the time he arrived in Mexico City, he had, like a latter-day Pied Piper, acquired a caravan of followers and was attracting tens of thousands of onlookers. Wallace was appalled at the state of Mexican agriculture; while it took an Iowa farmer ten hours work to produce ten bushels of corn, it took a Mexican two hundred. Back home, he made immediate contact with the Rockefeller Foundation and laid out a case for U.S. aid for development.[11]

The Rockefeller Foundation became the key institution in the new American engagement with global food issues. Established in 1913, the foundation's aim was to improve the lot of humanity through the application of science and technology. Established with a fortune made through the oil industry, it has always been close to the U.S. political establishment. A quotation from the director of the foundation's natural science program, Warren Weaver, encapsulates the rationale behind the agricultural programs supported by the U.S. government: "The problem of food has become one of the world's most acute and pressing problems; and directly or indirectly it is the cause of much of the world's present tension and unrest. . . . Agitators from Communist countries are making the most of the situation. The time

11. Culver and Hyde 2000.

is now ripe, in places possibly over-ripe, for sharing some of our technical knowledge with these people."[12]

Weaver's stress on "sharing" indicates the essentially collaborative approach taken by the foundation. The political flavor of the new U.S. involvement was set by President Truman in his inaugural speech in January 1949, in which he set out a four-point program. Point four stressed the importance of sharing technology with "peace-loving peoples" and made clear the distinction between America and the old European powers that he referred to as "the old imperialism—exploitation for foreign profit." Instead he promoted the idea of development through the "vigorous application of modern scientific and technical knowledge."[13] The prewar isolationism of the United States was over, and "development" was born. Plant breeding was to be at the heart of it.

The Rockefeller Foundation knew its limits, however. There was to be no challenge to existing economic, social, or political realities. David Hopper, an economist with the foundation, explained in a 1968 paper, "Strategy for the Conquest of Hunger," that it was necessary to "separate the goals of growth from the goals of social development and political participation. . . . [T]hese goals are not necessarily incompatible, but their joint pursuit in unitary action programs is incompatible with development of an effective strategy for abundance. To conquer hunger is a large task. To ensure social equity and opportunity is another large task. Each aim must be held separately and pursued by separate action. Where there are complementarities they should be exploited. . . . [I]f the pursuit of production is made subordinate to these aims (social and political), the dismal record of the past will not be altered."[14] This was at the very core of the U.S. philosophy—the idea that improved food supplies would result in improved economic growth, which would automatically lead to greater participation in the political process of all citizens of the country concerned. Economic inequality, social injustice, and political oppression and exclusion would not be tackled but would be eroded by the economic advances powered by a more productive agriculture. This was at best naïve, but realistically it would have been very difficult for the foundation to tackle head-on the inequalities present in the countries it worked in.

As the 1950s wore on, it became apparent that U.S. foreign policy was not

12. Quoted in Perkins 1997, 138.

13. Quoted in ibid., 145.

14. Quoted in Pearse 1980, 79.

prepared to make any connection between social justice and fighting Communism. The results of foreign policy's obsession with Communism were a disastrous series of alliances with a ragbag of dictatorships, often vicious. In such regimes, the Green Revolution never stood a chance of reaching whole populations.

In a nutshell, the programs that Truman's presidency initiated were an attempt to replicate America's success with hybrid corn throughout the developing world. A fundamental belief was that yield was all-important: the more that could be got out of one piece of land, the better. Yield would be improved through the introduction of American agricultural practices, in particular, new varieties produced through the application of scientific plant breeding. Key to this was technology transfer—the training of plant breeders and agricultural scientists in developing countries. The assumption was that the American way was best and that traditional farmers would only benefit from adopting American agricultural techniques (in other words, native skills and knowledge were worthless). This is a good place at which to point out that some developing countries made major progress with crop breeding quite independently of the American-backed development program. Sri Lanka, for example, achieved a 53 percent increase in rice yield between 1952–1953 and 1967–1958 through varieties produced entirely by its own research centers.[15]

IOWA FARM BOY MAKES GOOD IN MEXICO—NORMAN BORLAUG

These days, many development workers earn high wages and drive around the country in air-conditioned four-wheel drive vehicles. It was not so in the 1950s. Norman Borlaug, working on a wheat breeding program in Mexico, had to endure illness, bad food, and primitive living conditions. Conditions at the northern test plot in a remote area of Sonora province were particularly difficult; Borlaug had to sleep in a hayloft, cook over a camp fire, and borrow machinery from neighboring farmers.

Borlaug's name dominates any discussion of the Green Revolution. His achievements undoubtedly have been enormous, not just in the research work he has undertaken over a long career but also in his tireless promotion of the importance of plant breeding and in the training of the young. Every sphere of human life needs heroes and secular saints—and for plant breed-

15. Pearse 1980.

ers and agronomists it is Norman Borlaug who is the leading light in their pantheon. But like every great achiever, he is not without his critics.

Born in 1914 on a small farm in Iowa, Borlaug grew up surrounded by and fascinated by nature. During his first year in high school, the stock market crashed, and though he did not experience real poverty, he saw it around him and was greatly affected. His college career did not mark him out as somebody who in later years could have his biography titled *The Man Who Fed the World*—he failed the entrance exam for the University of Minnesota. He had a reputation for having a streak of impulsiveness and stubbornness—qualities that later in life may have a played a part in his achievements. Eventually he did graduate and in 1933 he started working with E. C. Stakman, one of the nation's most respected plant pathologists.

Elvin Charles Stakman (1885–1979) made waves with his thesis of 1913, which marked out a new direction in showing the ecological and genetic relationships between plants and pathogens. In 1941 he was named by the Rockefeller Foundation to head a team to look at developing new HYVs of wheat in Mexico, which inevitably would involve dealing with disease issues, in particular, wheat rust. It was Stakman who in 1943 recommended Borlaug for involvement with the Mexican Agricultural Program (MAP), although he had never worked on the key Mexican crops of corn, beans, or wheat, and had no knowledge of Spanish or indeed had ever traveled outside the United States. Personality appeared to have been the key issue: "he will not be defeated by difficulty and he burns with missionary zeal," observed Stakman.[16]

MAP had been established in 1943 by the Rockefeller Foundation and the Mexican government. This was only the second time that the foundation had been involved with agriculture, the first being in China in 1935. The background to the collaboration had been the left-wing administration of Lázaro Cárdenas, president from 1934 to 1940, who had instituted a major program of land reform and irrigation, and had nationalized foreign-owned oil companies, including those owned by the Rockefeller family. The Cárdenas regime had supported scientific plant breeding based on indigenous Mexican varieties as part of a strategy for improving the livelihoods of the entire rural population. Land reform, however, had not led to an improvement in productivity—a painful fact for egalitarians, reinforced by the experience of several other countries.[17] On leaving office in 1940, both Cárde-

16. Quoted in Hesser 2006, 34.
17. Discussed in detail in Sen 1974.

nas and America seemed to have agreed on a less radical successor, General Manuel Ávila Camacho, who was a great believer in modernization through industrialization and foreign investment—inevitably American. Camacho's policies are now familiar to us: encourage farmers to leave the countryside and work in new industries in the cities, have the remaining farmers extend the size of their holdings, invest in machinery rather than labor, and provide cheap food for the expanding urban workforce. The Mexican elite welcomed the MAP, hoping no doubt that agricultural "improvement" would divert peasant activists from trying to occupy large farms.

After initially working on fungal diseases, in 1945, Borlaug took over leadership of the wheat improvement program. The choice of wheat as the primary research subject was a clear indication of the political and social direction with which the Rockefeller Foundation had decided to work. Corn, along with beans, was the staple diet of the majority of the Mexican population, as it had been for thousands of years. The rising middle classes however, as rising middle classes do, wanted wheat—which had to be imported with the country's limited foreign currency reserves. The idea that Mexico should grow its own wheat was an attractive option, as was the logical corollary that the country could grow enough not only to be self-sufficient but to export, too. Indeed, Stakman and his team were reported to have been surprised by the strong desire of the Mexicans to focus on wheat. Wheat at the time was grown in two places in Mexico: on the large farms of the well-irrigated north and the central highlands, where farms were small and impoverished. Borlaug concluded that the farmers of the north would be the most receptive to new methods, but first he had to tackle wheat stem rust—a major disease problem.

From the start, Borlaug threw himself into his work, ignoring a variety of problems and privations. He was a great one for leading from the front, insisting that Mexican scientific colleagues get their hands dirty; in doing so, he came up against a culture where status depended on being seen to wear a clean shirt and getting others to do the physical work. Borlaug's response was to tell them that "the farmers have no respect for you. If you don't know how to do something properly yourself, how can you possibly advise them."[18]

Borlaug's work in Mexico involved three approaches. One was high-volume crossbreeding—the tedious and fiddly task of crossing thousands of wheat varieties to produce new combinations in the hope that an occasional

18. Quoted in Hesser 2006, 43.

one would turn out to be rust resistant. Another was to use short-strawed wheats that he obtained from Orville Vogel at Washington State University (see chapter 11) in an attempt to improve the plants' productivity—the harvest index. The third was his greatest and most controversial innovation—"shuttle breeding."

Shuttle breeding took advantage of the differences between the two main wheat-growing regions of Mexico, the northern Sonora province had a spring harvest, and the highlands had an autumn one. Until now, the accepted view had been that it was essential to breed a variety in the region in which it was to be grown so that it was as well adapted for that region as possible—an orthodoxy laid down in widely used textbooks by none other than H. K. Hayes at the University of Minnesota, Borlaug's alma mater. Flying in the face of his elders' advice, Borlaug proposed rushing the seeds of Sonora's harvest in April to be sown in the newly prepared fields of Toluca, in the south, which would then be harvested in October, their harvest then being sent north for a November sowing, and so on. Unable to see much good coming from this apparently harebrained project and much duplication of effort, his superiors, Stakman and George J. Harrar, told him to discontinue his work at Sonora. Borlaug's response was to hand in his resignation. But at the last minute the decision to close down the Sonora plot was rescinded, possibly because one of the neighboring farmers had written to Stakman and Harrar in support of Borlaug's work and implicitly criticized them for not giving him more assistance.

The result of Borlaug's apparent pig-headedness was that he did indeed start to produce two generations a year, effectively halving the usual ten-year process of getting a commercial release. By 1950, eight new varieties had been released, all of which had some advantage over existing Mexican wheats, particularly in rust resistance. The shuttle breeding approach also had consequences that challenged the rationale behind the orthodoxy of breeding for particular regions; the new wheats were remarkably tolerant of a variety of soil and climate conditions and to a wide range of pests and diseases; above all they were relatively insensitive to daylight. Normally a wheat variety will only flower and seed at a time dictated by day length; Borlaug's wheats were effectively being selected for daylight insensitivity because of being grown in regions two thousand kilometers apart and because one lot was harvested when days were getting longer, the other when they were getting shorter. This insensitivity had the effect of making them supremely adaptable to different regions around the world. Shuttle breeding soon became respectable, and has been taken over by CIMMYT (see below), to enable researchers to breed new wheats for a wide variety of purposes. The photoperiod insensitivity developed by the early wheats has continued to be a feature of CIMMYT wheats and is a major part of their success.[19] Shuttle breeding has also since proved suitable for India, with nurseries in the lower Himalayas shuttling seed down to the Nilgiri Hills of the south.[20]

The 1950s saw the release of more wheats, high yielding and disease resistant enough to make a major impact on Mexican agriculture. Borlaug knew, however, that more could be done and began to look for short-straw mutants, a search that resulted in his being put in touch with Vogel in 1953 (see chapter 11). Despite problems with 'Norin 10's' lack of resistance to disease in the fields of Mexico, Borlaug and his team had by 1962 two new semidwarf wheats: 'Penjamo 620' and 'Pitic 62.' Their release was premature—the decision to go commercial having been forced on Borlaug by a problem often faced by workers at research stations, the "leakage" of seed from the fields into those of neighboring farmers who had been helping themselves to free samples.

Central to Borlaug's success has been pragmatism, a "suck it up and see" approach to developing breeding techniques and an ability to know when to stop looking. "Perfection," he is reported to have said to a Mexican audience, "is a butterfly the academics chase and never catch. If we go on looking for the ideal wheat for Mexico, your countrymen will go on being hungry for a long time. We will have to do the best we can with what we have."[21] By 1960, the Rockefeller Foundation had concluded that its work in Mexico was completed. In order to keep the momentum up, the Mexican government worked with the foundation and the Ford Foundation to develop CIMMYT. Borlaug, although he continued to work with CIMMYT, became primarily a roving ambassador for HYVs. In this role, he has been immensely successful, a crusader and an effective spokesman for the work of wheat breeding with everyone from farmers to heads of state; much of this has been down to his ability to see the big picture—the relationships between science in the field and policy issues at national and international levels.

Wheat production in Mexico certainly took off, largely in the north in the provinces of Sinoloa and Sonora, where a new breed of commercial farmer developed vast fields of wheat on irrigated land. By 1958, Mexico had

19. Rajaram and Van Ginkel 2001. Borlaug 1983 is a paper by the man himself on shuttle-breeding.

20. Rao et al. 2001.

21. Quoted in Hesser, 2006, 46.

become a net exporter of wheat. The farmers of Sonora did not rely entirely on the new wheats and improved agricultural technology—they extended their holdings by illegally evicting smaller farmers, drawing more water from underground reserves than they were entitled to, and pressuring the government to guarantee them a high price for wheat. Increased food supplies came at a high environmental and social cost.

The irony of the increased wheat crop in Mexico was that this was the land that had developed corn, the world's most advanced and productive crop; subsistence farmers, who still depended on corn, had been effectively bypassed. They either had to struggle on, relying on the notably less successful corn breeding program, or leave their farms and go to work in the cities or to seek work in the United States, a process which has led to a great deal of immiseration and social dislocation.

In 1963, the ironic situation existed that Mexican wheat yields were among the highest in South America but corn yields, although greatly improved, had not kept pace. The assumption had been made that the American way was the best and so a strategy of producing F_1 hybrids was pursued, but poor farmers could not afford to buy more seed every year. Besides, their water supplies and soil fertility were often inadequate for hybrids. Many in the Mexican Ministry of Agriculture had recommended that open-pollinated varieties be the focus; although they were lower yielding, they would reach more farmers, especially the poor. The Rockefeller Foundation, however, ignored their advice.[22]

In 1943, leading U.S. corn expert Paul Mangelsdorf had been enlisted to help run a Mexican corn program. Starting to work with strains from America, Mangelsdorf and his colleagues soon realized that they would simply not perform in Mexican conditions, so instead started on a program to collect a range of Mexican varieties, inventory them, compare them in trial plots, and then finally inbreed among the better varieties. The aim was to produce some new synthetic varieties and modified hybrids that could be released as a short-term measure while working on what was regarded as more important—American-style double-cross hybrids. There was clearly an additional aim—to collect landraces that might be useful for U.S. agriculture.[23] In parts of Mexico, it is possible to grow two crops a year, which made the work of analyzing results in just a few weeks a demanding schedule, although with good organization, the time taken to produce new hy-

22. Dahlberg 1979; Hewitt de Alcantara 1976.

23. See technical note 32: Synthetics.

brids could be halved—indeed, by the end of 1947 ten double-cross hybrids were ready.

A central problem with hybrids was that only the wealthier and more commercially orientated farmers could afford to buy new seed every year. It soon became apparent that F_1 hybrids were not making the impact the Americans thought they should. So, in an effort to try to get more new varieties to small farmers, the U.S.-led team took a step back and put more effort into the synthetic varieties, which rapidly began to yield good results; by 1948, the country had had a record harvest, and for the first time since the revolution of 1910, Mexico did not need to import its staple crop. By the 1960s, over one-third of the corn acreage was down to HYVs, and total production was three times what it was before the program started, although still poor in many areas. Other South American countries were able to learn from Mexico's experience, and by the early 1980s, half the corn acreage on the continent was planted with HYVs. Yields were soon up by one-third over the early 1960s. Whatever mistakes the Rockefeller Foundation may have made initially, the stress on training Mexican scientists, in particular plant breeders, had paid off—as now the Mexicans were able to train others from elsewhere in Latin America. Mangelsdorf's contribution should not be forgotten either—the collection, conservation, and classification of Mexico's immense number of corn varieties was another valuable result of the foundation's intervention, one which potentially serves all interests, from the U.S.-owned multinational right down to the humblest tenant farmer.

Mexico's new wheat and corn varieties soon made it around the world, and although their impact in the Old World was never as dramatic as it had been in the New, they enabled many more mouths to be fed. China, otherwise cut off from the rest of the world by the regime of Mao Zedong, made particularly effective use of both its own high-yielding wheats and Mexican varieties. The photoperiod-insensitive Mexican wheats enabled expansion of the crop southward into areas where it not been grown before—largely as a second (winter) crop after rice. Fields which had often been left fallow could now produce.[24]

Borlaug's role in the Green Revolution has been recognized many times, but the crowning achievement was being awarded the Nobel Peace Prize in 1970—the first scientist to be awarded this prize. In doing so, the Nobel Committee recognized that:

24. Wortman 1975.

he has given the economists, the social planners, and the politicians a few decades in which to solve their problems, to introduce the family planning, the economic equalization, the social security, and the political liberty we must have in order to ensure everybody—not least the impoverished, undernourished and malnourished masses—their daily bread and thus a peaceful future.[25]

A FANTASY NO MORE—INDIA FEEDS ITSELF

The gardens around the government buildings in New Delhi, India's capital, are magnificent, planted in a style introduced by the country's former rulers, the British. Beds of colorful flowers are arranged in precise and regimented patterns and are tended with great care by an army of gardeners. But in 1967, the flowers around the official residence of Prime Minister Indira Gandhi (in office 1966–1977 and 1980–1984), were replaced—by Mexican wheat. It was the new Prime Minister's way of celebrating an early success—a bumper harvest.

As in Mexico, land reform, which had been instituted by the Nehru government in the 1950s, did little to improve productivity. For many in India, this was a failure of hope and of progressive politics. By the mid-1960s, the food situation in India was becoming increasingly troubled: there were serious failures of the monsoon, leading to a real concern that famine would return; there was tension with Pakistan; and an there was upsurge in Communist-instigated political violence. The United States was ready with food aid in the form of grain shipments, but these would only be forthcoming in exchange for a devaluation of the rupee. It is not surprising that C. Subramaniam, the Minister of Agriculture, in announcing the Fourth Five Year Plan for agriculture in August 1965, decided to go all-out for increasing yields through science and in state support for entrepreneurs.

The key figure in India's Green Revolution was Monkombu Sambasivan Swaminathan, born in 1925 in the southeastern state of Tamil Nadu. His father had been a doctor, active in a variety of health campaigns, in the freedom struggle against British rule, and in campaigns for the reform of the caste system. The younger Swaminathan was one of a rising tide of young Indians who had the opportunity to study abroad, at Wageningen University in the Netherlands and at the Plant Breeding Institute at Cambridge. On

25. Hesser 2006, 132.

returning to India, he worked at the Indian Agricultural Research Institute (IARI) helping to turn it into a major research station. In 1961 he began to work with the All India Coordinated Wheat Improvement Project, which was trying to coordinate various attempts at producing semidwarf wheats; realizing that the origin of some of the most hopeful experimental wheats at IARI were from Mexico, he invited Norman Borlaug to visit.

Borlaug arrived in February 1963—the first of many visits—and stayed for several months, touring wheat-producing areas. He came to the conclusion that Indian wheat could only be improved in the context of major changes in Indian agriculture itself, specifically, a better supply of fertilizer and more extensive irrigation. Later that year, one hundred kilograms each of the most successful Mexican wheats were shipped, along with samples of more than six hundred additional unreleased lines.

Even at an early stage, results were spectacular, with Mexican wheats yielding five or six times as much as local varieties. Indian scientists, however, were cautious, worried in particular about yellow rust. There were political problems too—rice was the main Indian grain, and questions were raised about the relevance of investing in a crop that was the diet of only a minority of the country—was the politically dominant *chapatti*-eating and Hindi-speaking north benefiting at the expense of the rice-eating Dravidian south? The sudden death of Nehru in May 1964 might also have threatened progress. His replacement as prime minister, Lal Bahadur Shastri (in office 1964–1966), saw food self-sufficiency as being an essential part of national security and national self-respect—not surprising when President Johnson had the country on a humiliating month-to-month arrangement for emergency food aid. Subramaniam, newly appointed to the agriculture portfolio, was determined to promote a science-orientated agenda against opposition from economic nationalists who opposed the level of foreign financial assistance required to fund the reforms, as well as Gandhian idealists who believed in small-scale agriculture and village-based economics.

Conditions in the country were sufficiently severe for a big risk to be taken; Mexican seed was to be imported and distributed among farmers before waiting for farther breeding. In July 1966, eighteen thousand tons of 'Lerma Rojo 64' wheat was shipped from Mexico to India, following a visit to Mexico by an Indian delegation, which had to arrange the purchase of the seed from individual farmers and cooperatives and oversee its cleaning and warehousing.

Borlaug's advice to the Indian breeders was that they should be "aggressive in deciding to multiply and distribute promising lines. Don't look for the perfect variety—or for a variety which will last for 15 years commercially, for you will never find it. If your breeding program is dynamic and aggressive, the lines entering increase will already be obsolete."[26]

Borlaug and Swaminathan's working together focused on the short-straw, high-yielding wheat varieties that Borlaug had developed in Mexico, making selections which would be suitable to grow in Indian conditions. In 1965, Swaminathan set up thousands of demonstration and test plots in northern India to show small-scale producers that they could make the new wheat flourish in their own fields. An approach that stressed farmer education played a vital role in the success of the new wheats, particularly since they would only perform with greatly increased levels of fertilizer.

Despite more monsoon failures, the farming year of 1966–1967 marked the turning point, with around 504,000 hectares planted with the new wheat. India's imports began to drop. Initially, one problem was simply finding places to store the extra grain—in 1968, schools had to be closed down to provide temporary warehousing. Over the first four seasons, the total wheat yield had increased from twelve million tons to twenty-three million. By 1974, India was effectively self-sufficient in all cereals, just six years after Paul Ehrlich had claimed that such an occurrence was a "fantasy."

As elsewhere, the Green Revolution was about integrating new varieties with the use of fertilizer and irrigation. However, around 70 percent of India's agricultural lands are rain-fed with no additional irrigation—so effectively the initial phase of the revolution passed them by. Later on in the 1960s, Indian scientists were able to make progress with a variety of dryland crops, notably kharif (i.e., autumn-sown) sorghum, the output of which doubled during the 1970s in Maharashtra state. As with wheat, harvest index was the key, the crop being transformed from the traditional tall, long-maturation varieties to relatively short ones with a quicker maturation time. In addition, they were bred to be more drought resistant. The irony was that human consumption of the crop started decreasing as an increasingly wealthy population began to turn to wheat-based products. New uses have been developed that reflect the fact that in one way or another the whole plant can be utilized: in cattle feed, or for the production of industrial starch, syrup, malt and malt-based foods; this in itself required its own research program.[27]

26. Quoted in Perkins 1997, 242.
27. Rao 1991.

FOOD FOR HALF OUR RACE—BOOSTING RICE YIELDS

Rice is perhaps the world's most important crop, with around half the human population dependent on it. Millennia ago it established its position through its amazing productivity, its suitability for double or even triple cropping in many tropical climates, and its nutritional qualities—cooked rice has a protein content of around 4–5 percent, whereas many tropical carbohydrate sources are very much lower in protein.

Inspired by Vogel's work with semidwarf wheat 'Norin 10' in 1949, the FAO launched the International Rice Commission, which established a project in Cuttack, in Orissa State, India, to work on what seemed like the most hopeful direction, the hybridization of the two main rice groups: *japonica* and *indica*. Results, however, were slow in coming and were not widely available until the 1960s. American researchers reached for the then fashionable weapon of first resort—the irradiation chamber. Sterile rice was the main result. The Rockefeller Foundation was more realistic—it went plant hunting with the result that their researchers found a semidwarf variety of Taiwanese origin, 'Taichung Native 1.'

Some thought that only an international effort could make much progress with rice, including J. George Harrar (1906–1982), who had been the director of MAP; he first proposed an international rice research center in the early 1950s. Learning from his experience in running MAP, he believed it must be truly international so that scientists could work together regardless of ethnic and linguistic backgrounds. But others saw only disadvantages, notably that every rice-growing country (and indeed region) had its own particular problems that were arguably best addressed by local research. The idea of the international center was consequently stalled, but by the late 1950s, the lack of progress being made by national bodies put the idea back onto the agenda. So in 1958, the Ford and the Rockefeller foundations agreed to cooperate on the building of just such a center. Pressure for results in rice breeding had been mounting throughout the decade as the physical limits of intensive rice cultivation had been reached in most rice-growing countries—the neat terraces of rice paddies climbing prodigiously steep hills make nice tourist photographs but were also a graphic reminder of just how severe the problem of space was becoming. Growth had to be intensified; yields had to be raised.

The IRRI opened at an eighty-hectare farm near Los Baños in the Philippines in 1962. The country was chosen partly for climatic reasons, partly to build on existing strengths in rice research, and possibly for reasons of

U.S. foreign policy—America and the Philippines were close allies. From the start, IRRI was a very international organization, yet its atmosphere and way of working were American, being described as a combination of "the research division of a transnational corporation, a military base, and a diplomatic enclave." IRRI expanded to become one of the world's leading agricultural research centers, but for some in the Philippines there was a feeling that it took too much away from the Philippines itself, reducing breeding efforts which were the most relevant for its own conditions.[28] Initially, plant breeding was the overwhelming priority for IRRI, with two targets: varieties with as wide an application as possible that would also yield best under optimal conditions.

Improving harvest index was key, with new varieties at 100 centimeters tall (rather than the 160–180 centimeters of most traditional varieties), with short, upright leaves that allowed sunlight to penetrate further, and that were only mildly photosensitive, enabling them to be planted at anytime of year. Little was known about rice genetics at the time, so one of IRRI's first appointments was a geneticist, Dr. Te Tzu Chang from Taiwan. Starting in 1962, using 'Peta,' a popular, traditional, tall variety from Indonesia and a Chinese ancestor of 'Taiching Native 1,' 'Dee-Geo-Woo-Gen' (DGWG), the IRRI team soon found out that the F_2 generation was tall to dwarf in a 3:1 ratio—a perfect Mendelian ratio. Dwarfism was clearly controlled by a single recessive gene. With further breeding, in fact, by the F_5 generation, the new variety was judged ready for testing.[29]

Named 'IR-8,' it had a harvest index of 1.0, as opposed to the 0.6 or 0.7 of traditional varieties, and a yield of up to 10 metric tons per hectare, compared to then developing country average of 2 and the Philippine average of 1. It also cropped in 120 days instead of the 150–180 days for most older varieties.[30] Unlike traditional rices, which when fed with nitrogen-rich fertilizer tended to fall over, IR-8 stood firm, but the fertilizer was necessary for results: with no extra feeding, in IRRI trials it only gave 5 tons per hectare—little more than 'Peta'—but at 120 kilograms of fertilizer per hectare, it gave 10 tons.[31]

IRRI was under great pressure for results. 'IR-8' had been distributed for

28. Anderson et al. 1991.
29. Hargrove and Coffman 2005.
30. Griffin 1974.
31. Hargrove and Coffman 2005.

testing in 1965, but its "escape" into local marketplaces forced IRRI's hand into releasing it in 1966. Many rice scientists criticized its early release; it has been suggested that there was political pressure from the United States because of the need for some good news from Asia—this was when America was deeply mired in the Vietnam War.[32] Philippine president Ferdinand Marcos also brought pressure to bear; he is reported to have flown into Los Baños by helicopter and ordered the new variety to be released as soon as possible, as he had an election promise to fulfill—to make the country self-sufficient in rice during his first term of office.

Like many rice varieties that break new genetic ground, 'IR-8' had cooking and taste problems: it broke easily in milling, it had an unattractive chalky appearance, and it tended to harden after cooking. "It scratches my throat," one young woman was reported to have told a researcher.[33] In general, early IRRI rice varieties were dogged by pest and disease problems: during 1968–1969, 'IR-8' was attacked by bacterial blight in Southeast Asia and then by tungro virus. In 1977, 'IR-36' was developed to be resistant to these two pathogens but then fell victim to new viruses: ragged stunt and wilted stunt. In addition, the new rices seemed more vulnerable to a number of insect pests. By 1971, floods and pests created so many problems in the Philippines that 250,000 tons of rice had to be imported, comparable to pre–Green Revolution imports.[34] Not surprisingly, once the issue of harvest index had been cracked, IRRI moved on to breeding for pest and disease resistance.

Breeding for pathogen resistance began not a moment too soon. During the 1980s, a British postgraduate student, Michael Loevinsohn, who was studying rice pests in the Philippines, became alarmed at the amounts of pesticide being used and the slapdash methods of application. He went on to look at local mortality records and found that during the years in which pesticide use had doubled, death rates from diagnosed pesticide poisoning had gone up by 250 percent. His work stimulated more research with the result that a virtual epidemic of pesticide poisoning was revealed.[35] Other research then showed that pesticide poisoning is a huge problem across the developing world, in contrast to industrialized countries, where it is com-

32. Anderson et al. 1991.
33. Hargrove and Coffman 2005, 38.
34. Palmer 1976a.
35. Conway 1997.

paratively rare.[36] Simply on public health grounds, the shift in research over the last few decades toward finding plant breeding solutions to pests and diseases rather than chemical ones has to be welcomed as a major advance. Some countries have also tried launching public health campaigns to reduce pesticide use; an IRRI-backed project in Vietnam in 2002 even won a national "Golden Rice Award" for its "No early insecticide spray" project.

Despite its problems, 'IR-8' made a huge impact, simply because of its ability to produce. One Indian farmer in Tamil Nadu was so impressed with the performance of the variety that he named his newborn son IR-8. In 2005, it was listed as one of the "top 50 inventions that have 'rocked the world' during the past half-century" by the U.S. magazine *Popular Mechanics*.[37] In India the success of the new varieties—and the media interest they created—did, however, lead to a false impression that Indian agriculture had been unchanging or stagnant in the past. One commentator wrote that "Indian agriculture has not been the hopelessly inert toad that is often presumed. Rather it has been more like a collection of sleepy frogs, each one of which makes an occasional spirited hop between long bouts of torpor."[38] Not only did Indians at all social levels innovate, but the British imperial authorities had contributed too; in particular, rice benefited from the breeding of new varieties, with some suggesting that as large a gain had been made in the immediate pre-Independence period as that achieved by the Green Revolution.

By 1980, however, only 25 percent of the world's rice fields were planted with the new varieties, the problem being that, as IRRI recognized, the new varieties were only suitable for irrigated areas, and given that they were short, areas that could be irrigated to a precise depth of water—too deep and the rice would drown. IRRI also acknowledged the pest and disease problems.[39] IRRI's second decade saw an approach which was much finer-grained in its approach to rice breeding, focusing more on identifying farmer's needs and problems and working more closely with national research organizations. Increasingly it saw itself as a catalyst and facilitator rather than a one-stop independent operator. A genetic evaluation and use (GEU) program was put

36. For information on pesticide problems, see work by the Pesticide Action Network (http://www.pan-international.org), which, while ideologically committed to the elimination of all chemical crop protection, is widely respected for its thorough research and reporting strengths. It does, however, take an anti-GM line, which many may find perverse.

37. IRRI 2007.

38. Baker 1984, 46.

39. IRRI 1980.

into place designed to enable breeders working for different conditions to draw on whatever knowledge and expertise was appropriate for them. Rice varieties for deep-water cultivation, rain-fed agriculture, high soil salinity, and drought-prone situations were among the new specialist research targets. Much of IRRI's aim now was not so much to increase yield, per se, but to help poorer farmers in more marginal areas. Breeding varieties with high protein content (up to 10 percent) was also part of the focus on the poor.[40]

NEW SEEDS, NEW HOPE, NEW PROBLEMS—THE GREEN REVOLUTION'S FIRST DECADE

The Green Revolution grew out of an attempt to radically improve people's lives through plant breeding. As we have seen, it only makes sense to see the new plant breeding in the context of a major change in how people in developing countries farmed. Like a Russian doll, these issues are enclosed within others; notably these changes in farming were part of a much bigger movement, now fashionably called "globalization." Inevitably, such major changes were not going to work out in everyone's best interests. There were going to be winners and losers, with some of the losers precisely the people who the Green Revolution set out to help.

In looking at the post–Green Revolution world, the obvious question is: why are so many people still so poor? And why is there so much malnutrition, even outright starvation? But given what we know about increased yields, we might also ask: what would the world have been like had there been no Green Revolution?

Something as momentous as the Green Revolution needs to be assessed; we need to draw up some kind of balance sheet. This is not just for academic historical interest or for political point scoring, but so that we can approach the issue of how to continue to fight hunger from a position of knowledge. The effects of the Green Revolution—agricultural, economic, social and political—have been many and varied. Trying to establish a balance sheet is actually a very complex business, as there were different effects in different countries, and in some cases, economic or social trends changed over time, even reversing direction. Looking at what might be termed negative effects—issues of increased poverty, social conflict, and environmental pollution—is also a way of taking into account various criticisms of the Green Revolution.

40. Ibid.

In dealing with criticisms of the Green Revolution, it is probably fair to surmise that the vast majority of plant scientists and development workers could probably find something they would have done differently if they had their time again. This is despite the tone of messianic fervor that sometimes comes across from those involved, a tone typified by the World Food Prize (WFP) organization, established by Norman Borlaug in 1987, to honor those who have achieved the most to improve the "quantity, quality, and availability" of food in the world. The WFP organization has acted as the cheerleader for the Green Revolution, putting its faith in science and policy making as the way forward.[41] While it appears to be highly effective at networking among those involved in maintaining the momentum of the Green Revolution and in keeping food issues high on government agendas, some question its closeness to corporate America. Its headquarters in Des Moines, Iowa, the heart of U.S. agribusiness, does little to reassure the Green Revolution's radical critics.

In contrast to the constructive critics, the radical critics of the Green Revolution are ideologically opposed to capitalism and the global market economy. Some are opposed to science and development in general. The radical critique of the Green Revolution is important to understand, as from it grew an important part of the opposition to biotechnology. Although perhaps few in number, they are very influential, with many of the media and articulate and highly politicized people who identify themselves as environmentalists and believers in social justice among them.

Here, we shall look at some of the effects and problems of the Green Revolution in attempting to draw up a balance sheet, particularly as it affects the ongoing development of plant breeding and then take a closer look at the ideological critics.

Who Grew the New Seeds? Issues of "Take-Up"

As we have already seen with hybrid corn (see chapter 10), new seeds are not taken up equally. One of the concerns with Green Revolution seeds was that they were not tried by the small and poor farmers—did the revolution leave them behind?

The evidence from many countries was that large farmers innovate, middle-sized imitate and the poor don't. One study in India showed that the average holding of innovating farmers was 90 percent bigger than noninnovators. Perhaps not surprisingly, innovators tended to have higher assets, be own-

41. See http://www.worldfoodprize.org.

ers rather than tenants, be more educated, and have more contact with extension services.[42]

Access to credit or subsidized seeds, fertilizer, or irrigation water, appears to be the key in getting poorer farmers to try new varieties. One Indian author reports how uptake was very dependent on the existence of government backed credit/subsidy programs, and that where such programs existed, uptake was good across all sizes of farms, with smaller farmers even taking the lead in some areas. In line with many other studies on small farms, research across India showed how small farmers often use inputs more intensively, and small should not be equated with backward or less efficient.[43] Credit is generally only extended to small farmers or the poor, if governments are proactive. In India this did happen, with the result that many smaller farmers were able to benefit. In Mexico the poor were failed by their government in this respect.[44] It is worth pointing out the political contexts: India during the 1950s and 1960s was a great deal more democratic than Mexico.

One Size Fits All—Adaptability

"IR8 was to tropical rices what the Model T Ford was to automobiles . . . a rugged variety that could go anywhere," trumpeted IRRI in a review of its second decade of research.[45] It is certainly true that one major difference between Green Revolution crop varieties and older ones is this adaptability—Green Revolution varieties tend to be far less sensitive to different environments than previous ones. In particular, the HYVs' relative insensitivity to day length allowed for cropping at different times of year, so allowing farmers to use land more efficiently. Combined with the short cropping period of many HYVs, this allowed many farmers to get two crops from a piece of land, whereas before they could only get one. This alone was a revolution, enabling great increases in productivity and family income. In northern India, rapidly maturing HYVs have enabled whole new cropping patterns to emerge involving wheat, rice, potatoes, pulses, and green manures—the land can produce three or even four times a year.[46]

In many cases, studies have shown how the new crops are generally in-

42. Griffin 1974.
43. Sen 1974.
44. Hewitt de Alcantara 1976.
45. IRRI 1980, 3.
46. Conway 1997.

sensitive to their environments—the genotype x environment interaction is only small, and that many HYVs are superior in all environments, at both high and low management regimes. Even at zero fertilizer input, many HYVs outperform traditional varieties. HYVs also use water more efficiently so that they produce more grain per liter of water than conventionals—which in fact might be expected from the measure of harvest index as the plant shifts resources from stems to seeds. However, this does not mean that they are more drought tolerant, and indeed many HYVs are more vulnerable to drought.[47]

One Indian scientist and dryland agriculture expert is of the opinion that "wide adaptation, adaptability and high yield are certainly not negatively correlated . . . this is not to be misunderstood as pleading for one variety for all areas or adaptability for all conditions."[48] Indeed, it became apparent very early on—by the end of the 1960s— that the "one size fits all" concept had been pushed too far; this was accepted as such by IRRI. The second phase of the Green Revolution involved breeders adapting HYVs much more to local conditions.

Environment-insensitive varieties worked much better with some crops than with others. In the case of rice, there were problems. Bangladesh, where, until the Green Revolution there were more than ten thousand different varieties, was an example. Such a vast range of landraces had evolved for different soil, water depth, and salinity conditions. In an area where flooding is common, it made sense for local communities to make fine distinctions between landraces for a variety of water depths, from as little as 5 centimeters to as much as 8 meters! The latter "floating rices" are also very important in Thailand and Vietnam—they crop during the wet season and are harvested by boat. Irrigation has been described as the "leading input" in such environments; without an ability to accurately and predictably control water levels there is little hope of being able to grow a rice variety at the correct water depth, and the extra expense of new seed and fertilizer would be wasted. In these regions, farmers who had a choice of varieties to grow at a range of water depths saw little point in growing only one and making labor intensive or expensive changes to their fields and irrigation systems in order to do so. Such farming systems are a classic example of crops being adapted to suit the environment; for them, the Green Revolution offered

47. Lipton 1989.
48. Rao 1991.

the opposite, the need to change the environment to suit the crop. Not surprisingly, penetration of HYVs into Bangladesh and much of Southeast Asia was slow.[49]

You Only Get Out What You Put in—Inputs and Their Costs

One of the most frequently voiced criticisms of the Green Revolution approach to improving food security has been that it involves farmers having to invest considerable sums in fertilizers, irrigation and pesticides. It could be argued that the first two come with the territory—that you cannot have HYVs without more inputs. Several decades later, we know that organic production methods can, in the tropics, make a substantial impact in reducing, even eliminating fertilizer applications, but this approach requires a major and very extensive effort put into training and extension services.[50] Increased costs are inevitably a large part of the Green Revolution—which brings us back to the key issue of how much farmers, especially small farmers, are supported by financial institutions, government, and NGOs in order to make the necessary investment. In the case of irrigation, this often requires political decisions made at a high level and highlights how a successful program of agricultural improvement requires a high level of coordination among many professional groups—all too difficult to achieve in many developing countries with poorly coordinated and often corrupt institutions. The results are usually that large irrigation projects almost inevitably benefit large landowners.

The reputation of HYVs has been established on performance at research stations—in optimum conditions. Achieving such ideal growing conditions is inevitably almost impossible for anyone else. It has been suggested that on average, the difference between optimum and realistically achievable yields have been as high as 25 percent for rice and 50 percent for wheat.[51] However, the second decade of the Green Revolution saw much more widespread testing of new varieties and the expansion of the whole field of participatory breeding has brought breeders face to face much more with the reality of life down on the farm (see chapter 14).

As with hybrid corn in the American Midwest, HYVs are generally much better at extracting and using nutrients and so are more likely to exhaust

49. Boyce 1987.
50. Gypmanteseri 2003.
51. Dahlberg 1979.

the soil. Poor farmers, who are only used to or able to think short-term may then engage in "soil mining," resulting in soil exhaustion, or alternatively, higher yields may cause them to abandon traditional practices which preserved fertility. These are both issues about soil husbandry, and can only be addressed through education; the growing organic movement and the dissemination of its more evidence-based ideas through the wider agricultural and research community has begun to play a more important role at the turn of the century.

Security or Plenty—The Conundrum of Yield and Reliability

"I met Borlaug once, I thought he was a maniac, all he was interested in was yields, nothing else," reported a former FAO employee.[52] Of all the unintended effects of the Green Revolution it is the trade-off between higher yields and crop reliability (and therefore food security) which has caused perhaps the greatest amount of alienation of both farmers and activists.

Norman Simmonds recognized this, writing once that "successful plant breeding tends to narrow the genetic base of a crop."[53] Landraces, being genetically diverse, include a range of genotypes with different levels of adaptation to environmental stresses and pathogens, whereas modern varieties, being genetically more or less uniform, put all the eggs in one basket. The choice is between predictable low yields and less predictable high yields—for many of the world's poor, the clear choice is the first, as security depends on year-to-year reliability. Being unable to risk taking the opportunity to reap the rewards of a bumper harvest is another example of the "poverty trap" long recognized by economists.

Small farmers have gotten in debt often because they borrowed money at high rates of interest to plant HYVs that, when they did not deliver, ruined them financially, often with the effect of them becoming landless laborers or forcing them off the land altogether. It is this, perhaps this more than anything else, that has damaged the reputation of the Green Revolution. In India, this problem is inextricably linked with the notorious figure of the village moneylender—and is part and parcel of a system of caste designed to keep the poor in their place.

Some NGOs and activists have made considerable political capital out of this problem; one positive response has been the establishment of community seed banks (see chapter 2) that enable farmers who have failed with

52. Anon. [former FAO Schwartz employee] 2006.
53. Simmonds 1991, 323.

HYVs or who do not wish to take the risk to have access to traditional varieties and landraces. In many areas, these had almost disappeared as they were displaced by HYVs. In some cases, Green Revolution varieties have been rejected, or indeed replaced, after failing. One story from the Terai region of Uttar Pradesh, India, tells of a farmer called Inder Singh who continued to grow traditional rice varieties after his neighbors started growing HYVs. Irrigation lowered the water table, and Inder Singh found that his neighbors came back to him for the more drought-resistant varieties; one in particular was named after him, 'Indarasan,' which eventually displaced HYVs on many local farms.[54]

Since the 1960s, the problem of reliability has been addressed by the increasing finesse of plant breeding in and for the developing world, particularly by the greater emphasis placed on drought-, pest-, and disease resistant crops from the 1970s onward.

Another way around the narrowing of the genetic base is to use "multilines," where several cultivars are deliberately mixed—basically a modern take on the landrace. This was tried in Kenya in the 1960s, where the important wheat crop was facing particularly bad problems with rust. A number of cultivars were maintained as pure lines but were mixed for sale to farmers, with up to ten cultivars per multiline crop. Since levels of rust resistance vary, when one cultivar was observed to suffer, it was removed and another one substituted. Canadian "advisors" who arrived in the early 1970s completely failed to understand this innovative, low-tech technique and terminated the program. Other attempts at developing this approach have also stumbled on the failure of those involved in variety evaluation, testing, and registration to be able to adapt their procedures to cope with multilines.[55] Now, with plant patents becoming increasingly important, there are problems with registering multilines, as the seed bought by farmers is not composed of a single patentable variety.

A Revolution Only for Some—Regional Inequalities

Farmers in areas marginal for agriculture were generally left out by the first phase of the Green Revolution; the particular demands created by their environments were too specialized to warrant spending limited research funds on. The result was that areas already marginal and poor became even more so. It might be argued that the whole of Africa fell into this category.

54. Singh and Prakash 1985.
55. Leakey 2007.

Africa is agriculturally very different than Asia: irrigated lands are very limited, and there is an immense range of climate and soil zones that necessitate very highly targeted breeding efforts. Soils are often very poor, and there are particular pest and disease problems. Political instability has also played a part—for example, the Rockefeller Foundation had a strong presence in Uganda until the increasing chaos under Idi Amin's dictatorship in the early 1970s forced all foreigners to leave and most institutions to close down.[56] It is only within the last few years that a concerted new effort to address an African Green Revolution has been made, relying on a variety of approaches, including biotechnology.[57]

One of the factors in Asia that has deepened regional disparity has been the split between irrigated areas and those areas where agriculture is dependent on rainfall, often upland regions. The latter have, in any case, a more varied range of environmental conditions, so making the breeding of appropriate HYVs a slow task. In many cases, dry or upland regions are inhabited and farmed by minority ethnic communities, who are already disadvantaged and marginalized.

Research over the last decade of the twentieth century has begun to target regional inequalities much more deliberately. A key issue is often education, simply having enough trained agronomists, plant breeders, and geneticists to go around. The more varied a terrain, the more research effort is needed to produce new crops for a patchwork of different conditions. Participatory breeding has made an impact; the 1960s approach of the scientist parachuting in with new varieties bred with no reference to local farmer knowledge is now firmly in the past. A great many breeding programs now emphasize local involvement.

A Balanced Diet? Protein Questions

It has sometimes been claimed that one deleterious effect of the Green Revolution has been a diminution in the quality and quantity of protein available. In many developing countries with largely vegetarian diets, an adequate quantity of protein is obtained by balancing different vegetable

56. Ibid.

57. The biggest effort is that being made by the Bill and Melinda Gates Foundation, an institution self-consciously following in the footsteps of the Rockefeller. See http://www.gatesfoundation.org/GlobalDevelopment/Agriculture/RelatedInfo/AfricanFarmers.htm.

proteins, largely through a diet including both grains and pulses, such as beans, lentils, and in India, *channa*, a pulse closely related to the familiar chickpea (both are forms of *Cicer arietinum*). It has been argued that HYV grain has displaced these other crops.

In some areas, pulse and other "protein crops" have declined, whereas in others they actually increased during the 1960s and 1970s.[58] In any case, several studies report that overall protein quantity has increased because of the greater production of wheat and rice and because many of the new wheats have a higher protein content (16–17 percent as opposed to the 9–12 percent of older varieties).[59]

It is important to remember that overall nutrition was improved by the new crops simply because they hugely increased availability and reduced the price of basic foodstuffs; farming populations who were able to increase their income through growing HYVs were also able to afford to eat better. In addition, a number of studies of specific communities clearly show markedly increased nutrition.[60]

Who Benefits? Questions of Social Justice and Equity

It is an iron fact of history that not everyone benefits from change. In the case of the Green Revolution arguments, opinions and studies have ranged backward and forward on this issue. A major historical process affecting many very different countries and geographical regions would be highly unlikely to cause the same economic and social effects everywhere, and there is a clear danger of extrapolating from the study of one area to the entire process.

Lower food prices were one of the main impacts with positive benefit to the poor all over Asia—whether they were urban or rural.[61] Indeed, urban dwellers probably benefited unequivocally, as they were removed from more direct contact with the changes the Green Revolution set in motion.

Rural populations involved in farming have been affected in a variety of different, and sometimes contradictory, ways. One of the most important has been that of scale. It was the belief of many during the 1960s that the Green Revolution was "scale neutral"—its costs and benefits affected farmers in direct proportion to their size of plot. Seed and fertilizer are indeed

58. Sneep and Hendriksen 1979; Palmer 1976a; Evans 1997.
59. Dahlberg 1979; Evans 1997.
60. See, e.g., Pinstrup-Andersen and Jaramillo 1991.
61. David and Otsuka 1994.

infinitely divisible, but other inputs may not be, notably costs of irrigation (canal building, pumps, etc.) and machinery. Since the Green Revolution was part of a wider process of agricultural modernization, frequently involving mechanization, it is difficult to really test this proposition; in general, though, it appears as if larger farmers did benefit more, but many small farmers did well, too, at least if they took the risk of trying the new seeds when they first became available.

Some studies appear to show that, in India at least, there is no relationship between size of plot and yield, others show that the Green Revolution has widened the gap between rich and poor farmers, partly because of economies of scale and partly because large farms generally started using the new seeds earlier[62] Higher productivity of the new varieties also appears to have reduced one of the advantages that small farms used to have—their greater yields.[63] Government subsidy and taxation policies generally benefit larger farmers, too.[64]

Much of the debate over the benefits of the Green Revolution concern labor—whether a greater demand for labor has been created, so benefiting the landless poor, or whether it has been decreased, so worsening poverty. This is a complex issue because of the parallel increase in mechanization in many developing countries and a growing move to the cities. Initial worries over increased inequality in employment and income have often shown to be misplaced. In Thailand, for example, HYV adoption appears to have increased labor income.[65] A Bangladesh study showed that HYV adoption increases the number of tenancy agreements as opposed to sharecropping. This is beneficial for smaller farmers, and there were also beneficial effects for labor, both the availability of work and higher incomes.[66] Where regions benefited very differently from the introduction of HYVs, disparities appear to have been ironed out to some extent by migration, with workers moving on a seasonal basis to areas where agriculture demanded more labor.[67] The introduction of irrigation often appears to be the key to absorbing landless labor, as higher intensity farming tends to employ more.[68] However, there

62. Sneep and Hendriksen 1979.
63. Nulty 1972; Bhalla and Chadha 1983.
64. Khan 1975.
65. Isvilanonda and Wattanutchariya 1994; Palmer 1976.
66. Hossain et al. 1994.
67. David and Otsuka 1994, Jatileksono 1994.
68. Sen 1974.

are also examples of landlords appropriating the gains made by the new crops by increasing rents for tenants or not passing on gains to laborers in increased wages.

Looking back on the Green Revolution, it does appear as if early studies of its effects appeared to show increasing inequality, evictions of tenants, and unnecessary mechanization. Later studies generally show a more positive picture, as more small farmers took up the new seeds and a wider range of crop varieties became available. There was also a positive "knock-on" effect in rural economies, which took time to emerge—the greater productivity stimulated a range of nonagricultural activities in the countryside, as now-wealthier farmers spent money locally.[69] Unfortunately, though, some initial negative assessments of the Green Revolution have created a myth of surprising durability: that it helped only the rich and made things worse for the poor.

In considering social and economic impacts, it is vitally important that the effects of the technology of the Green Revolution and its social and political contexts are considered separately; too many critics have thrown the proverbial baby out with the bath water by condemning the Green Revolution on the basis of it being carried out by oppressive or exploitative regimes. The agricultural changes of the 1960s onward experienced in many countries were and are part of a wholesale change in societies and economies—the process of "creative destruction" so typical of the way in which capitalism transforms societies, to use the phrase of political economist Joseph Schumpeter.[70] For those interested in social justice, the question is: has the economic and social policy context within which the Green Revolution been implemented and benefited all members of society or simply an elite bent on gaining a Western consumer lifestyle as soon as possible? Asian countries varied in the egalitarianism of their Green Revolutions. The Mexican experience, though, is a powerful example of how agricultural change fails a large proportion of the population, basically because Green Revolution policies were not brought to the mass rural population. Instead, the countryside was used as a source of capital for industrial development and a place to grow cheap food for a new urban proletariat. Small farmers in Mexico received paltry help from their government, with little cheap credit, poor extension services, police suppression of rural unions,

69. Hazell and Ramasamy 1991.
70. Schumpeter 1976.

and many instances of loss of rights to land.[71] It should come as no surprise that opposition to modern crop varieties, and GM in particular, should be stronger in Mexico than in any other emerging country; people who have been so badly served by agricultural change in the past can hardly be expected to think very positively about the prospect of yet more change.

A Revolution in Attitudes? Social Impacts

As with judging the economic impact of the Green Revolution, the effect on society and politics presents a complex and at times contradictory picture. Establishing causal links between the new varieties and changes in class structure, migration, and regional conflict is particularly difficult—especially when so much else is changing.

In many cases, the advent of the new crops, the increased investment they involved, and the greater income they brought to rural communities has been one of the most dramatic changes in many years. This is particularly the case in very traditional societies where relationships between social groups are deeply established, very rigid, and in many cases very oppressive.

Demands for increased labor that puts more money into the hands of poor, landless communities has, it has been suggested, led in India, to low-caste or outcaste laborers gaining in social confidence in their dealings with landlords and employers and improving their status. On the other hand, this may have led to employers then trying to replace them with machines, leading to unemployment and increased migration with disruptive effects on families as a result. Successful farmers have tended to "pyramid" their gains, as they invest in more machinery, irrigation, and land. In many cases, this represents their leaving behind the complex web of feudal relationships they grew up with and joining new, more commercial networks based on local towns—"the transition from feudalism to capitalism." Feudalism involves high-status individuals being responsible for low-status ones, for feeding them in famine, for example. With the breakdown of old feudal structures, such responsibilities will no longer be felt, traditional cultures of deference will break down, and society will become more fluid, unpredictable, and unstable.[72] Traditional landowning elites have also been partly displaced by once–socially inferior small landowners who have improved their financial position through good farming. Newly wealthy farmers may well want to get their machinery mended locally but will not necessarily spend

71. Hewitt de Alcantara 1976.

72. Frankel 1971; Harriss 1977a, 1977b.

all their newfound gains in the neighborhood; one Indian commentator rather crossly remarked that "having reached a plateau in farm investment," rich Punjabi farmers "seem to be squandering their surpluses on conspicuous consumption, including purchase of jeeps, cars and television sets, excessive indulgence in alcoholism, and demonstrative expenditure on social ceremonies, etc."[73]

It has even been suggested that the Green Revolution has had a major impact on the religiously founded set of attitudes that have traditionally underpinned life in rural India. The new seeds and the technology that goes with them, the theory goes, has eroded traditional ideas about the futility of human action and the possibility of altering the course of one's destiny. This is a truly radical idea for popular Hinduism, where traditionally someone's place in society was defined very clearly. The new thinking may have encouraged a variety of rural political movements in particular Marxist ideologies, such as the Naxalite phenomenon, which involved small-scale guerilla warfare in some parts of rural India, notably Bengal, Bihar, Orissa, and Kerala.[74] Others disagree, claiming that most violent disputes over land have been in areas where agriculture has remained stagnant, and that violence has been least in the wheat belt where more labor has been employed and incomes raised.[75]

Others in India have suggested that the political Right rather than the Left has benefited from the social changes brought about by the Green Revolution. Since many low-caste farmers have benefited from change, the disturbance to the rural order has contributed to a feeling that the *Kali Yuga* (the time of Kali) has arrived—the time when the divinely sanctioned order is overturned. This has possibly stimulated the rise of a reactionary and xenophobic Hindu nationalism.

The role of women in the Green Revolution has not been systematically researched. Clearly many women's lives were improved by the greater availability of cheaper food, but the changes associated with growing and processing crops in some cases have resulted in marginalization. Women in traditional societies work almost entirely within a domestic sphere, so the commercialization of agriculture might be expected to have reduced their role or ability to earn income, especially in societies where social mores restrict their working outside the home. One example of undesirable

73. Bhalla and Chadha 1983, 161.
74. Frankel 1971.
75. Sen 1974.

change comes from Indonesia; where traditionally women have harvested rice by cutting it about 15 centimeters below the panicles of grain with a knife know as the *ani ani*, earning a proportion of the grain as payment. This method was ideal for landraces that matured at different rates. HYV rice varieties, however, all tend to produce at once and so need to be harvested simultaneously; the rice is shorter, too, making the *ani ani* method less practicable. During the 1960s and 1970s, men took over the harvesting, using sickles, thereby depriving women of a source of work, income, and pride.[76]

Green Slime, Sick Farmers, and Salty Soil—Negative Environmental Impacts

Problems with pesticide use in developing countries have already been noted. Further problems arose with the use of nitrogen-rich fertilizers causing pollution of water supplies with nitrates, and algal blooms wiping out fish and other animal life in rivers. Continued use of chemical fertilizers also damaged soil structure over many years. Irrigation may have hugely increased yields, but it often brought its own problems, notably salt deposit build-up in soils and unsustainable use of limited water resources.

The higher cropping densities and lush growth of HYVs made attack by fungal and bacterial pathogens more likely. Farmers were tempted to grow the same productive crop more frequently, breaking with traditional crop rotation systems that reduced the possibility of pathogens from infecting a crop grown on the same land. The pests and diseases that were always present with landraces, but that rarely became a problem, now multiplied more rapidly and on occasion did more damage than they would have done previously.

The environmental side effects of the Green Revolution were never as intensively researched as the economic ones or perhaps as widely acknowledged by those in mainstream agricultural development, leaving the field open for those with a more fundamentalist critique. Many of these problems might be more accurately described as the side effects of changes in agriculture rather than simply that of using new seeds. As with all societies with new technologies, many problems are the result of inexperience—large numbers of people with little appropriate knowledge of how to use synthetic fertilizers or pesticides, together with unregulated markets, corrupt sales practices, and inadequate extension services.

76. Conway 1997.

BEYOND YIELD—THE REVOLUTION MARCHES ON

Since making early breakthroughs with harvest index and linking new highly productive varieties with increased fertility, the Green Revolution approach has segued into the main body of plant breeding. So have the genes. As noted elsewhere (see chapter 2), one of the most interesting effects of the post–Green Revolution period has been how HYVs have crossed with landraces to produce new landraces with many of the desirable characteristics of the HYV. This has happened not only with vigorously out-crossing corn but also with rice.[77]

HYVs have also been bred into national varieties, as in Thailand where people were initially very cautious with HYVs, owing to the country's status as a major producer of high-quality and distinctly flavored rice. Dwarf strains were also less useful to them, as much of the rice growing area was liable to floods. Eventually, Thai government breeders undertook the crossing of IRRI-bred HYVs with local strains to produce varieties better suited to local conditions and that maintained the flavor for which Thai rice is world famous.

Once breeders had cracked the main issue of harvest index and improving yields, the second decade of the Green Revolution saw a variety of different goals in crop improvement; these often changed owing to the need to respond to different factors. Pests were the main barrier to the adoption of HYV rices in the Philippines, Indonesia, and Malaysia; in the latter two countries, breeders tended to keep to national improved varieties but introduced genes from IRRI varieties. Breeding upland rice varieties responsive to fertilizer was a priority until 1974, when fertilizer costs rocketed owing to the oil crises resulting from the Arab-Israeli conflict of the year before. In general, plant breeding has became more nuanced, more attuned to developing varieties for particular environments and more "pro-poor" so that HYVs were more robust and reliable in the face of pathogens and challenging environments. The early 1970s saw a particular focus on moisture stress tolerance in IRRI's work, along with a similar focus in sorghum and millet breeding, with 'CSH-1' sorghum producing spectacular yields in drought-prone parts of India in the late 1970s. The latter was one of several bred by a team headed by N. G. P. Rao in Hyderabad, the result of The All India Coordinated Sorghum Improvement Project, founded in

77. Boyce 1987.

1968.[78] Dependence on costly and sometimes damaging fertilizer usage was also tackled, with research by the International Centre for Tropical Agriculture (CIAT) explicitly aiming at low nutrient input agriculture.

Pest and disease resistance has continued to be of huge importance in plant breeding. Breeding for resistance is a constant struggle; most pests and diseases are capable of adapting, so the breeder and the pest are involved in a constant "arms race." Many resistant varieties can only be expected to be commercially viable for ten to fifteen years. The constant need to be one step ahead of pathogens makes the availability of landrace and wild species originated genes a vital resource for breeders.[79]

By the late 1970s and 1980s, HYVs were more and more likely to outperform traditional varieties even under unfavorable conditions and were outwitting some of the worst pathogens. A particularly interesting story concerns witchweed, the parasite that is the "strongest biological constraint to crop production,"[80] according to Ethiopian agronomist Gebisa Ejeta, who, while working with the Sudan office off the International Crops Research Institute for the Semi-Arid Tropics (ICRISAT), developed 'Hageen Dura-1,' the first commercial sorghum hybrid in sub-Saharan Africa. Drought-tolerant and yielding as much as 150 percent more than traditional varieties, it still fell victim to witchweed. After taking up a post at Purdue University, Ejeta joined a team to work on the parasite.

A Zambian colleague had noticed that sorghum produces a compound, sorgolactone, which stimulates germination of witchweed seeds. Another colleague developed a simple test for sorgolactone production, leading to the discovery that one varietal line had a recessive gene which greatly reduced its production. Breeding this into new varieties then proved relatively straightforward. Introduced in 1995, the new varieties are now grown across Africa's dry belt. The new sorghum even spread across hostile borders and throughout Sudan despite civil war.[81]

Research over the final decades of the twentieth century tackled a progressively wider range of crops. The Green Revolution started in the 1950s and 1960s with the basic sources of carbohydrate: wheat, rice, and corn. Crops that fed smaller numbers of people, who incidentally tended to be poorer and more marginalized, were dealt with next: sorghum and millets.

78. National Research Centre for Sorghum 2001.

79. See technical note 33: Horizontal and Vertical Resistance.

80. Mann 1997, 1040.

81. Mann 1997.

Other crops came later still, including the protein-rich crops upon which many of the world's poor depend for a protein-balanced diet, notably pulses. An example is the Pan Africa Bean Research Alliance (PABRA), established in 1996 with the support of a variety of donor country development agencies. Ten years later, PABRA had released 245 new varieties across 18 countries and had greatly increased the numbers of researchers working on beans. New varieties included traits such as resistance to the highly damaging bean root rot fungus, a tendency to climb—which enables higher planting densities, an ability to grow on poorer soils, and beans that need shorter cooking times, allowing for saving on firewood for cooking.[82]

Indian agronomists also talk about the "Yellow Revolution" of the 1980s and 1990s, yellow for the flowers of sunflower and mustard, oilseeds vital for both cooking and industry. Three hundred new varieties were bred by institutions working under the umbrella of the Indian Council of Agricultural Research, leading to huge increases in production; from 1985–1986 to 1994–1995, groundnuts went up by 27 percent, mustard and rapeseed by 111 percent, and soybeans by 450 percent. In many cases India went from a net importer to an exporter.[83]

NOT GREEN AND NOT A REVOLUTION? CRITICAL VOICES

Just as the world of politics has reformists and revolutionaries, so does the world of agricultural development—there are those involved in the business who believe that problems can be solved through the constant self-correcting mechanism of science and the market, and those who reject the entire system.

Frances Moore Lappé and Joseph Collins, of the Institute for Food and Development Policy, an NGO based in California, have been at the forefront of a radical critique of food and development policy since the 1970s. Lappé wrote *Diet for a Small Planet* in 1971—it was very influential at the time, introducing a radical critique of global food issues, as well introducing many readers to the idea of a vegetarian or reduced-meat diet in order to make the world's limited supply of agricultural land go further. It also included some very good recipes—I still make their poppy seed cake regularly, some thirty years later. In 1977, with Joseph Collins, Lappé produced another very influential book, *Food First: Beyond the Myth of Scarcity.*

82. PABRA 2006.
83. ICAR 2003.

The Green Revolution, they say, "was a choice *not* to start developing seeds better able to stand drought or pests, *not* to concentrate first on improving traditional methods of increasing yields, such as mixed cropping . . . *not* to concentrate on reinforcing the balanced, traditional diets of grain plus legumes."[84] There is a reluctance to believe that the Green Revolution actually fed more people, and there are predictions of famine in the future as disease sweeps genetically uniform crops or some other disaster strikes. No one since the 1970s can deny that the Green Revolution has produced more food, but radical critics argue that simply relying on producing more food will not solve world hunger. This will only happen when unjust social, economic, and political structures are changed. Food, they say, is "plentiful," as do many others in the global justice movement; indeed, "there is enough food in the world" has become something of a mantra; the problem, they believe, lies in unjust social systems that stop it being equitably distributed. HYVs tend to be referred to sarcastically as "miracle seeds," and the role of plant breeding is implicitly denied.[85] There is no denying that social justice would indeed help to feed more people, but what is distinctive about the arguments advanced here is that there seems to be an unnecessary "either/or" inserted into the analysis, as if social justice on its own would make scientific advance unnecessary, and that scientific plant breeding is not needed. Many would go further and argue that scientific plant breeding is part of a plot for world domination by U.S. capitalism.

Indian activist Vandana Shiva belongs to this camp. The Green Revolution, she argues, is part of a sociopolitical strategy aimed at "pacifying" the poor "not through redistributive justice but through economic growth," and at ensuring dependency on the west and on multinational corporations.[86] Despite having started her career in science (nuclear physics), Shiva is now very antiscience, speaking of "the exaggerated sense of modern science's power to control nature and society."[87] Asian agricultural systems had nothing to learn from outside, she argues, quoting Sir Albert Howard (1873–1947), "The agricultural practices of the orient have passed the supreme test, they are almost as permanent as those of the primeval forest, of the prairie, or of the ocean."[88] Any vegetation ecologist will of course now

84. Lappé and Collins 1977, 114.
85. Rosset 2000.
86. Shiva 1991, 52.
87. Ibid., 23.
88. Quoted in ibid., 25.

tell you that there is nothing primeval or unchanging about any of these environments.

Who was Sir Albert Howard? He was an Imperial British agriculture expert who worked in India from 1905 to 1924 and became a convert to traditional farming technology, believing that it could not be improved upon; he went home to become an early campaigner for organic agriculture. What was he doing in India? He and his first wife Gabrielle were breeding new wheat varieties.[89] Unlike most of the wives of British imperial civil servants whose work was almost entirely taken up with organizing large teams of servants, Gabrielle Howard spent a great deal of time doing the intricate work of crossbreeding wheats, working under a parasol "to the astonishment of the ladies of the Station, who prophesied either a complete breakdown in household arrangements or at least sunstroke from so many hours spent in the field."[90] The Howards had taken the decision to use native Indian wheats as much as possible; they sifted through landraces to isolate pure lines and crossbred to improve yields and rust resistance, coming up with some fifty varieties, all which included 'Pusa' in the name. The new wheats made a huge impact on Indian agriculture and were widely used for breeding elsewhere. The message for today is that a belief in the worth of traditional agricultural systems can go hand in hand with scientific plant breeding. Sadly, this message seems to have been lost on Vandana Shiva and almost the entire alternative agriculture movement.

In her 1991 book, *The Violence of the Green Revolution*, Shiva maintains that the Green Revolution set off a spiral of social conflict. Between 1970 and 1980, the number of land holdings in Punjab state fell by around 25 percent as a result of poorer farmers leaving the land, often, according to Shiva, because they were unable to afford the higher costs of inputs needed by the new crops. She goes on to claim that increasing indebtedness in Punjab during the early 1980s led to agitation by farmers over the costs of agricultural inputs, which contributed to the destabilization of the state by Sikh separatists, culminating in the attack by the Indian army on the Golden Temple in Amritsar in 1984 and the assassination of Prime Minister Indira Gandhi. She tries to blame the Green Revolution for more or less everything that went wrong in Punjab during this time: alcoholism, smoking, drug addiction, pornography, and violence against women. In particular, she sees the Green Revolution as a "cultural strategy" replacing "traditional peasant values of

89. Rao et al. 2001.
90. L. E. Howard 1953, 21.

co-operation with competition, of prudent living with conspicuous consumption, of soil and crop husbandry with the calculus of subsidies, profits and remunerative prices."[91]

Shiva's critique of the Green Revolution centers on two main issues: she argues the replacement of diversity of crops with uniformity, as well as the substitution of the internal resources of the farm with inputs that have to be bought in (fertilizers, pesticides, seeds, etc.). In the case of the latter issue, she is putting forward an argument that has been a constant in the alternative agriculture movement: that of self-sufficiency. Her argument is very much that traditional societies managed very well through their self-sufficiency and recycling of nutrients, and that entry into the market place inevitably brings with it social and ecological disintegration. The seed in particular becomes the focus of her discourse, and that of many others in the this movement, as the repository of deep symbolism; indeed, it becomes a quasispiritual entity. In particular it becomes a block to the introduction of market economics to the world of the farm, for the seed (at least in the case of grains and pulses) is not just the end result of one's labors—an item of food—but it is also the means to start the next year's crop. It is both present sustenance and future crop.

The fact that traditional farmers can save their seed and start again next season presents a clear block to the interests of commercial seed suppliers. "The seed," explains Shiva, "has therefore to be transformed materially if a market for seed has to be created . . . . [M]odern plant breeding is primarily an attempt to remove this biological obstacle to the market in seed."[92] The marketing of F_1 hybrid seed that has to be sown every year clearly gives private sector seed producers an open door, yet Shiva appears to object to *any* "corporate seed" that puts control over seed and breeding beyond the control of farmers themselves, even if it is open-pollinated.

By effectively forcing farmers to become part of the marketplace—by buying seeds, fertilizers, pesticides, even water, from outside the village—the Green Revolution, so its radical critics allege, breaks open the tight and self-contained world of the traditional village and integrates its inhabitants into a global world where everything is a commodity. The implication of the critics is that this is, by its very nature, a bad thing. Those who argue for science-led development are entitled to ask: how is agriculture meant to progress in order to feed millions more hungry mouths, in particular,

91. Shiva 1991, 185.

92. Ibid., 242. This argument is also developed by Kloppenburg (1988).

those increasing millions who are moving to the cities? Radical critics of the Green Revolution such as Shiva tend to ignore the fact that many, indeed most, traditional societies were far more integrated into extensive market-based systems than is often supposed and that many traditional societies were extremely rigid, offering a life of poverty and ceaseless hard work for the vast majority; it is for this reason that many of those millions are leaving the country for the city—they want the possibility of freedom from centuries of class- and ethnically based repression, the chance to be part of a labor market that offers minimal options rather than none at all, and the opportunity to better themselves and their families through education and entrepreneurial activity. Interestingly, critics like Shiva rarely discuss the deeply oppressive nature of many traditional societies; instead, the traditional village becomes an idealized golden age. Any mention of India's own "peculiar institution"—caste—is strangely absent from Shiva's discourse.

Market-led interventions into traditional societies clearly generate social dislocation, but in democratic societies, this disruption is most likely to be minimized, and in any case, the costs of not engaging with the global market economy are stagnation and continued poverty. India provides two good examples here—of a country whose economy stagnated until it engaged with the global economy, and where governments that do not listen to the rural poor get thrown out (as happened to Chandrababu Naidoo, chief minister of Andhra Pradesh from 1995 to 2004; he lost an election when he proposed to drastically reduce the numbers of farmers by state diktat).

MALTHUS DEFIED! CONCLUSIONS

"Sustainable agriculture" is now regarded as an essential goal. Back in the 1970s the idea of being able to dramatically increase food production while at the same time respecting a broadly environmentalist agenda was new and challenging. One of the people who developed the idea of sustainable agriculture was Professor Gordon Conway, a British agricultural ecologist. He was also a pioneer in developing integrated pest management (IPM), a system of crop protection which minimizes the use of chemical pesticides. Given this background, his thoughts on the Green Revolution should carry weight:

> Without the Green Revolution the numbers of poor and hungry today would be far greater. [In 1962,] according to FAO, there were about a billion people in the developing countries who did not get enough to

eat, equivalent to 50% of the population, compared to under 20% today [1997]. If the proportion had remained unchanged the hungry would be in excess of 2 billion . . . . The achievement of the Green Revolution was to deliver annual increases in food production which more than kept pace with population growth.[93]

Although small farmers have often not benefited as much from the Green Revolution as have larger ones, this is not universally true. Landless laborers may not have benefited greatly either, but neither have they been cast off in vast numbers. However, there has been little overall apparent effect in reducing poverty. The Green Revolution may succeeded admirably in its first goal, that of putting more food in more mouths, and in more than keeping pace with population growth, but it has not been notably successful in its second goal, that of kick-starting the economic growth that was meant to guarantee political stability. An example of this can be seen only a few hundred kilometers from IRRI headquarters in the Philippines, where fundamentalist Muslim separatist guerilla forces are still battling the government. The hope that the Green Revolution would fuel economic take-off was at best naive and at worst cynical, given how the elites who control government in developing countries tend to identify economic growth with industry and urbanization, rather than development in rural areas.

Population growth has undeniably hugely reduced the benefits accrued to rural areas by the Green Revolution and raised the amount of unskilled and landless labor. Fundamentally, though, Nobel prize–winning economist Amartya Sen argues that poor people do not need extra food, as such, but extra *entitlement*, that is, money. As the numbers increase who do not own or have access to land, they will become increasingly dependent upon earned income. Sen's argument has a bearing on the debate between those who argue that a shortage of food will be solved by making more food available (essentially the yield-focused Green Revolution) and those who argue that only redistributive justice will make any difference (who tend to be critical of the Green Revolution). Quite apart from whether the political changes which would ensure effective redistribution of wealth are likely to happen, the fact is that many countries are so poor that even if all the wealth in them were to be distributed fairly, they would still be terribly poor. Redistribution of wealth may be just and fair, and helps to iron out extremes of poverty, but there is no substitute for economic growth to raise living stan-

93. Conway 1997, 44.

dards—and the ability to buy food. Indeed, most workers in the field recognize that social justice and economic growth go hand in hand; it is no coincidence that the two countries that achieved major increases in agricultural productivity early in the twentieth century were Japan and Taiwan, both lands of small farms that are *relatively* egalitarian. And the opposite needs to be borne in mind—Mexico, where vast numbers have had to go north to seek work and where social divisions are massive because of a failure to link economic growth with social justice, in particular, a failure to bring the benefit of the Green Revolution to the rural poor. For agriculturally based economies, economic growth can perhaps be most easily achieved by investing in agriculture, so we are back to "Green Revolution politics" again and—of course—plant breeding! But plant breeding that benefits as many people as possible, and which in particular targets the poor, is a practical and moral imperative.

SUMMARY

The era of the Green Revolution represents the globalization of plant breeding during the 1950s and 1960s, as techniques developed in the industrialized world were spread throughout much of the poorer, so-called developing, world. The key characteristic of the new crop varieties created by the Green Revolution was high yield—vital in feeding rapidly expanding populations. The key plant trait, at least among the all-important grains, was a tendency toward dwarfism, so plants did not put energy into stems but into grain. Varieties themselves tended toward the global—initially the outcome of one of the key technical innovations of the era, shuttle breeding, whereby two generations of a crop can be bred in different locations in one year, which helped to ensure that varieties performed better over a wider range of environments than traditional ones.

The Green Revolution resulted in greatly increased grain production, which undoubtedly saved several countries from famine, and enabled some to become grain exporters rather than importers. Plant breeding may have been at the core, but fundamentally, the Green Revolution needs to be seen more holistically, as an all-round agricultural revolution involving introducing synthetic fertilizers, irrigation, pesticides, and modern techniques, along with new varieties; the role of credit and other financial mechanisms for poorer farmers should not be forgotten either.

Perhaps the best way of looking at the Green Revolution is to see it as the beginning of the modernization and globalization of many rural areas of South America and Asia. This was quite deliberate, as part of a geopolitical strategy led by the United States, to prevent the spread of Communist insurgency through economic improvement. The process of modernization of economies and societies is inevitably disruptive, however, especially in countries where powerful elites have prevented democratic accountability. As a result, the Green Revolution has tended to become a scapegoat for criticisms of the wider process of development, which has often obscured its benefits. As time has gone on, many lessons have been learned and applied, in particular, a new generation of plant breeders in the developing world have begun to improve a much wider range of crops and are focusing on issues other than yield, in particular, pest and disease resistance.

THIRTEEN ❦ **ORNAMENT** ❦ FURNISHING OUR GARDENS

It is a spring Saturday afternoon down at the garden center. The place is heaving with people, whole families who have come out to stock up with seed and plants for their gardens; think about big new expenditures on patios, ponds, and fountains; buy presents; or just have an afternoon out. Here is a good opportunity to assess the wide range of plants available to the citizens of an industrialized country at the beginning of the twenty-first century and to consider their origins.

Racks of seeds with colorful packets illustrating successful care of their contents tempt those with vegetable gardens or who like growing flowers from seed. Many of the varieties for sale are the result of decades of intensive breeding and genetic research, some of it specifically for the home gardener, but often a spinoff from work done for the agricultural or commercial nursery sectors. Bedding plants, still small and undistinguished looking, are lined up in a variety of plastic trays, hints of their bright futures given only by accompanying notices. A few are already in full flower, mostly hardy pansies and violas whose cheerful faces scorn cold weather and make a good stop-gap plant until the summer beauties perform; these particular plants have been selected and bred for centuries but have recently benefited from the growing sophistication of breeding programs. Farther out, beyond the shelter of the greenhouses, are beds of containerized shrubs and perennials, bamboos, climbers, roses, and trees. Many of the shrubs and perennials do not look too different from their wild ancestors; little tufts of the grass *Stipa tenuissima* could sit unrecognized next to their fellows on the edge of the New Mexico desert, while only a botanist would probably notice that a *Perovksia atriplicifolia* 'Blue Spire,' magically transported to be alongside its ancestors in the mountains of Pakistan, was not the wild species. Out amidst the fruit trees, old and young mingle—not just people, but trees. Apple 'Bramley's Seedling' commemorates the Vicar of Bramley in Nottinghamshire, England, who first picked it out as the best of a batch of seedlings in 1809, while a modern variety may be the result of a whole textbook of breeding technologies.

"An endless search for novelty" is perhaps the best way to sum up the aim of those who are involved in breeding new ornamental plants for gardens and public spaces. Throughout history, whenever people have had the leisure time and income to stop worrying about spending all their working hours producing edible crops, they have turned to growing a few plants for decoration, often picking out natural oddities: plants with variegated leaves, double flowers, or flowers a different color from normal. Once gardening became a popular hobby and the nursery trade an industry, this search for novelty became a business.

The ornamental plant breeding business presents a very different picture to the agricultural one—although not at the "top end" of the business, where multinational corporations invest in laboratory techniques of great sophistication to produce varieties for the cut flower, seed, and bedding plant industries. The difference lies in the enormous and steadily increasing range of plant species and varieties sold by smaller businesses and amateur growers. If plant breeding can be thought of as a historical progression from unconscious selection through conscious selection, then onto hybridizing and laboratory-based biotechnology, then every stage of history is represented in ornamental horticulture—all are alive and thriving and keeping people in business.

The history of ornamental plant breeding is a peculiarly difficult one to research. Unlike plant introduction, where the writing of biographies of "the great plant hunters" has become practically a subgenre in its own right in the world of gardening literature, that of plant breeders is almost minimal. The plant hunters were dashing romantic figures whose lives of scrambling over river and mountain, escaping avalanches and bandits, in the search for new plants for horticulture, lend themselves to endless retelling. But what happened to plant introductions when they got back home is another story, and one rarely heard. Breeders, usually in the nursery trade or amateurs, are often shy and retiring figures, rarely interested in publicity. With their eyes firmly fixed on the future, they often failed to keep breeding records or destroyed them when they were no longer of use. Something in the personality of a good breeder makes them strangely unfitted for the art of writing up the story of their lives—and they have attracted few biographers.

What was once said about one plant could have been said about many others: "the silence of all reference books on the subject of the breeding of the hybrid calceolarias would suggest that it is regarded as a shameful secret to which no reference should be made."[1] The history of ornamental plant breeding is also a repetitive one, as the same story can be told again and again,

1. Gorer 1970, 178.

with just locations and plant and personal names being changed. What is most interesting are general trends, in particular, the cultural and social historical aspects of the story. It is the intention to cover this whole subject in a future book, but for now, we must be content with an outline.

THE SHOW'S THE THING—
THE FLORISTS AND THEIR FLOWERS

Flowers have been a focus for plant selection and breeding efforts in those cultures that have developed a "flower culture" and a nursery trade, primarily European, Islamic, Hindu, and East Asian. What we know of the Aztecs also indicates that they had a strong interest in flowers and commercial horticulture. Cut flowers have been fundamental to the human love of ornamentation since the earliest times; in particular, they have played an important part in religious worship. English churches today still have "flower rotas" for filling the vases, Hindu temples in India are nearly always accompanied by sellers of marigold and tuberose garlands for offering to the gods, while Islam's austerity is softened by the joyous casting of rose petals over the tombs of saints. Beautiful foliage has been a more minority interest—except in Japan, where a sophisticated aesthetic has predisposed people to look with favor on a pretty leaf or imposing stem.

The origins of ornamental gardening, as opposed to simply an artistic interest in flowers, seems to lie in the selection and cultivation of flowers and plants that stood out as distinct or unusual. People have always been attracted by curiosities and diversions from the norm, and in the preindustrial era when the vast majority of people lived in the countryside and were intimately familiar with it and its plants—a great many of which would have had medicinal or other uses—anything odd would have been quickly spotted. Wild plants with double flowers, unusual coloring, or variegation might be dug up and moved to a garden, which otherwise would have been used entirely for growing vegetables and herbs. Those with wealth and leisure time would have begun to collect such plants, propagate them, swap them with others who were interested, and commission artists to record them in paint and ink. Thus began ornamental plant breeding.

Double roses have been known since 300 BCE, with Theophrastus mentioning a variety with a hundred petals. Peonies have been collected in China from very early times, varying in color and the number of petals; over ninety varieties of tree peony (*Paeonia suffruticosa*) were known by the eleventh century. In the Islamic world, roses have long been popular, joined in the Ottoman Empire by the tulip (*Tulipa* species) and the ranunculus. Trade brought these latter two to Europe, where they joined the array of what came to be known as "florist's flowers"—plants which were grown by dedicated amateur growers known as florists. The florists grew plants simply for the love of the flower and often to exhibit at public shows. *Hortus Eystettensis* (see chapter 3) records most of what was known in seventeenth-century Europe; among ornamental flowers it includes seventeen varieties of daffodil, fifty tulips, and twenty-three pinks and carnations.

Interestingly, it was not the wealthy and well-connected who brought flowering plants into their first phase of glory, but men (and a few women) of the shopkeeper and skilled artisan class, particularly in Flanders and what is now the Netherlands, during the sixteenth century. It was this culture of seeking both perfection and variation in flowers that later gave rise to the extraordinary outburst known as "tulipomania," but also to the development of florist culture in Britain. Fleeing persecution (many were Protestant), weavers came from Flanders to England, where they settled and started a movement which was to last a good two centuries.

Exhibiting plants at shows, largely for the benefit of other growers, was a key part of florist culture. In most cases, shows were competitive. Until the nineteenth century, eight categories were recognized in British florist shows: tulip, carnation, pink, auricula, hyacinth (cultivars of *Hyacinthus orientalis*), polyanthus, anemone,[2] and ranunculus.[3] Most were derived from occasional mutations or sports and were vegetatively propagated. Occasionally, however, someone would sow seed, and a whole new level of variation would be opened up. Florists were accustomed to taking the weakest plants and latest germinators due to the belief that they had the best flowers.[4]

Florists were organized into societies, which held competitive shows at public houses, accompanied by dinners, known as florists' feasts. Prizes were generally given for the best plants—often valuable but utilitarian: silverware and kitchen utensils. Florists' societies generally flourished in industrial towns: in Paisley, Scotland, weavers grew pinks; in Sheffield, metalworkers grew auriculas. Such gentle pursuits were at great variance with the lives of much of the urban population at the time. In 1837, the curator of Sheffield Botanic Garden noted that "Horticultural and Floricultural Exhibitions . . . have within these few years been working a change in the

2. *Anemone coronaria* cultivars.

3. *Ranunculus asiaticus* cultivars.

4. Laidlow 1848.

tastes and recreational pursuits of the inhabitants of this densely populated island"; he was happy that such pursuits were replacing "bear-baiting, bull-baiting, a cock-fight, a dog-fight or mayhap two animals in human form similarly engaged."[5] During the late nineteenth century, the florists' societies generally disappeared and were replaced by more general gardening associations; their shows gave way to the bigger and more wide-ranging horticultural shows we are familiar with today. Some florists' societies are still going strong, however, the two oldest being The Ancient Society of York Florists, founded in 1768, and The Paisley Florist Society, founded in 1782.

A very characteristic feature of florist culture was the way flowers were categorized; tulips were classed as *roses*, *bizarres*, or *bybloemens*, depending on the distribution of ground colors and markings; auriculas were categorized as *doubles*, *stripes*, and *selfs*. Over time, different categories went in and out of fashion. Prizes were given for plants that came nearest to recognizing an ideal, the form of which would have been well-known among the growers, as is illustrated by one advertisement for a show from around 1780, headed "the perfection of the carnation, auricula, and polyanthus as laid down by the Society of Florists, at Leicester."[6]

Show rules have at times been a straitjacket on creativity. By the 1840s, the strict rules for the pansy (cultivars of *Viola x wittrockiana*), which permitted no variation in the search for the perfect expression of the character prescribed for a particular class, were stifling innovation. A letter to the *Gardener's Chronicle* of April 5, 1862, expressed the opinion that "the show pansies seem to have nearly run their length, owing to the want of variety among them, for though the so-called novelties are still introduced in considerable numbers, . . . yet as far as the public can appreciate, they present year after year little more than repetitions of stereotyped forms. . . . [W]e have not infrequently seen stands of 24 blooms, in which 8 or 10 were virtually alike."[7] The situation was not unlike that of the American corn shows (see chapter 10) and could be turned to advantage. John Salter (1798–1874), a cheesemonger turned nurseryman of Hammersmith in London, and later of Versailles, France, persisted with groups of plants for years even though they were not particularly popular with the public or show judges, specifically, Fancy Pansies and Japanese chrysanthemums. He introduced Fancy Pansies from France and Belgium, where breeders were not limited by show

5. Quoted in Elliott 2001, 173–74.
6. Duthie 1984, 19 (from the figure caption).
7. Quoted in Gorer 1970, 157.

rules but bred for size and color rather than conformity or regularity—and it was these that became the ancestors of the modern bedding pansy.

Some of the florists' flowers ended up as major commercial plants: tulip breeding and growing is now a huge industry, still dominated by the Dutch; violas/pansies and pinks are both popular, with considerable industry investment in exploring their genetics for the pot and bedding plant industry; the auricula, however, is little more than a cult plant for a dedicated few; and the ranunculus has fallen back into obscurity, with the extinction of nearly all its cultivars.

It is in Japan that florist culture has been raised to its greatest level of development. Developing during the later part of the Tokugawa era (1603–1868), Japanese growers have concentrated on around thirty plant genera, picking out for cultivation naturally occurring forms different from the norm, and in some genera breeding them more actively. In some cases, the plants have gone on to become of major commercial importance, such as *Iris ensata*, flowering quince (*Chaenomeles japonica*), and wisteria (*Wisteria sinensis* and *W. floribunda*); others have stayed as obscure hobby plants—of which the strangest must be the very primitive fern relative *Psilotum nudum*.

Members of the *samurai* (warrior) and *daimyo* (feudal landowner) classes were the main growers, although in some cases, the common people also participated. Often, interest in particular flowers would stimulate the formation of societies, very similar in many ways to those of the British florists. Even the humble dandelion (*Taraxacum* species) was grown with what has been described as a "rage" of contests and prizes during its late-nineteenth-century heyday. The real explosion of interest occurred during the Meiji period (1868–1912), when, with the dissolution of the old feudal order, dispossessed *samurai* and *daimyo* had to turn to new ways of making a living—and new ways to enjoy their now more modest incomes.

Beautiful and intricate foliage has been a focus of interest as well as flowers—perhaps no more so than in the case of *Asarum* species, which enjoyed a series of booms during the late nineteenth and early twentieth centuries, with some plants selling for huge sums. Of the flowers, the water-loving *Iris ensata* and the morning glory (*Ipomaea nil*) have particularly interesting histories. The iris was the symbol of the Higo clan, members of which during late Tokugawa times were particularly fond of raising new varieties. Clan members had their own societies of growers, with strict rules, such as the offering of new varieties to other members before offering them to others, even members of one's own family.

Morning glories were introduced from China around 800 BCE and have long been a particular favorite of amateur breeders from all social classes, as well as a very popular subject for artwork. During the Tokugawa period, procedures for their selection became very advanced. The plants are annual and can only be propagated from seed, which appears to have stimulated considerable research into genetics. Particularly prized were *demono* plants with bizarre leaf and flower shapes—these were sterile, however, and the genes coding for these characters recessive; "demono" actually means "segregating." Breeders corresponded extensively, and there is clear evidence from publications that they had reached the same conclusions as did Mendel a hundred years later.[8]

TULIP MADNESS AND THE ORIGINS OF CAPITALISM

If there is one episode in plant breeding history that everyone knows about, it is the extraordinary outburst of financial speculation in Holland during the 1630s—tulipomania. In fact, breeding was a relatively small part of the story—a history of virology might have just as great a claim. Tulips had been introduced into northern Europe some one hundred years before from the Ottoman Empire (Turkey), along with hyacinths, which also attracted the attention of speculators. Although still theoretically under Spanish rule, the Dutch had been building up an extremely successful economy, largely through trade. Many of the key aspects of modern capitalism were invented in Amsterdam; for example, the city had the world's first full-time stock exchange. Given the country's geography, investment in land (the investment of choice for *nouveaux riches* the world over) was difficult—so people had to seek other routes to grow their money. Tulips were one such speculative investment; they became status symbols for the newly rich merchant and financier classes, which stimulated both a rise in prices and efforts to breed ever more exquisite blooms. The bulbs that were most sought after were those infected with the tulip-breaking *potyvirus*, which caused a pattern of elaborate streaking in the petals. As far as the breeder was concerned, a tulip was only as good as its infection, which, since there was no understanding of either genetics or the existence of viruses, had to be left entirely to chance. Desperate measures were tried, with one grower tying together half a red tulip bulb and half a white in an attempt to get striped flowers.

As the love of tulips grew and grew, so did the prices, with particularly fine bulbs regularly reaching hundreds of florins, at a time when a pig might cost

8. Yamanaka 2000, 2004.

only thirty. In 1636, a 'Semper Augustus' bulb sold for six thousand florins.[9] Given the inventiveness of Amsterdam's financiers, it is not surprising that a futures market developed—in bulbs not even planted. With tulips being traded on stock markets and the less-wealthy members of society joining in, a speculative bubble soon grew. When traders could no longer get the prices they wanted for their bulbs, the bubble burst and many were ruined.

The opinions of historians and economists on tulipomania are mixed, some claiming that the whole business was exaggerated out of all proportion, others that investors were behaving more or less rationally. For our purposes, though, tulip fever marks a threshold—the first time in history that ornamental plants and the diversity among them become an important commodity, attracting the attention not just of connoisseurs but of society as a whole. Other countries were not immune—the French had experienced a similar, but less overwhelming, burst of enthusiasm for the bulb in the first two decades of the century. In Turkey, a history of breeding had resulted in many fine interspecific hybrids well before their export to Europe, but during the reign of Sultan Ahmed III (1703–1730), the Ottomans imported millions of the bulbs from the Dutch, stimulating a frenzy of breeding among merchants. The sultan became obsessed by them and held extravagant tulip festivals, the costs of which contributed to fomenting a military rebellion that ended in his abdication.

PASSION AND OBSESSION— NURSERYMEN BREEDERS AND GENTLEMAN AMATEURS

Thomas Fairchild of London, creator of the first hybrid known to be a cross (see chapter 4), was one of many nursery owners who, during the seventeenth and eighteenth centuries, began to cluster around major towns and cities. Exploiting a growing market for plants, they were constantly on the look out for the new or the bizarre. Their clients were the primarily the new middle classes or lower gentry, people with money to spend and some leisure time. The nineteenth century saw the nursery trade continue to expand; railways allowed plants to be sent from nurseries at one end of a country to another, and trade became more geographically dispersed. Nurseries and amateur growers saw what was for a time almost an exponential growth in new plant introductions as explorers, plant hunters, sailors, botanists, and

9. See technical note 34: The Importance of the Cultivar.

commercial collectors poured seed and plants into the countries of western Europe and North America. With so much to play with, nursery owners and amateurs soon found species that crossed easily, either through nature or design. For many, plants became a passion, and it is not surprising that progress in understanding genetics was driven during the eighteenth and nineteenth centuries as much as by interest in ornamental plants as in edible or useful crops.

Broadly speaking, there were three groups of ornamental plant breeders at the beginning of the nineteenth century and four by the end: amateurs (usually "gentlemen"), men of some means and more importantly some leisure; their gardeners; nursery owners; and increasingly, the seed companies. A word needs to be said about the gardeners. A head gardener was a man of considerable authority in the English or French country house and might become nationally famous in his own right; those who were interested in breeding and had the support of their employers would often have had superb facilities and a wide range of plants with which to work. New varieties grown by noncommercial gardeners often ended up being given to nurseries to propagate—whether money changed hands—and how much—is usually unknown.

At the simplest level, plants were brought into cultivation from foreign lands, usually in the form of seed. Many ornamental shrubs, trees, and herbaceous perennials were then propagated vegetatively, which meant that identical clones were widely distributed. Plants that bloomed early and had good-quality flowers and foliage would tend to be picked out for propagation. The next step was to take a plant that really stood out from its fellows, give it a name, and propagate it as a selection or cultivar. Giving plants names was an opportunity to promote one's business, commemorate a dead relative or horticultural notable, be obsequious to an employer or benefactor, flatter a pretty girl, or be patriotic. The subject of the social history of variety naming would make a very good—and quite entertaining—book. The makings of selections has not always been honest, particularly where there is very little difference between the selection and the typical form of the species; this may be done now for reasons of claiming plant breeder's rights. It can also be done simply for marketing purposes—the goldenrod *Solidago rugosa* will not sell as well as *Solidago rugosa* 'Fireworks.'

Selection from seed collections became particularly important as the tide of introductions to Europe and North America reached its crescendo at the turn of the and twentieth century. This was the high point of the "plant hunter" in Asia, intrepid individuals who, doubling as explorers (and sometimes as spies), set out for regions that no European had ever entered. The eastern foothills of the Himalayas were particularly fertile territory, with an immensely rich flora which had been evolving with little break since the tertiary period and enough mountains and other breaks in terrain to ensure that plenty of genetic isolation—and hence, speciation—had taken place. The plant hunters were funded by wealthy collectors, primarily in Britain, but also in France, Belgium, and the United States; magnolias, rhododendrons, camellias, bamboos,[10] and maples (*Acer* species) flourished particularly well in the mild British climate. Everything has conspired to give the plant hunters of the late nineteenth and early twentieth centuries a high profile in horticultural history: spectacular plants in some of the grandest gardens of the time, the ability of many of the plant hunters (e.g., Frank Kingdom Ward [1885–1958] and Ernest Henry Wilson [1876–1930]) to write prolifically and vividly, and the romantic nature of the terrain they covered.

Preliminary selection was in fact often carried out by the plant hunters themselves—they would often try to identify particularly good-looking plants from which to select seed or would collect a species from high altitudes to ensure that the hardiest forms would be chosen. In a great many cases, selection of one or two good plants, either by plant hunters or by nurserymen, resulted in clones, which then became *the species* as it was represented in cultivation.

In the case of some genera, the ease with which species crossed and the fact that their hybrids were fertile resulted in a rapid proliferation of new varieties. Such was the case with the southern African pelargonium.[11] Pelargoniums succeeded in European greenhouses with a vengeance. A great many species were in cultivation by the 1820s, and hybrids soon started appearing; Loudon's *Greenhouse Companion* of 1824 suggested that nosegays of pelargonium flowers be purchased and suspended over plants in order to create new varieties: "nothing forms a more pleasing gardening amusement for the ladies of the family than saving and sowing these seeds." Not surprisingly, the genus was declared to be "botanical chaos" by the influential garden writer Charles M'Intosh in 1838; another writer in *The Florist* declared that

10. Botanically, bamboos are members of the tribe Bambusea, in the subfamily Bambusoideae, within the grass family Poaceae.

11. Popularly known as "geraniums" after their original naming by Linnaeus, now rejected. This causes endless confusion with the true genus *Geranium*, which is related but herbaceous and frost-hardy.

"the opinion in some quarters is that they have been brought to the height of perfection and no farther improvement is possible."[12]

Botanical chaos on a somewhat grander scale was the result of growing rhododendron species together. The ease with which new plants could be created made this an ideal subject for dabbling. Not surprisingly, the more disciplined gentleman-gardeners tended to look askance at what some of their fellows were doing; "burn the lot" is what John Charles Williams is reported to have advised Lionel de Rothschild when he heard about his efforts, before lecturing him about Mendelian genetics.

J. C. Williams (1861–1939) was one of the great patrons of plant collecting as well as a consummate breeder of shrubs and daffodils; with several properties scattered around the mild and wet southwestern peninsula of England, he was also in an excellent position to trial plants. His name has been immortalized in *Camellia x williamsii.* Camellias had been popular in Victorian times, but in northern Europe they were mostly conservatory plants. Williams's stroke of genius was to combine *C. saluensis,* collected by George Forrest (whom Williams had sponsored) in China in 1917 and 1919 with *C. japonica.* The latter was well-established in cultivation through the importation of Chinese and Japanese varieties over the previous century but had performed poorly in most of Britain because of low winter light levels. *C. saluensis* seemed to confer hardiness on its progeny, was able to flower in low light, and had a trait that was a great advantage for a breeder, that of flowering as young as two years from seed.[13]

Largely named and promoted after his death, the new camellias were a great success, hardy and free flowering—launching the plant as a popular garden shrub rather than one for the conservatory-owning classes. One was 'J. C. Williams,' which produces so many flowers its growth rate is reduced. Williams was a successful breeder of other genera; several of his rhododendrons are still commercially important, such as 'Blue Tit' and 'Yellow Hammer,' while most of his daffodils he did not launch commercially. With rhododendrons, his approach was always to breed between species, believing that this gave much better results than crossing hybrids. He was undoubtedly a methodical hybridizer, but unfortunately, he lost all the key records when his suitcase was stolen on a train journey in 1923. This is what the family maintains; others in the camellia world have hinted that this was a story concocted to hide the fact that he did not keep written records. J. C.'s cousin,

12. Quotations from Kingsbury 1991a, 299–300.

13. Trehane 1998.

P. D. Williams, was a noted breeder of daffodils, an important part of Cornwall's cut-flower industry. He produced several varieties that have stood the test of time, including 'Carlton' in 1927, which still covers many acres of the county as a cut-flower crop; however, he certainly kept no records and his reputation is as having an intuitive, almost mystical, approach to breeding.

Many early-nineteenth-century hybrids occurred naturally, or if they were made artificially, records of their production and parentage have been lost; fear of being accused of meddling with creation may have put off their creators from claiming credit. The first British rhododendron hybrid is known from 1814, a cross between a Turkish form of *Rhododendron ponticum* and the American *R. periclymenoides.* Many others soon followed, for the genus is what gardeners, with a dash of prurience, like to call "promiscuous"—seedlings popping up around the borders were often dug up and transplanted to nursery beds in the hope that something better than the parent might result. It not infrequently did.

Rhododendrons provide a good case study of the importance of interspecific hybridization in horticulture. Pioneer growers in the nineteenth century found themselves in a frustrating situation—the most sumptuous flowers were produced on plants with somewhat tricky cultural requirements, but the hardiest and most robust plants were never going to win any of the prizes at a flower show. The beginnings of a solution appeared in May 1826, when flower heads of *R. arboreum* were packed into boxes of damp moss in Edinburgh and sent by rail to Hampshire in southern England to be used to supply pollen to a plant that was itself a cross between *R. ponticum* and the American *R. catawbiense.* The Himalayas met Turkey and eastern North America, and immense trusses of scarlet met two less outstanding brands of lilac purple; crucially, though, the American plant was very hardy and the Turkish immensely vigorous. The success of the result, named 'Altaclarense' started off a boom in rhododendron hybridizing that has never ended. Apparently some 1,800 seedlings were raised from the cross and widely distributed. It requires only a moment's reflection to realize that they were probably all genetically different. It was only several decades later that growers really began to appreciate that when making a cross, the progeny are going to display considerable variation and that a name is not going to describe what we today call a cultivar, but rather a hybrid swarm.

Nurseries and amateurs were attracted to hybridizing "rhodos" because they are easy, they flower when relatively young, and the results are rarely an outright disaster. The result is at least six thousand named hybrids, plus innumerable others which have lost their names or were never commercially propagated.

The occasional breakthrough made waves. Waterers was one of the great rhododendron nurseries of southern England; it was located in Surrey, where the wealthy can live near London but not in it, and where acidic sandy soil makes for rhododendron heaven. John Waterer (1826–1893) inherited the nursery from his father in 1868 and in the family tradition, hybridized rhododendrons. One day he noticed a particularly good plant in a stock bed but then found that it had disappeared. Since the workers were allowed to take plants home for their own gardens, he went for a walk around the local cottages and soon found the plant. After recovering it, he named it 'Pink Pearl.' It was then launched commercially by John's son Gomer (1867–1945) in 1896. It won an RHS First Class Certificate in 1900 and was a best-seller by 1902. 'Pink Pearl' had a winning name (which always helps) and a massive head of pink flowers that faded to white as they aged in the attractive manner of apple blossom. It was, moreover, a very good garden plant. It was blowsy enough to have been dubbed "the barmaid at the hunt ball" by one gardener, but the market by this time was made up by the rising middle classes, not aristocratic connoisseurs—customers who wanted bright color and lots of it. It proved an unstoppable seller, a plant to help popularize the whole genus.

Hybridizing has always attracted people with an obsessive streak. If commercial breeders manage to combine this with either being good in business or having a good manager, then success was often assured. Amateurs or those outside the nursery business can produce a vast amount of new material, but unless this gets taken up by the trade, it inevitably becomes dispersed after the hybridizer has departed to explore the gene pool of the Elysian Fields. One in the first category was Victor Lemoine (1823–1911) whose professional life spanned a period when hybridization was not only becoming established and adventurous, but also had an air of pioneering bravado. In such times it is possible for one person not only to make a great impression but also to lay down rails along which the future itself will travel—technological development always has a strong component of inertia, the past dictating the limits of the future. Lemoine was not only addicted to hybridizing any plant he came across, he was very good at it. In doing so, he did much to establish certain genera as garden plants, which others then returned to in order to breed further.

Lemoine came from a long line of horticultural professionals; having trained in Belgium, he set up on his own at Nancy in France in 1850. He then started on a life of sixty years of frenetic crossing. He started with gladioli and begonias, moved onto *Potentilla*, *Portulacea*, and *Streptocarpus*, and then to on to pelargoniums, chrysanthemums, clematis, phlox, and many more. *Many more*. Some of his most productive work was done in the last fifteen years of his life, with delphiniums, penstemons, hydrangeas, honeysuckles (*Lonicera* species), *Weigelia*, *Astilbe*, *Pyrethrum*, and others. And still more. But it was his transformation of the pretty *Syringa vulgaris* into something more—the sumptuous lilac—for which history will remember him. Laden with heavy flower clusters, its penetrating scent drifting through the still air of a day in early summer, the lilac has become a vital part of the early summer garden. In 1843 he bought a double-flowered sport from another grower; most of the lilacs he then bred were selections from seedlings grown from this plant or its descendents—only ten were the result of deliberate crosses. Madame Lemoine had better eyesight, so she would stand at the top of a ladder and carry out the actual cross-pollination. Lemoine's lilacs—among them, 'Belle de Nancy,' 'Mme. Abel Chatenay,' and 'Princesse Clementine'—all encapsulate what the French have done best in breeding—they are decadent, flouncy, and voluptuous. Some idea of his contribution to horticulture is gained from knowing that he was the first foreigner to receive the Victoria Medal of Horticulture of the RHS.

At the opposite scale are those who work either as amateurs or who hybridize as an adjunct to another job in horticulture. Almost as a random example, Albert Victor Pike, was a highly regarded head gardener at several leading private British gardens during the 1950s and 1960s and was a writer for garden magazines. According to his daughter Cynthia, after whom one of his best known introductions is named—an almost frost-hardy *Abutilon*—"hybridizing was an absolute obsession . . . he was like a kid with a meccano set." He understood Mendelian genetics, but like a lot of hybridizers, possibly like Lemoine, he was primarily interested in plants, not genetics or perfection. Hybridizers seem to fall into two groups: grazers and perfectionists. Pike was clearly a grazer, experimenting with very different genera and constantly moving on—his hybrids of *Lachenalia* and *Ixia* (both bulbs), *Anchusa* 'Royal Blue' (herbaceous), and buddleias (shrubs) received awards or ended up in commerce. It is interesting to note that as well as the one named for this daughter, Pike raised another very good *Abutilon*, also still in commerce, 'Kentish Belle.' Both were raised from seeds from the same capsule—an approach that would astonish commercial breeders, used to sowing thousands of seeds in order to get one worthwhile result.[14] Perfectionists stick with a limited number of genera and cross generation after generation—for which a knowledge of Mendelian genetics is vital; grazers, however, rarely breed beyond an F_1 generation. Our gardens and public spaces have benefited from both approaches.

Sometimes it is chance that brings the right set of genes to the attention of the right person in the right place at the right time. More often, however, it is intuition based on experience and real plant knowledge that launches a good new variety. Fuchsia hybridization began in the 1820s, based on only two species: *Fuchsia coccinea* and *F. magellanica*; as the century advanced the number of cultivars increased hugely, with most breeding being done between cultivars with little input from other species. Lemoine introduced over 390 during an 80-year period, an average of 5 a year. A Mr. Rundle of Edinburgh produced only three plants during the 1880s, all of which were still rated as among the best a century later: 'Mr. Rundle,' 'Mrs. Rundle,' and 'Duchess of Edinburgh.' The Duchess is no longer in cultivation, but the humbler Mr. and Mrs. Rundle are still commercially available in Britain.

A ROSE IS A ROSE IS A ROSE— BUT THE GENES ARE NEVER THE SAME

Roses are the favorite flower of the Christian and the Muslim worlds: varying hugely in color (but famously never blue), shape, and scent, and are loaded with cultural significance and symbolism. The ancestral rose, *Rosa gallica*, may have been pink, but it threw up genetic combinations for red forms early and was stable enough to be still very important in rose breeding. It dominated the gene pool of the rose until the nineteenth century. One of the biggest rose gardens famously belonged to Napoleon's Empress Joséphine (Joséphine de Beauharnais, 1763–1814), who began her collection at Malmaison in 1804, which, by 1814, encompassed every type known. The conquering armies of her husband sent back any interesting roses they discovered—hostilities between Britain and France were actually suspended on occasion to allow for the transportation of roses, and nurseryman John Kennedy was given a passport to travel between the two countries during the Napoleonic Wars. Joséphine's patronage encouraged breeders and encouraged a flurry of activity in the early nineteenth century, with one cata-

14. Pike 2007.

log (Desportes) listing over 2,500 varieties in 1829. Her head gardener, André Dupont, may even have made attempts at artificial pollination.

"Old" roses flower only once every year, in early summer. The Chinese *R. chinensis*, however, carries genes for repeat flowering, so its introduction brought about a revolution in rose breeding. However, the repeat flowering habit was recessive, which limited the breeding possibilities in the early nineteenth century, in addition to which hybrids between this species and others were usually sterile. However, a fertile cross was achieved in France in the 1830s and gave rise to the hybrid perpetual group. As well as flowering only once, "old" roses were also only available on a spectrum from crimson to white. It was not until the 1820s that the first yellow rose species were introduced to Europe and the United States, but it took many decades before successful hybrids were created. The French Joseph Pernet-Ducher (1859–1928), regarded by many as the greatest rose breeder of all time, created the first yellow roses; his hybrid tea class 'Soleil d'Or' of 1900 was the first "real" yellow, as opposed to the wishy-washy yellows of previous attempts. It—and he—has been blamed for the introduction of a susceptibility to black spot, as yellow roses seem to have a genetic susceptibility to the disease, but in fact the disease was not unknown before.

Importations of roses from China were frequent during the nineteenth century; often they arrived on cargo ships bearing tea, which led to the roses which were bred from them to be dubbed "tea" roses. Many had a touch of yellow, were well scented, and, unlike the flat blooms of traditional European roses, had a slight point in the center of the flower. This point gave each flower a very different character to what people were used to—and made the flower look particularly attractive just as it was about to burst out—a bud that spoke of promise as well as beauty. In 1867, a rose breeder in the south of France, Jean-Baptiste Guillot (1803–1882), discovered a particularly good flower on a plant in a batch of seedlings and felt confident enough in its future to call it after his country—'la France.' It was the first of the group that has dominated rose growing and breeding ever since—the 'Hybrid Teas.' Guillot confessed he did not know what the parentage was, as M. T. Masters, editor of *Gardeners' Chronicle*, in writing about the records kept by rose breeders at the time, noted that "too often it is to be feared that the pedigrees given are at best mere guess work."[15]

It was an Englishman, Henry Bennett (1823–1890), a farmer and cattle breeder, who really developed the hybrid teas as a class, crossing tea roses

15. Quoted in Wylie 1954, 565.

with the hardier hybrid perpetuals. Bennett, perhaps because he was a cattle breeder, was a systematic hybridizer and kept thorough records. He was also good at marketing his varieties and is generally seen as the founder of the class. One of the roses he bred was named 'Jean Sisley' for a French breeder who had advocated artificial pollination but had been ignored by colleagues. Meanwhile, once-blooming roses were being rapidly discarded in France; leading breeder Victor Verdier (1803–1893) jettisoned all his once-blooming varieties in 1838 in favor of *remontants*, those that flowered all summer. Out went all the old gallica, alba, damask, moss, and centifolia classes, and with them some of the best scents of all.

A GARDEN IN A PACKET—THE FLOWER SEED AND BULB TRADE

The seed trade may have started out primarily with vegetables and forage crops, but flower seeds were soon added—answering a growing desire for homeowners, the lady of the house in particular, for a splash of color among the cabbages and a source of cut flowers to cheer the house. As the trade grew in importance through the nineteenth century, some companies began to specialize in flowers; selection soon followed, and by the end of the century, the seed companies were involved in increasingly complex breeding programs (see chapters 5 and 6).

The flower seed trade is overwhelmingly about annuals, plants whose life cycle is completed within a year, giving color within a few months of sowing. At first these were all "hardy annuals," the sort that can be scattered on the ground in spring and more or less look after themselves. As exploration and plant hunting brought more exotic species to Europe and North America, another category appeared, that of the "half-hardy annual," species that were frost-tender, but which could be raised in greenhouses, and then, when all danger of frost was passed, planted out to flower until they were killed by the first frosts of autumn. Clearly, only species that produced plentiful seed were suitable for such methods, but with relatively little effort and a little help from the new technology of greenhouses, large numbers of plants—and dramatic effects—could be produced. During the 1850s, a new art form was born—"bedding out."

Bedding out involved laying out a pattern on the ground and then filling it with colorful flowering plants, generally just for the summer. As time went on, the patterns got more complex and color schemes more sophisticated—or garish, depending on the point of view. Writers on gardening had a range of diverse opinions: one fulminated against mixed colors—"avoid streaky colors as you would star-shaped beds," another advocated doing just this, "let all the nurserymen in the country be laid siege to for variegated scarlet geraniums, to make 'shot silk' beds . . . for if we do not strike while the iron is hot, the half of us may forget the thing altogether before another season comes round."[16] Fashion clearly played a big role, and fashion's current style has always been of major importance in deciding breeding priorities in ornamental plant breeding. By the late 1860s, bedding out morphed into carpet bedding, which was more precise still, with tiny foliage plants and succulents as well as flowering plants used to create extremely detailed designs—useful for heraldic crests or spelling out names. The French excelled in this field, and carried on making elaborate plantings long after such efforts became almost a joke in Britain. Among his other achievements, John Salter was a great breeder of carpet bedding, although his virtually black-leaved *Ajuga reptans* 'Purpurascens' has been lost, despite being a form of a very common European wildflower.

Bedding out is best thought of as "painting with plants"; just as artists have sought an ever wider palette, so have gardeners, which meant that the seed companies made great efforts to produce different colors from one species and ensure that seed strains could be developed which consistently produced a particular color. It almost goes without saying that large and long-lived flowers were selected for. Another important characteristic was height, or rather lack of it, for bedding out plants need to be short and compact.

Fashions in how plants were used changed rapidly throughout the nineteenth and twentieth centuries. The relationship between planting design and the plant palette available is a complex one—but there is little doubt that the availability of particular plants, made possible through introductions from abroad, had a major impact on planting styles. Within a particular planting style, designers wanted as much variation in color as possible, and often form and visual texture, too. The nurseries and the public wanted variety as well, particularly in color. Color was a great obsession in the nineteenth century, partly a result of advances in chemistry, which allowed for improvements in the range of colors available as dyes for fabrics, artists' materials, ceramics, and stained glass for church windows. Color theory—how we perceive color and how colors look together—was consequently of great interest. There is an additional factor behind the furious pace of variety development; it was a sign of progress. Progress was one of the dominant

16. Quoted in Elliott 1986, 13.

themes of the century; advances in science and engineering were improving human life so visibly, and improvement of the plant kingdom was part and parcel of this. That new varieties appeared in the seed catalogs and in the nurseries every year was yet another visible manifestation of progress.

Plants propagated from seed could be improved by processes similar to those used in crop breeding, methods that the seed companies began to use extensively in the latter part of the nineteenth century. Simple selection was always the beginning of this process, and in many cases the end of it. Reverend Wilks of Shirley in Surrey, who had been the secretary of the RHS's hybridization conferences at the turn of the century (see chapter 6), was a keen amateur plant breeder. His fame lives on in the Shirley poppies, a strain of the wild European field poppy (*Papaver rhoeas*). He started the strain when, in 1880, he noticed a poppy in a cornfield with a white edge to the petals, marked it, and later collected the seed. He discarded all the seedlings that came up with red petals or the normal dark blotch at the base of each petal, keeping and sowing seed from any that showed variation from the normal, ending up with a strain that showed a range of color from white through pink to crimson.[17]

Bulbs and other plants that grow from underground storage organs that can be dried are a bit like seeds—easily transportable and easily confused if someone mislays a label. Given their ability to produce color and growth as near to instantly as it is possible to get in the natural world, they have leant themselves to the mail-order business. They have also proved very successful as bedding plants for spring displays. A key development was the turning of the tulip from a florist's flower into something that could be mass-produced. This was achieved by the Dutch firm of Krelage, who in 1885 bought up the stock of a Belgian nursery that was closing down. Krelage took the best tulips and in 1886 launched them as 'Darwins,' named for England's scientific genius. The Darwins are somewhat military in bearing, making them ideal for dotting color into spring borders—bright orbs of clear color hovering above a low layer of blue forget-me-nots (*Myosotis* species) or red and yellow wallflowers (*Erysimum* cultivars). By the 1920s, the London parks were being planted with over two hundred varieties—all British grown. Garden guru and color theorist Gertrude Jekyll (1843–1932) loved them too. That tulips made an impact as bedding plants stimulated breeding work on other bulbs. The bulb that British breeders went to work on was the daffodil, up to that point, grown mainly as a cut flower. Daffodil breeding, however, was

17. Anon 1923.

utterly unlike tulip breeding. Whereas tulip breeding has been dominated by commercial nurseries for over a century, daffodil breeding was for most of the nineteenth and much of the twentieth centuries the work of gifted amateurs or individuals who dedicated their lives to breeding work. The flower was a favorite with English and Scottish country gentlemen and aristocrats, among the Scots one of the greatest was Major Ian Brodie (1868–1943), or to give him his full title, The Brodie of Brodie. Between 1899 and 1942, the Brodie raised hundreds of new varieties in the grounds of his Highland castle, his trial plots organized with military precision. By the end of the twentieth century, some twenty-four thousand daffodil cultivars had been recognized and placed on the International Daffodil Register—maintained by the RHS.

PLUS ÇA CHANGE, PLUS C'EST LA MÊME CHOSE—THE TWENTIETH CENTURY'S ENDLESS SEARCH FOR NOVELTY

Compared to agriculture, the world of ornamental horticulture is a conservative one; if a gardener and an arable farmer were both sent back in a time machine to 1890, the gardener would fit very quickly into the world of wooden greenhouses, clay pots, and coal-fired stoves, whereas the farmer would be much more disorientated. The point is not just that farming has become so hugely mechanized but that the functions of arable farming have changed from grain, fodder, and vegetable production to include growing for industrial chemicals, specific ingredients, oils, and so forth. The function of horticulture has remained more or less the same, and consequently the role of the breeder has not actually changed that much. The seed, bulb, and nursery trades are growing for amateur gardeners, landscape professionals, and companies and local governments involved in beautifying public space. There are only two really new areas: one is that of amenity landscaping, where from the 1950s on there has been a demand for low-maintenance ground-cover shrubs and narrow trees for use in urban public space; the other is the cut-flower trade. Houseplant breeding too might be regarded as virtually an invention of the twentieth century.

Whereas once flowers tended to be dual-purpose—they could be marketed as garden plants or for the cut-flower trade; there is now a definite divergence. The cut-flower trade grew enormously during the last half of the twentieth century and now attracts a large proportion of the investment that goes into ornamental breeding. With the expansion of the export-orientated cut-flower trade into warmer regions of the world—Colombia,

Kenya, Rwanda, Israel, and South Africa—genus after genus is being sucked into a showy new world, the secrets of their genomes unraveled, the limits of their hybridization probed, and every aesthetic outcome explored.

Breeding for gardening has kept more or less the same goals it had at the end of the nineteenth century: bigger, longer-lasting flowers in as many colors as possible, with compactness often an important goal. Breeding for gardens tends to concentrate resources on the remunerative end, so that bedding plants like petunias, Busy Lizzies, (*Impatiens* species), African and French marigolds (*Tagetes* species), salvias, and primulas get a lot of investment—after all, they only last a few months, and the customer is soon back to buy more. Perennial garden plants and shrubs, which can live for decades, get a lot less investment. A gradual extension of plant breeders' right systems, starting with the US Plant Patent Act passed in 1930, has ensured that more breeding gets done, but a vast amount of plant improvement is still very "old fashioned." As a general rule the less commercially important a plant is, the simpler will be the breeding technology. With genera relatively new to cultivation, the most basic selection and pre-Mendelian crossing can produce dramatic results. The hellebores (*Helleborus* species) are an example. Murky flowers that everyone would ignore if they bloomed in June, their appearance in late winter is something of a boon to gardeners in the temperate zone. Their mysterious grays, pinks, and maroons have a sophisticated cachet to which is added their origins—much of the best genes were in plants collected on the outbreak of war in the Balkans in the 1980s and 1990s. More good genes are now probably growing on top of landmines. I once met a renowned and commercially successful breeder of hellebores, so inevitably I asked him about his methods and mentioned Mendel: "Who's Mendel?" he asked.

What is important in ornamental horticulture is novelty—something to keep the customers coming back year after year. Nothing too new, however. Gardeners, professional and amateur, feel safe sticking to a limited number of genera. The key flowering shrub genera today were already well-established and well-hybridized over a hundred years ago; few have joined their number. Writing in 1975, garden historian Richard Gorer expressed the opinion that "the hybridists appear to have gone on breeding the same plants that have been popular for so long . . . they seem to lack enterprise."[18] New versions of familiar plants sell well. The world of herbaceous perennials has been more dynamic, many of the most popular genera today, *Geranium,*

18. Gorer 1975, 170.

Euphorbia, and *Hosta*, were little known, and certainly little bred, before the Second World War.

The dahlia offers an intriguing question. With a remarkably wide range of colors as intense as any flower can offer and some truly extraordinary shapes, such as the almost spherical *pompom* group, the dahlia's habit of dying back to a tuber that can be dried and sent cheaply by post has long made it a bulb merchant's favorite. It has also been a favorite of both grand borders and tiny working class gardens. What is most extraordinary about it is the apparent speed of its apparent development. Six species were introduced to Europe between 1789 and 1804, but by 1806 the Berlin Botanic Gardens was able to have fifty-five cultivars in flower; by 1830, some thousand cultivars were known, with nearly all the colors with which we are now familiar. That they are determined cross-pollinators and their high level of polyploidy increases their level of variability, despite the limited range of chemicals involved in flower color. However the real reason as to why European breeders were able to make such progress in such a short period of time may well be that the work had already been done for them—the dahlia had been a favorite flower of the Aztecs.[19]

Modern varieties may include an enormous amount of complexity—the 'Queen Elizabeth' rose, introduced by veteran breeder and grower Harry Wheatcroft (1898–1977) to celebrate the coronation of Queen Elizabeth II in 1953, was of American origin, bred by Dr. Walter E. Lammerts (1904–1996), "known in the United States as the father of scientific rose-breeding." Wheatcroft researched its background and found that it was the result of no less than thirty-nine different varieties.[20] Wheatcroft's life combined both breeding and importation—in both he showed flair; his marketing was that of a showman, helped by what must have been the world's best-known mutton-chop whiskers. Nursery owners and plant breeders are generally quiet, unassuming people, but not Wheatcroft, who was courageous but sometimes naive in his loud espousal of left-wing views, his dress sense, and his private life—all were flamboyant. Lammerts was flamboyant too in a curious way; his approach to breeding may have been scientific, but he was also a Creationist, and in 1964 he became the first president of the Creation Research Society.

The ornamental plant business has always had a small but significant

19. Crane and Lawrence 1952.
20. Wheatcroft 1959, 1.

whiff of skullduggery about it—many small nursery owners, and even curators of botanic gardens, will tell of good new plants mysteriously disappearing overnight, only to appear in the catalog of a major company a few years later. The fact that the trade attracts both dedicated individuals with little head for business and the odd ruthless entrepreneur with little respect for the truth tends to result in a field of both lambs and wolves. Burpees was about to introduce double-hybrid nasturtiums in 1934, when someone stole a considerable quantity of seeds from a field—company boss David Burpee was luckily a pioneer in seeking publicity and turned what could have been a major blow into free advertising in the form of newspaper headlines. Armed guards looked after the next crop.

Getting your customers involved in the tiresome business of plant selection and driving novelty forward is always a good ploy, a technique still occasionally used today. In 1954, Burpees offered a $10,000 prize to the first gardener who could provide seeds for a white marigold, something that had frustrated company breeders. During the next two decades, more than eighty thousand customers sent in seeds for testing. None were white enough or big enough to satisfy Burpees' demands. By 1975, the company itself had come up with several possible strains, but in setting the results before a panel, an amateur gardener, Alice Vonk, the widow of an Iowa farmer, won the prize.[21]

"Colchiploidy" may have not have achieved as much in crop breeding as some had hoped it might (see chapter 11), but in the world of the garden where appearance is often more important than substance, it has proved very useful indeed. Burpees was among the first to use colchicine to breed ornamentals, with the launch of their marigold 'Tetra' in 1940, produced in conjunction with the New York Agricultural Experimental Station. 'Super Tetra' varieties of *Antirrhinum* (popularly known as "snapdragons") soon followed, as well as a scarlet zinnia with blooms up to seven inches across—zinnias had also been bred initially by the Aztecs. Colchicine-induced changes have been of major importance in a great many other genera too. One of the first ornamental plant breeders to experiment with colchicine was Edward Steichen (1879–1973), an American breeder who raised delphiniums and irises; his delphinium program involved five gardeners growing up to a hundred thousand seedlings per year. Plant breeding was a sideline, however—Steichen's "day job" was as a photographer, painter, and gallery curator.

21. Burpee n.d.

Plant breeding, in his opinion, was an art form, flowers a material to form and sculpt.[22] In 1936, the Museum of Modern Art in New York City held an exhibition of his delphiniums. Among his circle was Georgia O'Keeffe (1887–1986) perhaps the most famous painter of flowers of the twentieth century.

Radiation breeding has proved of some use, mostly for species used in the cut flower trade (see chapter 11). In the breeding of garden plants, both radiation and polyploidy have been particularly useful to make the "color breaks" so keenly ought by breeders. Color breaks are commercially important as they offer the familiar in a new guise and are well known for attracting the interest of everyone involved in the garden world. Radiation breeding has also proved useful for other factors. Although long since grown, the southern African perennial *Streptocarpus* was revolutionized by X-ray treatment at Britain's John Innes Institute, which produced a photo-insensitive plant that was able to flower for much longer. This breakthrough enabled the "strep" as it is popularly known, to be much more widely grown as a houseplant.[23] Once incorporated into the gene pool in cultivation, the origin of such breaks is often forgotten.

An interesting subject for consideration is the complex relationship between three factors: number of varieties, genetic complexity/diversity, and variety distribution. The rose is certainly a leader in all three, having been derived from several species, with enormous genetic complexity, and very familiar. The iris, specifically the so-called bearded iris, is almost the opposite. Perhaps genetically more complex than any other garden flower, bearded irises are widely grown but are not of major commercial importance, nor are they used much in the cut flower industry; yet there are a vast number of varieties (approximately 2,500 commercially available in Britain). They are overwhelmingly a connoisseur plant. The daylily tells a similar story of rampant genetic diversity, but combines both public familiarity and private obsession.

The Englishman George Yeld (1845–1938), by profession a schoolteacher, was the first to hybridize daylilies (*Hemerocallis* species), starting with 'Apricot' in 1892, followed by nurseryman Amos Perry (1869–1953), a very prolific hybridist of many genera, most of which were lost when his land was requisitioned for farming during the Second World War.[24] The daylily's

22. Mahan 2007.
23. Davies and Hedley 1975.
24. Bourne 2007.

real climb to fame started with the American Arlow Burdette Stout (1876–1957). As a child, Stout had wondered why his mother's 'Europa' (*Hemerocallis fulva*) never produced seed, a question he remembered all his adult life. In 1929, at the age of 53, he discovered it was a sterile triploid. As a botanist and professional plant breeder, Stout worked with the New York State Agricultural Experimental Station, producing 175 new seedless grapes and hybrid poplars for forestry, plus various fruits, Chinese cabbages (*Brassica rapa subsp. pekinensis*), and numerous ornamentals. Renowned for being straight-laced, humorless, and direct, Stout was one of those who was clearly happier in the company of his plants than with his fellow humans. Obsessed with his hobby of daylily breeding, he would dive into his collection whenever he could. His hobby led to the introduction of eighty-three high quality hybrids; his greatest achievement was to use *H. fulva rosea* in a systematic breeding program; yellow is the usual daylily color, but *rosea*'s genes introduced a new spectrum of pinks to reds.[25]

Daylilies are easy to cross, making them ideal for amateur breeding. Practically any seedling from a cross looks good, and like proud parents, it seems that many breeders have been unable to rate one of their children any higher than another, let alone to objectively judge them against others. A vast array of cultivars has been the result—around fifty-six thousand registered with the American Hemerocallis Society by early 2007. Founded in 1946, the society's membership of over ten thousand forms a vibrant and lively community, and perhaps more than any other specialist plant society, involves its members in breeding, including laboratory-based techniques. Colchicine treatment gives good results with daylilies, as tetraploid plants have petals whose colors have an extraordinary sense of visual depth, which allows for the expression of very intense colors. One use of colchicine has been the conversion of old cultivars into tetraploids, resulting in their having stronger stems and better and more robust flower shapes.[26]

Even more frenzied hybridization has taken place in the orchid world. The first orchid was deliberately crossed at the Veitch nurseries in England in 1856—the company was not only the recipient of many new introductions but also the pioneer in hybridizing many genera. Orchids at the time were in the process of becoming the fashionable flower for the wealthy of Europe and North America to grow in the newly perfected hothouse. It was soon discovered that orchids cross very easily, even between genera. Since

25. Apps 1986; Thomas 1986.
26. Grenfell 1998.

then, their extraordinary genetic flexibility has led to the creation of ever more complex crosses—a vast number of which never become commercialized but remain the playthings of hobby breeders. Some new intergeneric crosses are composed from four genera. All need new names—from 1988 to 1990, the registration of new hybrid genus names included *Nornahamamaotoara*, *Georgeblackara*, *Charlieara*, and *Yonezawaara*. If ever a plant has been completely redesigned and remade by human intervention it has to be the orchid.[27]

Cultivar names are all important for marketing, something which canny nurserymen realized a long time ago. Jean-Nicolas Lémon (1817–1895) was an early-nineteenth-century grower and breeder of irises entirely through random natural crossing. Lémon's masterstroke was his naming of irises to capture people's attention. A familiarity with the classics resulted in a score of Greek and Roman gods, goddesses and heroes having their names attached to irises. Popular novelists, and indeed their novels, were memorialized too. More popular still was to name varieties after the heroes or heroines of popular novels or opera.

The endless search for novelty was established in the nineteenth century but became formalized with motor manufacturer Henry Ford's concept of "this year's model." Annual shows make this almost a necessity. The result is a great many plants that differ little from their predecessors. Sometimes the plant is just a peg on which to hang a name, as when charities team up with rose nurseries to introduce a plant named after them at a major show, such as Chelsea. This is an extension of a tradition of company sponsorship established as long ago as 1912 with 'The Daily Mail' rose, raised by the French firm of Pernet-Duchet.[28]

Genera with large numbers of cultivars pose a problem with naming as the range increases; whimsy turns into absurdity, as clearly illustrated with daylilies—a selection from the top of the alphabet for cultivars registered during 1998–1999 reveals: 'A Babbling Baboon's Bouncing Babies,' 'A Bodacious Pattern,' 'Alice's Day Off,' 'Art Gallery Explosion,' 'Blue-eyed Frog,' and 'Bogg's Wharf.'

The shrinking size of gardens in urban areas and the perversely opposite growth in interest in gardening have come together to stimulate development of compact-growing cultivars, but with surprisingly little outright miniaturization. 'Patio' clematis, which grow no bigger than two me-

27. Rentoul 1991.
28. Wheatcroft 1959.

ters and flower profusely, are one example of a group which has been successfully miniaturized. Interestingly, Raymond Evison, the breeder responsible, did not resort to any modern hi-tech trickery to achieve this; he went back to surviving nineteenth-century clematis hybrids, finding not only a dwarfing gene but also one for axillary as opposed to terminal bud production, so plants flower all the way up the stem. The origin of these nineteenth-century hybrids may lie farther back, in Tokugawa-era breeding, as many of the original introductions of large-flowered clematis were from Japan and were notably small growing. Evison's clematis are a powerful reminder of just how important it is that old ornamental cultivars be preserved—and indeed "national collections" of old cultivars are increasingly being organized.[29]

Rhododendrons, however, tend to be big, and when they are not in flower, funereal and dull. There are plenty of dwarf species, wiry little shrubs at home on hilltops across Europe and Asia—but they tend to lack garden impact. What the nursery industry needed was something in the middle. *R. yakushimanum* was introduced from the Japanese island of Yakushima: here was a dream set of genes, a species with wonderful, tidy dark foliage and a habit so compact that in open conditions the plant can form a perfect hemisphere. A Waterers employee, Percy Wiseman (1890–1970), had the job of crossing it with other dwarfs, and so during the 1960s, the nursery released a steady stream of what came to be known as the "yak" hybrids, named, for popular appeal, after the Seven Dwarfs: 'Sleepy,' 'Dopey,' 'Grumpy,' and so on.

Increasing adaptability to average environmental conditions is another mark of twentieth-century breeding, most clearly appreciated in orchids. Breeders have helped take orchids out of the hothouses of the wealthy into the average urban home with work on the cool-growing *Cymbidium* genus during the 1950s and 1960s—best known through their robust and long-lasting flowers being widely used in corsages, and then with *Miltonia* and *Phalaenopsis* during the last quarter of the century. *Phalaenopsis*, a genus of south East Asian rainforest origin, can flower more or less nonstop and relish the same conditions of temperature and humidity as humans—making them ideal for the centrally heated home. They and modern propagation technology have finally democratized the most aristocratic of flowers.

29. Evison 1998; Chesshire 2007. Britain has the most comprehensive range of National Plant Collections, which are overseen by the National Council for the Conservation of Plants and Gardens (http://www.nccpg.com).

THE PERFECTION OF THE UNNATURAL—ORNAMENTAL PLANT BREEDING AS ART AND FASHION

A walk around a flower show is a chance to appreciate the wide range of what the human race considers beautiful in its long history of domesticating plant life. Flowers predominate, and illustrate in their extraordinary diversity of color the human desire to achieve as wide a range of hue as possible. In form, however, there seems to be a definite convergence, certain shapes appearing again and again in dissimilar and unrelated plants; in particular it is clear that we (or most of us) love doubles.

It is clear that plant breeding is an art form; Luther Burbank clearly thought so, "we have learned that plant life is as plastic as clay in the hands of the artist. Plants can be more readily molded into more beautiful forms and colors than any sculptor can hope to equal."[30] However, neither the art world nor the horticultural world seem very keen on finding common ground. Oregon-based artist George Gessert is one of the few who has tried to find and create links, arguing that ornamental plant and animal breeding is a "genetic folk art." Gessert breeds a number of genera but has a particular love of Pacific Coast Iris[31]; as part of his work he documents breeding projects, and has presented collections of his hybrids at galleries as installations. He has also invited audiences to help him select plants, the corollary being that rejected plants are sometimes disposed of—in this way he encourages people to think about the interaction between people and plants and the dark side of our relationship to genetics, such as eugenics.

Much of what is on the stands at the flower show would never survive in the wild, doubles in particular. The double is at the core of the ornamental plant aesthetic; with its stamens turned into petals, unlike naturally double flowers (water lilies, cacti), the artificial double is often sterile or unattractive or inaccessible to bees, which makes its genetic survival completely dependent on us. Its dependency and artificiality is perhaps part of its appeal—here is a living thing that clearly bears our imprint, an alchemical amalgam of nature and culture. Like the profile of a dachshund or the curls of a poodle, doubles read "domestication." Ever since the ancient Greeks noticed double roses and the Chinese peonies, double flowers have been sought after and bred.

30. Quoted in Gessert 1993b, 205w.

31. A group of hybrids descended from closely related species of the North American west coast, including *I. douglasiana* and *I. innominata*.

There are different categories of double: formal doubles with a regular arrangement of petals, informal semidoubles, where a central boss of stamens is generally visible, and informal doubles, where a tight mass of petals forms a flower that tends towards the spherical and where any relationship with the natural shape of the flower is lost. The latter does seem particularly popular—petal-packed doubles tend to dominate among roses, carnations, camellias, peonies, and marigolds, but also crop up among a great many other genera. Additionally, there are those plants which have been bred to produce a "mop head" whereby large numbers of small individual flowers are packed into a single head—hortensia hydrangeas are a good example. Time and time again, it seems, people want flowers to form rounded, almost ball-like shapes. Generations of selection and hybridizing are used to suppress the natural flower shape.

Another common goal of breeding is ruffled petals. Plants that cannot be reduced to a formless ball of petals are bred so that they do the next best thing—open shapes with ruffled or wavy-edged petals, such as many modern daylilies and Louisiana irises.[32] It seems that in this breeding style, the natural, and often very elegant, shape of the flower is lost; this seems particularly strange in the case of irises, whose shape is very distinctive, indeed, unique. In some cases it seems almost perverse, as in the "split corona" daffodils—where the central corona is split and splayed out to form a pseudo-double effect. Presumably there is a market for the things, but I have never met anyone whose reaction to these particular daffodils is anything other than sheer hatred. Orchids, which almost uniquely among flowering plants, do not seem to throw up doubles, get a subtle variation on this treatment, most clearly seen in hybrids of the genus *Cattleya*, with prominent lips edged with ruffles. Flowers given the doubling and ruffling treatment create an effect which is like the flouncy petticoats of cancan dancers at the Moulin Rouge. It has already been noted how French breeders have been notably successful, or possibly determined, at creating whole ranges of plants which have undergone "petticoatisation"—there is a thesis here for somebody exploring issues of culture and maybe even sexuality! The flower as mistress?

Breeders of ornamental plants may have a particular obsession, but their work and passion will only become well-known if they either create a mar-

32. These are moisture-loving species bred originally from three species of the southeastern United States: *Iris brevicaulis*, *I. fulva*, and *I. nelsonii*. They are interesting in that they are amongst the very few ornamental plant genera whose breeding history is entirely American, with no Old-World input.

ket or give their market what it wants. Markets are very much influenced by fashion. When a plant is in fashion, seed company and nursery breeders tend to try to produce new varieties as often as possible, simultaneously helping satisfy the demand for novelty and creating an expectation of more novelty. The results tend to be a lot of varieties with little real distinctiveness, which then disappear as the plant species or genus goes out of fashion. It is rare, however, for entire categories to face extinction: the florist's ranunculus was greatly reduced, but is now being built up again by the nursery trade, John Salter's pyrethrum daisies have now been almost entirely lost, and *Abronia*, or sand verbenas—nineteenth-century bedding plants, have almost totally disappeared. In all cases these were short-lived plants with very particular cultural requirements. Other loses have been of varieties bred in conservatories and greenhouses—nineteenth-century gardeners grew many selections and hybrids of the sweetly scented climber *Bouvardia* under glass, along with a profusion of *Epacris* hybrids, an Australian heather-like plant with masses of tiny tubular flowers. Both are now virtually unknown in cultivation.[33]

Sometimes it is difficult to understand why a plant became popular in the first place—why did the *Epiphyllum* cactus become so popular in Germany in the 1920s and 1930s? Proliferating into a wide range of varieties; the Nazi leadership all had one named after them, but with two weeks of glory a year and fifty of looking like a piece of giant seaweed, it is no surprise that it is now rare as a house plant.

Popularity, and therefore the pressure to breed and to diversify both plant form and gene pool, is to some extent led by practical issues, but largely by cultural factors. An example of the former is the desire for low-maintenance plantings as the cost of hiring labor to maintain public landscapes and private gardens has risen from the 1950s onward—the result was a growth of interest in ground-cover conifers, shrubs and heathers, and a flurry of selection work. Cultural factors often involve a large dose of social snobbery; whereas lower-income gardeners are relatively predictable in their desires for compact and colorful plants, higher-income consumers are less so. More aware of their social standing and the need to both conform to an accepted standard and to be seen to be displaying a small amount of originality, higher-income gardeners tend to be a fickle public. A good example is provided by the dahlia. In Britain for much of the twentieth century, this exuberantly colorful flower was largely the preserve of working-class gardeners; anyone with an eye to their social standing saw it as irrepressibly vulgar.

33. Kingsbury 1991a, 1991b.

But once garden guru Christopher Lloyd (1921–2006), the doyen of upper-middle class British garden society, decided he liked dahlias and planted them in his garden, the popularity of the plant in Britain soared. Gardeners in the United States were heavily influenced by Lloyd, too. Increased sales soon laid to a flurry of breeding activity. The groundwork for the dahlia renaissance had actually been led (in Britain at any rate) by the 'Bishop of Llandaff,' not the cleric, but a highly distinctive dahlia of that name—deep red leaves and single scarlet flowers, which for some unaccountable reason (the name?) was socially acceptable. Needless to say, the bishop was soon used in breeding efforts, and a range of "Bishop's children" and even "Bishop's bastards" began to circulate among the horticulturally well-connected.

Within a genus, fashion too plays a part. The Ottomans liked tulips with attenuated, almost dagger-like, petals, whereas seventeenth-century Europeans preferred bowl-shaped flowers; among roses, early-nineteenth-century French hybrids were "quartered"—densely packed with petals in a bowl-like flower; shapes became looser as the century wore on, and with the arrival of the hybrid tea, the "perfect" rose shape for a long time was this class's distinct pointed bloom. By the later twentieth century, however, there were complaints that these roses, indeed, all modern roses, had no scent. Harry Wheatcroft was scathing, that roses were now scentless was a claim that he described as being "regularly trotted out by people who should know better . . . a perennial complaint . . . like sighs for the good old days."[34] His sneer was remarkably prescient, for it warned us that gardening was about to enter into a period where nostalgia was about to become very important indeed. It might have been aimed at Englishman Graham Stuart Thomas (1909–2003), who in 1955 had published *The Old Shrub Roses* and had embarked on a mission to rescue varieties in danger of extinction. The "old roses" then began a long climb to popularity, largely through high-income trendsetters, many associated with the British aristocracy, for whom nostalgia is second nature. That many were often poor garden plants, usually only flowered once and were available in only a limited range of colors, did limit their popularity. Until, in the dialectical process so often seen in ornamental plant breeding, a British rose grower, David Austin, hit upon the idea of crossing them with modern roses. The results, released from the 1970s onward, were dubbed "English roses" and combine the best of old and new.

It is clear from the preceding discussion that opinions about the aesthetic merits of ornamental plants vary and that opinions are often linked to social

34. Wheatcroft 1959, 205.

position. Colorful summer bedding plants have always tended to attract the strongest accusations of vulgarity, a reflection perhaps of their continued commercial success. This success, and the major investments put into breeding new strains and periodic attempts to widen the gene pool, have been more or less continuous since their initial impact in mid-nineteenth-century Britain and France. During the nineteenth century they also attracted some sustained opposition, expressed in both the gardening journals and in fiction.

Dislike of attempts to improve nature have been noted earlier (see chapter 4). The seventeenth-century metaphysical poet Andrew Marvell (1621–1678) foresaw many nineteenth-century objections to hybrid exotics in his poem "The Mower, Against Gardens." He regarded "improved" flower varieties and foreign introductions as artificial and immoral; the double pink he likened to hypocrisy of the mind, while colored tulips were compared to women with painted faces (by implication, "loose"). The poem stands as a condemnation of just about every contemporary human intervention in the supposedly unsullied world of God's creation, and is worth quoting in full:

LUXURIOUS man, to bring his vice in use,
Did after him the world seduce,
And from the fields the flowers and plants allure,
Where Nature was most plain and pure.
He first enclosed within the gardens square
A dead and standing pool of air,
And a more luscious earth for them did knead,
Which stupefied them while it fed.
The pink grew then as double as his mind ;
The nutriment did change the kind.
With strange perfumes he did the roses taint ;
And flowers themselves were taught to paint.
The tulip white did for complexion seek,
And learned to interline its cheek ;
Its onion root they then so high did hold,
That one was for a meadow sold :
Another world was searched through oceans new,
To find the marvel of Peru ;
And yet these rarities might be allowed

To man, that sovereign thing and proud,
Had he not dealt between the bark and tree,
Forbidden mixtures there to see.
No plant now knew the stock from which it came ;
He grafts upon the wild the tame,
That the uncertain and adulterate fruit
Might put the palate in dispute.
His green seraglio has its eunuchs too,
Lest any tyrant him outdo ;
And in the cherry he does Nature vex,
To procreate without a sex.
'Tis all enforced, the fountain and the grot,
While the sweet fields do lie forgot,
Where willing Nature does to all dispense
A wild and fragrant innocence ;
And fauns and fairies do the meadows till
More by their presence than their skill.
Their statues polished by some ancient hand,
May to adorn the gardens stand ;
But, howsoe'er the figures do excel,
The Gods themselves with us do dwell.[35]

Marvell's opinions were echoed by John Ruskin (1819–1900), writing two hundred years later, "a garden is an ugly thing, even when best managed: it is an assembly of unfortunate beings, pampered and bloated above their natural size, stewed and heated into diseased growth; corrupted by evil communication into speckled and inharmonious colors."[36] Ruskin's comments on bedding plants have been echoed ever since their introduction in the 1850s: that they were without scent, artificial, and vulgar. Their opponents instead praised the "traditional" hardy annuals to be found in the gardens of simple country people. There was a definite moral overtone; "old" annuals symbolized unpretentious old-fashioned moral values, "new" ones were brash and associated with industry and the greedy new middle classes. "Old" flowers were associated with a disappearing social order, "new" ones with its replacement by a new world of uncertainty and new values. "Artifi-

35. Marvell 1681.
36. Quoted in Gessert 1996, 294–95.

cial" was a term often used to describe bedding exotica, which referred to both the way it was produced (in heated greenhouses) and its hybrid origins. Ruskin fulminated against the products of "vicious science."[37]

By the twentieth century, opposition to bedding plants had largely died down, but a new generation of sophisticated gardeners were tackling questions of planting design and had to form opinions about plants—and a growing gardening public wanted to know what they thought; apart from anything else, they wanted to be seen doing "the right thing." The immensely influential Gertrude Jekyll rejected most hybrids as aesthetically inferior, although she was not against them per se. Double roses she thought beautiful, whereas she considered the double canterbury bells (*Campanula medium*) "confused and disfigured" and the double Madonna lily (*Lilium candidum*) "wretched and misshapen."[38] For the aristocratic Vita Sackville-West (1892–1962), opinions on double flowers were grounded in pure snobbery—old roses and old peonies were fine, because they were from old aristocratic civilizations, most others she treated as bourgeois upstarts. None of these writers or indeed anybody else has ever argued for a standard of aesthetics for plants which is anything other than a set of judgments made a priori. Historically, the standards of beauty used to grade auriculas, irises, roses, and everything else are simply stated—never argued for or justified.

Romantic unease about bedding plants is echoed today in a number of movements within horticulture. In the United States, "grandmother's" or "heirloom" plants are immensely popular, as are old vegetable varieties, and are often linked to organic or "natural" gardening—it is possible that this need to look back to simpler times is a common phenomena to be found in times of social and technological change. The past is idealized and romanticized, its poverty, oppression, and lack of opportunity edited out. Given that the vast majority of the population cannot or do not wish to opt out of modern capitalist society, symbolic alternatives are sought—the growing of grandmother plants in the garden is one such. Many of the plants sold as "heirloom" varieties today may well have been plants that nineteenth-century romantics objected to; today's romantics' love of them perhaps says more about their attitude to modernity than about floral aesthetics. The fact is that today, the garden is largely seen as a place in which to retreat from stress and turmoil—a place in which to keep the modern world at bay. In any case, variety improvement and running a business never get in the way

37. Quoted in Waters 1988, 120.

38. Gessert 1996.

of strict definitions and some dealers in heirloom plants have also seen fit to create new varieties too, calling them "created heirlooms."

Less directly oppositional and more creative, is the move toward using either native plants or "naturalistic planting."[39] Varieties used are either natural species or selections from natural species. However, the breeding dialectic can be seen here at work as well—in seeking "naturalness" pioneer natural garden practitioners use wild species as opposed to hybrids; however, the aesthetics of wild species inevitably do not satisfy a great many gardeners, and so a process of improvement inevitably begins. The synthesis is that a great many new plants, mostly herbaceous perennials, but also grasses and sedges (*Carex* species), are currently being brought into the world of plant breeding. The 1990s onward have been an immensely fertile time for this kind of ornamental plant breeding, mostly using simple selection techniques. The message is clear, we cannot leave nature alone.

SUMMARY

Plant selection for ornament is immensely old, but closely linked to a society having a certain level of disposable income. A fashion for tulips in the seventeenth-century Netherlands, resulting in a period of manic speculation, was a turning point—from then on there was increasing investment in plant breeding for ornament. Much early hybridizing work and explorations of heredity was undertaken using ornamental plants. Selection and breeding work expanded enormously in the nineteenth century, using as raw materials the vast number of species being introduced into Europe and North America. New areas of horticulture demanded selection and breeding work, notably public bedding out and the needs of expanding urban populations with small gardens.

Ornamental plant breeding in the twentieth century continued to work very largely with many of the genera included in late nineteenth century breeding programs. Rates of investment and technological input have reflected the commercial importance of different plant genera; in addition to much hi-tech breeding, there is also a thriving specialist nursery trade which uses traditional methods with less commercially important species. Aesthetics and ease of cultivation have tended to drive ornamental plant breeding. Fashion has always been very important, but certain flower groups, colors, and shapes do seem to be always popular.

39. Kingsbury 2003.

FOURTEEN ❦ OWNERSHIP AND DIVERSITY ❦ ISSUES OF PROPERTY RIGHTS OVER PLANT GENETIC RESOURCES

In the summer of 1968, corn farmers in Illinois and Iowa started to notice a new disease affecting their crops—lesions were gradually spreading to kill whole leaves. The same happened to a few more farmers the following year, but in the autumn of 1970, the disease, southern corn blight (*Helminthosporium maydis*), was well and truly on the rampage, fed by a particularly wet season. Field after field in the eastern and southern states were flattened as weakened plants collapsed. "There was a sense of panic," reported Stan Jensen, a retired breeder of the crop. The Chicago Board of Trade reported record sales as commodity traders frantically tried to stockpile all the corn they could. By the end of the year, the national yield was down by around 40 percent. Jensen who was employed by Pioneer at the time, recalled, "they [Pioneer] and other seed companies were threatened by class actions which, if they had been won, would have bankrupted them."[1]

Those concerned with genetic diversity were not surprised. In 1948, Minnesota farmers could choose from over six hundred different varieties of corn; in 1969, 71 percent of total U.S. corn acreage was planted with six hybrids. At the same time, 96 percent of the peas produced in America were from two varieties, and 69 percent of the sweet potatoes were one variety.[2] The Achilles heel of crops at the time was often a narrow genetic base. Hosts and parasites (all pathogens can be described as parasites) are in a constant state of balance, which in nature is constantly shifting as both evolve—the host to resist and the parasite to exploit. If a parasite establishes in a field of a crop its progeny will, through natural selection, tend to be those which can exploit the weaknesses of the crop; if this crop is almost genetically identical, these weaknesses are exposed to an onslaught, which is potentially much greater than that to which a genetically diverse landrace is exposed. Now imagine what happens when a whole country grows only a few varieties of a crop.

Such an onslaught happened when southern corn blight hit the United

1. Jensen 2007.
2. Kloppenburg 1988.

States. Corn breeding had become dependent upon crosses made with female parents which carried a factor for male sterility in the cytoplasm—essential to enable controlled cross breeding to take place (see chapter 11). Unfortunately the male-sterility factor in the Texas CMS strain used was linked to a susceptibility to particular strains of the southern corn fungus. Filipino scientists had spotted this vulnerability some years before, but their warnings had not reached the United States. These strains of the disease then multiplied rapidly when faced with acre upon acre of corn of genetically identical corn. It is to the credit of the U.S. plant breeding and corn industries that substitute male-sterile varieties could be rapidly mobilized to breed new, more resistant varieties. In most cases, these were preexisting varieties, but, in a similar scheme to Norman Borlaug's shuttle breeding of wheat in Mexico, the seed companies were able to exploit the reversal of the seasons south of the equator, growing seed crops in South America. There was also a move to go back to detasseling rather than using male-sterile variants. It is impressive how quickly the problem was solved, part of the reason being that private companies were in a good position to react, as they had much better stocks of seed of different varieties than public institutions.[3]

The story of southern corn blight brings together several of the key issues in plant breeding history since the Second World War—issues concerned essentially with *diversity* and *ownership*. Crop plants have traditionally exhibited huge diversity, the result of a wealth of genetic material developed by farmers over thousands of years, but who did this material belong to? And what would happen to it if it were displaced by modern varieties? Crop biodiversity, its evaluation, preservation, and use, has been an important issue for much of the twentieth century and has been much written about; here we will simply look at the issue from the point of view of the plant breeder. Another issue that has been extensively written about is the ownership of varieties in general, in particular, the very controversial area of patenting plant varieties; indeed this has been one of the most fiercely disputed areas in plant technology. Ownership of varieties is closely linked to the question of who does breeding: farmers, publicly funded bodies such as universities or state experimental farms, private companies, or international agencies? Again, the intention here is only to give a broad outline of issues covered in greater detail elsewhere.

3. R. Murphy 2006.

USE IT, SAVE IT, OR LOSE IT! GENE BANKS

During the 1930s, the government of Turkey funded an agronomist, Mirza Gökgöl, to make a huge collection of Turkish wheats, with more than thirty-six thousand accessions. Turkey's range of wheat races was so vast he considered that the country would never need to look anywhere else for genetic material. Tragically, his collection was abandoned and then destroyed in the 1960s.[4]

It was to prevent any more disasters such as this that the creation of seed banks under international control became a priority in the 1970s. The Green Revolution may have fed many more people, but it came at a price for the crops that farmers used to grow before they turned to HYVs. As the compact and productive new varieties spread, traditional varieties and landraces disappeared with remarkable and frightening speed. This was happening at a time when breeders were increasingly aware of the importance of these "uneconomic" varieties—for among them could be the genes for a wide range of highly desirable qualities: pest and disease resistance, drought tolerance, dwarfism, higher vitamin content, and so on. In 1965, in response to an increasing level of concern over disappearing landraces, the FAO established the Crop and Ecology and Genetic Resources Unit to promote the formation of seed banks and other propagating material, in what has become to be known as gene banks.[5] The next stage was the formation of the International Board for Plant Genetic Resources (IBPGR) in 1974 within the framework of the Consultative Group on International Research (CGIAR), which over the next ten years guided rescue collecting worldwide. A great deal of material was gathered, with IBPGR helping to provide common protocols for collecting and storing material and recording the accessions.

The idea of collecting varieties of useful plants was not new, Wahlberg had done it in Moravia in the late eighteenth century (see chapter 4), Tschermak-Seysenegg had assembled collections of crops in Austria, and Vavilov had made it his life's work. Collecting and classifying corn varieties had also been part of the Rockefeller Foundation's brief in Mexico—a wise move on the part of the Foundation and its collaborators, as it had resulted in the securing of genes simultaneously with the development of new crops. In the United States the first seed banks had only been established in the 1940s.

The 1920s and 1930s saw a growing interest in the collection of landraces,

4. Braun et al. 2001.

5. See technical note 35: Gene Banks.

as Mendelian genetics directed the attention of breeders towards sources of a wide range of characteristics, as well as genes that might be potentially useful but remained hidden by being recessive. As late as the 1920s, major new crops had resulted—soy, for example, valuable for the U.S. South as a replacement for cotton that was increasingly being affected by disease. Starting in 1909, the USDA had imported a small number of varieties from China, Japan, and India, followed by collectors being sent to Asia during 1929–1931, some going as far north as Sakhalin, and gathering over three thousand soy varieties.

Controversy has always dogged the plant genetic resources movement. Points of concern revolve around ownership and control of the material included, and the uses to which it is put. Politics can interfere, as when the U.S. government embargoes the supply of seed to countries (such as Syria or Cuba) with which it has political differences. More fundamental, however, are attitudes to the loss of landraces in their regions of origin. This issue was flagged up early, by Dr. Carl Sauer, who had worked with the Rockefeller Foundation on its Mexican Agricultural Program. In 1941, he warned that:

> A good aggressive bunch of American agronomists and plant breeders could ruin the native resources for good and all by pushing their American stocks. And Mexican agriculture cannot be pointed toward standardization on a few commercial types without upsetting native culture and economy hopelessly. The example of Iowa is about the most dangerous of all for Mexico. Unless the Americans understand that, they'd better keep out of this country entirely.[6]

The warnings of Sauer and those of his colleague, botanist Edgar Anderson, were angrily rejected by Paul Mangelsdorf, one of the three plant scientists sent into Mexico to make the original survey for the Rockefeller foundation, and arguably the very sort of "good, aggressive plant breeder" who so alarmed Sauer. At a meeting of foundation officers, he attacked the arguments of Sauer and Anderson:

> If the [Rockefeller Foundation] program does not succeed, it will not only have represented a colossal waste of money, but will probably have done the Mexicans more harm than good. If it does "succeed it will mean the disappearance of many ancient Mexican varieties of corn and other crops and perhaps the destruction of many picturesque folk ways, which

6. Quoted by Kloppenburg 1988, 162.

are of great interest to the anthropologist." In other words, to both Anderson and Sauer, Mexico is a kind of glorified anthill which they are in the process of studying. They resent any attempt to "improve" the ants. They much prefer to study them as they now are.[7]

The establishment of the IBPGR was itself the outcome of a tug of war over gene banks which had been going on for some years, with much political rhetoric being let loose on both sides. Many developing nations wanted to keep the gene banks under the loose control of the FAO. The industrialized nations, led by United States, wanted a more centralized system, independent of the political pressures that might be exerted by FAO members. In the end a compromise was agreed, that an independent organization, the IBPGR, would be set up at FAO headquarters, but independent of it and linked to the CGIAR. Since the latter was the body, closely linked to the World Bank, which oversaw the work of all the international crop development agencies (CIMMYT, IRRI, etc.), critics claimed that this was an attempt by the industrialized nations, driven by their corporations, to ensure that world seed resources would be made available for commercial plant breeding. The deal did indeed have the fingerprints of U.S. interests all over it—not only did the World Bank play a crucial role, its director at the time was Robert McNamara, who simultaneously ran the Ford Foundation and had been Secretary of Defense under President Kennedy. In 2006, IBPGR followed corporate fashion and rebranded itself—as Bioversity International.

Some gene banks took over existing collections and often rejuvenated neglected ones, but sometimes only after considerable political pressure. The most contested example was in India in what is now the state of Chhattisgarh (formerly part of Madhya Pradesh). It is worth noting that the state is poor and marginalized, with a high population of tribal people, and during the mid-2000s, it had the worst Naxalite insurgency problem in the country. It was proposed that IRRI take over the rice collection held at the Central Rice Research Institute in Cuttack, much of it built up over the years by its director, Dr. R. H. Richharia, renowned as an independently minded researcher. Richharia had already been engaged in a long-running dispute with IRRI—he and several colleagues believed that IRRI varieties were infested with viruses that they wanted to keep out of central India. In 1966, he was removed after he resisted handing over the institute's collection of rice

7. Quoted in ibid., 302.

seed to IRRI, and for his opposition to the introduction of IRRI-bred HYVs. Several other scientists who disagreed with the IRRI approach were forcibly retired. Richharia went on to the Madhya Pradesh Rice Research Institute (MPRRI) at Raipur, were he worked with indigenous varieties that he claimed were yielding as much as some of IRRI's. He also discovered some new dwarf varieties and developed a method of clonal propagation for mass-producing them; the result of which he believed was that he could produce rice better for local conditions than IRRI could. The MPRRI refused to share their new material with IRRI, who then apparently, with the connivance of the World Bank (responsible for funding) engineered the closing down of MPRRI. Richharia had all his notes and research material confiscated on both the occasions he was sacked.

The story continued, and indeed it is highly likely that we have not heard the last of Dr. Richharia's rice collection. In 2003 the Swiss-based company Syngenta agreed to a deal with the Indira Gandhi Krishi Vishwavidyalaya, an agricultural university in Chhattisgarh, which now held the seed bank, that would have given them access to the repository of 22,972 varieties of rice germplasm. A number of organizations came together to launch the Chhattisgarh Seed Satyagrah,[8] and protests spread rapidly across the state, forcing the company and university to call off the deal.

The Richharia affair highlights one of the tensions between the priorities of organizations working on a global scale and institutions that are locally rooted and believe that control is best kept at a local level. For many involved in the opposition to corporate involvement in agriculture, the key word is "sovereignty," which encapsulates the idea that communities have rights over the genetic resources that they have built up through generations of community labor and wisdom. Demands that communities be paid for the collection of their plant varieties are usually closely linked to long histories of exploitation and marginalization.[9] Agriculturally and economically successful communities, and those not politically marginalized, have shown far less interest in declaring sovereignty over their historic varieties, or indeed in preserving them.

Anger by marginalized people over what they see as the theft of their resources feeds radical critics of globalization in both the developing and the industrialized world. This critique can be summarized as follows: mod-

8. *Satyagraha* means nonviolent resistance.

9. See Pistorius and Van Wijk 1999 for an example of their work with the Mapuche people in Chile.

ern plant breeding destroys its own genetic base by eliminating landraces, it is a south-to-north "gene drain," and multinational companies based in rich countries are taking control of the breeding industry through patenting plants and preventing farmers from saving their own seed of commercially available varieties. However, there has been a contradiction in the stance taken by those who have stood up for the rights of growers of traditional landraces. During the 1990s they were demanding that such crops were the common heritage of humanity, and so should be included in seed banks protected by international statute. By the early 2000s there was a shift towards claiming "ownership" of these crop varieties by the communities which grew them, and claims of "biopiracy" being made against the CGIAR agencies.

One constructive outcome of anticorporate political activism has been the construction of grassroots gene banks and seed distribution systems. All that a small-scale gene bank needs is a fridge, some airtight plastic boxes, a package of silica gel dehydrating agent (easily found in the packaging of electronic equipment), and some envelopes. There is no doubt something very empowering for communities about the ability to achieve long-term seed storage for heritage varieties, but in the final analysis it is the uses to which the seed will be put that will be their test. Merely conserving the old without making an attempt to adapt and change in order to face the future, risks tying a simple and imaginative idea to the limitations of gesture politics.[10]

FOR THE PUBLIC GOOD—PLANT BREEDING BY THE TAXPAYER

Our survey of plant breeding history has revealed a variety of types of ownership of the institutions that have undertaken crop improvement. Sweden had a strong initial private input, which led to the formation of the Svalöf Institute, which then became a part of a state system. In Britain, too, private breeders and seed companies did much of the important work in the late nineteenth and early twentieth centuries. Private and amateur breeders were particularly strong in the Netherlands, and private companies were crucial for wheat breeding in Poland. It is somewhat ironic that publicly funded breeding was strongest in what we are used to thinking of as the homeland of private enterprise—the United States, but circumstances were totally different—farmers and growers were facing vast new acreages with

10. See AScribe Newswire 2002 for an example from Mexico.

unfamiliar climates and required considerable support, and the state recognized its responsibilities—farming had to work if the young nation was to feed itself and build up an export trade.

Over time, the balance between public and private breeding has shifted. The aim here is to give a broad outline of these movements, the reasons for them, and how they affected breeding. It is important to recognize though, that breeding has not just been carried out by privately owned corporations or government-owned or funded bodies such as state experimental farms or universities. Since the end of World War II, international agencies have played a major role in breeding new crops for the developing world—the role of these will be considered later. There is a minor role played by NGOs, which could, if the political will were there, play a much larger role in the future in developing countries. Finally there is the "one-man band" (although as likely as not to be a woman)—individual breeders, important in the earlier part of our story but now almost an extinct species in agriculture in the industrialized world. In ornamental horticulture they are still very much alive and play an economically important role (see chapter 13). In the developing world individual breeders still do have a role, although often unnoticed and unappreciated beyond their local communities.[11] As a general rule, small-scale breeding cannot be expected to remain viable once a crop achieves commercial importance. Once gold is spotted, large-scale commercial interests inevitably move in—the result may be greater resources available for breeding programs, but risk taking or daring innovation becomes much less likely.

Recognizing the importance of agriculture to the settlement and hence to the success of the young nation, the U.S. government took the key role in plant introduction and breeding until well after World War II. Introduction was organized centrally through the U.S. Patent Office, with free seeds of new crops and varieties made available to anyone who wanted to try them. U.S. consuls and the navy had a particular role in plant collecting, Commander Perry's gunboats, for example, brought back material from Japan, Java, Mauritius, and South Africa. Generating new varieties was initially in the hands of farmers, but with the Morrill and Hatch acts (see chapter 6) publicly funded bodies took control of breeding.

From 1910 to 1925, breeders' outlooks were changed by the enthusiastic

11. See http://www.masipag.org/breeding.htm for one of the few examples of "community breeding," from the Philippines. One suspects that there are many more, though unpublicized.

adoption of Mendelian ideas. Seeing plants less as whole organisms than in terms of collections of genes, the idea developed that specific traits could be imported, or sought out. This had an impact on USDA-backed plant collection, as K. A. Ryerson, a USDA plant explorer noted, "species and varieties which in themselves have little or no intrinsic value become of first importance if they possess certain desirable characteristics which may be transmitted through breeding."[12] Plant hunting was no longer about species, but genes. This resulted in far greater efforts to collect a wide variety of landraces and an increasing stress on provenance—the exact geographical origin of collections. Until the 1930s, new varieties did little to raise production; instead breeding was closely linked to introduction, new introductions needing selection work in order to find varieties which would flourish in the conditions of the United States. Railroad companies, whose silver trails tied together the new nation, played a role in promoting agriculture, too; a special feature of the period of 1900–1920 was the demonstration trains, such as the "seed corn specials" in Iowa, which among other things promoted new corn varieties.

By the mid-nineteenth century, seed companies were becoming more visible and more sophisticated in their marketing but still catered to a clientele largely composed of private gardeners and market garden growers of vegetables. As urban areas expanded, so did the ring of market gardens (or truck farms) around each metropolis; increasingly, these growers bought seed from seed companies rather than saving their own. Mislabeling and adulteration were common, leading the USDA to start testing in 1905. Those companies, such as Burpees, who did engage in breeding, rather than selling existing varieties or contracting out the production of existing varieties to farmers, concentrated on vegetables; there was little corporate impact on larger scale crops until the era of F_1 hybridization began to change the rules of the game from the 1930s on. In fact, the seed companies were effectively marginalized from much research and production by the extent, quality, and reputation of the state experimental stations—whose seed bags were denoted by a blue tag. Commercial seedsmen were not happy with the fact that growers trusted "blue tag" seed above all others. They also objected to the generous government distribution of free seed of new introductions.

The exploitation of F_1 hybrids may have strengthened the corporate breeding sector in the long–term, but the post–World War II period was one of a state-directed industrialization of agriculture in the United States

12. Quoted in Kloppenburg 1988, 80.

and western Europe, with governments becoming deeply involved in its organization and high levels of government money being plowed into research. These developments need to be seen in the context of this period being the high water mark of state involvement in the economies of the West—social democratic policies in western Europe and the clear success of Roosevelt's New Deal in the United States led to mainstream politicians everywhere becoming strongly attracted to the concept of planning.

In Britain, there had been a flurry of establishment of plant breeding institutions between 1910 and 1921, again at a time of state intervention in the economy; even so, only 40 percent of the funds for plant breeding in Britain came from public sources during this period. Little was produced during the 1930s, possibly because the breeding institutions had lost touch with farmers, and became too academic. The Plant Breeding Institute (PBI) had been well funded by the taxpayers—and maybe it was because of this that it could afford to ignore its constituency. Its products were sometimes failures—two wheats, 'Steadfast' and 'Holdfast,' were rejected by the milling industry. Postwar state intervention though saw a major boost to the PBI at Cambridge, its lands increased from 21 to 380 acres. New investment sometimes took a long time to yield results, but by the 1970s, PBI-bred wheats dominated British production; their yields were unprecedented, allowing the country to grow more of its own wheat than ever before. By the 1960s potatoes bred by the Scottish PBI slowly displaced the myriad varieties originally bred in the "potato boom" of the early years of the century.

What had happened in Britain had been an effective nationalization of the plant-breeding industry. Government investment had been for reasons of import substitution. The dominance of the state sector had quite possibly prevented private companies from developing, with private crop breeding and seed production largely limited to the vegetable sector and old established companies like Suttons, Dobies, and Carters either breeding themselves or by buying the rights from independent small breeders or even from individuals.

In the Netherlands, an institute of plant breeding had been set up in Wageningen in 1912, which went on to become one of the world's leading such institutions, concentrating on cereals, fodder, and potatoes. Small seed firms, hobby breeders, and farmer cooperatives played a very important part, however, possibly because there was some level of legal variety protection. Potatoes were a particular success, and have continued to be so—many of the best varieties of the latter quarter of the twentieth century were Dutch: 'Estima,' 'Record,' 'Wilja,' and 'Marfona.' One man's name is particularly linked to the potato, Geert Veenhuizen (1857–1930), who started in the northern region of Groningen, breeding for both food and industry. Apparently he originally thought of the idea of breeding potatoes after reading about the crossing of bromeliads in a gardening magazine (see chapter 13). He created dozens of varieties, some of which, including 'Eigenheimer' and 'Paul Kruger,' were exported all over world. The centenary of his birth on November 18, 1957, was celebrated with an illustrated thirty-two page book distributed to schools and colleges all over the country—if only now a plant breeder could be held in such high esteem by his countrymen!

In addition to Wageningen, an additional state-supported institution was established in the postwar period: the Foundation for Agricultural Plant Breeding (Stichting voor Plantenveredeling; SVP), founded in 1948 to support and advise private breeders. Its staff distributed breeding material and conducted research that was regarded as too financially risky for private breeders. But only private breeders were allowed to produce and market varieties gained from its work.

WHOEVER OWNS THE SEED OWNS THE CROP—PRIVATE AND CORPORATE BREEDING

F_1 hybrid seed, which made growers come back to the seed merchant every year, created what was almost a point of no return for the industry; from that point on the division would grow between their work and that of the publicly funded university and state experimental farms. From 1934 to 1944, seed corn sales went from almost nothing to $70 million, and from 1940 to 1950, they tripled. Corn yields doubled in the twenty years after 1935.[13] Not surprisingly the seed companies did very well, but what was their relationship with the public bodies from which they had drawn the genetic material which made their success possible?

This question was succinctly answered by none other than Henry A. Wallace, the complex, baffling, but undeniably honorable man who had launched the hybrid seed business, acting as a midwife to what some might see as one of the most barefaced rackets that capitalism has launched onto the world, but who had also been the highest placed radical of the Left that the United States had ever had. In 1955, Wallace addressed a meeting of the hybrid corn division of the American Seed Trade Association; he summed up what has become the essence of the division of labor between public and

13. Kloppenburg 1988.

private in genetics and plant breeding—that universities and government-owned and funded research stations do pure research while private companies produce the goods, the varieties that the nation's farmers would sow. He presented the division of labor as an example of cooperation, but it has never been this simple or benign.

During the 1920s and 1930s three seed companies: Funk Brothers, DeKalb Agricultural Association, and Pfister Hybrid Seed, all began to develop seed programs that appeared to duplicate much of what the state experimental stations were doing. It was in the production of hybrids that the real distinction between public and private was made, as the experimental stations continued to promote traditional methods of variety improvement, largely selection. With hybrids, the companies began to rival publicly funded bodies as sources of innovation and scientific authority, Funk Brothers especially appeared to mimic a state experimental station: they undertook trials, held yield contests, and published the results of their extensive scientific work. Relationships between the two sectors were close, with frequent contact between company directors and leading publicly funded researchers. One of the closest relationships was that which developed between the Funk family and James Holbert; as we have seen, this started with that classic exercise in bonding—the conversation that goes into the small hours (see chapter 10). Funk Brothers must have gained an enormous amount of prestige and profit from the links to the federal government and the University of Illinois that Holbert provided.

As hybrids took over the fields of the United States during the 1930s and 1940s, the standing of the companies in the eyes of much of the farming community grew too; that of the colleges and experimental stations did not necessarily decline, but they were no longer seen as the sources of all agricultural wisdom. Whereas once anyone who traveled the roads selling seeds from place to place was liable to be seen as no one more than a peddler or possibly a fraudster with a pack full of mustard seed impersonating cabbage and cauliflower, now the traveling seed "rep" increasingly came to be seen as having scientifically based knowledge. Because his knowledge was seen as being rooted in science, his company's products gained credibility.

As hybrid corn took over, it became increasingly difficult for state farms to compete, as their open-pollinated pedigree lines could not produce as much as the hybrids. At taxpayers' expense, any inbred lines they produced were snapped up and used by the companies—leading to them gradually dropping out of this work. By 1958 the USDA had stopped inbred line development altogether. Wallace's recognition of a "division of labor" between public and private has been largely adhered to by industry representatives ever since. To be blunt, the public sector does the donkeywork, the private earns the money; the public sector effectively subsidizes the private. Basic research increasingly became the province of universities and state farms, covering anything that could not be turned into a marketable commodity: making the advances in genetics research, trial plot design, statistical analysis, and new breeding techniques. Public sector workers did the unremunerative and often laborious work with landraces, separating out those characteristics regarded as desirable from those that were not, to produce elite lines which could then be used in the private sector. The companies applied the research and produced commodities which earned them income.

During the 1950s, pressure grew from the seed industry for public funding to limit itself to basic research. The companies were particularly annoyed by state experimental stations publishing lists of recommended varieties; such decisions, they thought, should be left to the consumers—and the companies' advertising departments. The division of labor that Wallace outlined was gradually becoming entrenched.

Developing highly remunerative crops was not the only route to commercial success. Sometimes breeders just walked out of a public body and set up on their own. One such was Jack Dunnet, who had worked at the Scottish Plant Breeding Station. His decision to go it alone had a long history. Disease resistance was the main issue the station had worked on since almost the beginning of its existence, arguably at the cost of work on yield—indeed, yield statistics were not kept for many years. In 1969, Norman Simmonds, then station director, published a paper in which he stated that potato breeders were failing to keep up with cereal breeders in the improvement of yield. This was, he said, because of their narrow genetic base, still too constrained by the heritage of the limited range of clones of 'Andigena' varieties that had been imported in the sixteenth century. Simmonds had grown an entire collection of Andigena varieties in a field; having obtained them from the British Commonwealth Potato Collection, and with the help of the bees hugely expanded the genetic base of the crop. Selection, he maintained, could be started again from scratch on this basis. Not everyone agreed, among them Dunnett, who eventually resigned and in 1980 went private, setting up Caithness Potatoes and concentrating on the development of pedigree lines. He could also, he said, "wash my own flowerpots, think about potatoes, and listen to opera at the same time." Since one of Dunnett's varieties, 'Nadine,' was by 2000 selling more than any variety introduced from a British state-owned breeding station for over sixteen years, it seemed that he was vindicated.[14]

Since publicly funded breeding regards the varieties it breeds as part of a global commons, there is no need for commercial secrecy—but as soon breeding began to go private, researchers began to clam up. Breeders in the United States began to become secretive about their inbred lines, and in early 1940s allegations started cropping up that some companies were marketing publicly bred varieties as their own creations under new names. Cooperation between state and company began to break down, with some state experimental farms beginning to curtail contacts and some states instituting registration requirements for new commercial varieties. The new companies also began to do what had been common practice among seed companies dealing with vegetables for a long time—taking someone else's variety, doing some minimal breeding to make it slightly different, and the remarketing it. W. R. Little, a seed department manager for the Tennessee Farmers Cooperative, complained in 1958 that what many large companies were doing was "'borrowing'" what has been developed by USDA and experiment station plant breeders, "adding a little private stock in some instances, slapping a fancy label on it, mapping out a Madison Avenue advertising program for it, and putting it on the market."[15]

The tipping point where private overtook public breeding was the 1980s. Politically, this was a time when the neoliberal economics of Ronald Reagan and Margaret Thatcher held sway over the world's most powerful economies. It was also the time when companies who dominated the agrochemical market, for example, Du Pont and Monsanto, judged the seed and plant breeding market to be profitable enough to try to take over. With the growing emergence of biotechnology as a possible future for the industry, these companies' R&D departments were also in a strong position to undertake the (very expensive) research that would usher in a new era in crop breeding. The agrochemical companies were also under pressure, both economic and political, to reduce the use of environmentally damaging pesticides; biotechnology offered an alternative to chemicals, as it adds a whole new suite of tools to breeders working on pest and disease resistance, specifically, the opportunity to transfer single genes responsible for resistance between species. An example is the very successful Bt cotton, where a bacterial gene

14. Dunnet 2000, 11.

15. Quoted in Kloppenburg 1988, 136.

responsible for killing caterpillars is inserted into the crop's genome. This particular technology, transgenics, popularly known as genetic modification, also offered enormously exciting new opportunities for specialty crops, including crops bred for specific chemical contents for nutritional, medical, or industrial reasons. The fact that breeding was set to become ever more dependent on laboratory work not only coincided with a wholesale corporate takeover but also appeared to give it a rationale.

Private-sector R&D investment in agriculture in the United States increased from $2 billion in 1970 to $4.2 billion in 1996, with public funding staying stagnant at around $2.5 billion since 1978.[16] From being a fragmented industry, seed production and plant breeding became highly concentrated, but as part of mega-companies who increasingly became vertically integrated. By 1980, eight companies had 72 percent of the seed market in corn[17]; by 1997, Du Pont had 40 percent and Monsanto 20 percent.[18] It was not just agrochemical companies who wanted to invest in seeds and breeding. That there was money in those fields was illustrated by the alacrity with which companies who had previously showed no interest in agriculture were ready to invest—the Swedish car company Volvo bought forty-seven seed companies. These mega-companies typically form a highly complex network of subsidiaries, their plant breeding and genetics often the result of collaboration with particular universities. Working with academic institutions enables them to have access to publicly funded science, the global commons of academic research and very often seed banks. That chemical companies such as Du Pont and Monsanto have bought up plant breeding resources has, in the eyes of some, resulted in a cynical manipulation of breeding programs. Rather than breed fungus-resistant cereals, for example, it has been suggested that they would make more money selling fungicide to farmers—so breeding for resistance is not always vigorously pursued.[19]

Parallel with this corporate takeover has been a wholesale abandonment of plant breeding in the public sphere. In Britain, this took the form of effectively privatizing the PBI, selling it in 1987 to corporate giant Unilever, who then sold it to Monsanto in 1998. This was not for lack of success—at the time of its sell-off, around 90 percent of the wheat grown in Britain

16. Schimmelpfennig et al. 2004.
17. Kloppenburg 1988.
18. Schimmelpfennig et al. 2004.
19. D. Murphy 2007a.

was from PBI-bred varieties.[20] PBI joined the coal, telecommunications, gas, water, and electricity industries in being sold off to private shareholders—"selling the family silver," in the words of the policy's numerous opponents. Given the notorious inefficiency of publicly run enterprises, it is worth noting that on the whole public plant breeding has not only produced an incalculable quantity and quality of new material, but also done so with relatively little cost. Cost-effectiveness could not have been a reason for privatization; the payback on plant breeding may be long-term but it is not expensive; it has been calculated that the cost to the taxpayer of running state-funded plant breeding bodies in Britain at the time of their closure was only 1–2 percent of the entire cost of agricultural research, a modest £6 million.[21] The fact is that investment in plant breeding research can reap enormous rewards, which must have been a major factor in the multinational rush to buy up and invest in seed companies.

To conclude, there are two sets of issues around the privatization of plant breeding: one is the benefit of public versus private ownership; the other is the increasing monopoly of breeding by a small number of very large multinational companies.

Public and Private

ARGUMENTS FOR PUBLICLY FUNDED BREEDING

- Public spending can stimulate private research—if public bodies drop out of breeding altogether, there is a danger of a gulf developing between fundamental research and day-to-day breeding.
- Market forces can distort the research agenda. As a British government study put it: "There is now a risk that plant breeding research will be focused on purely economic goals, at the expense of social and environmental objectives"[22]; minor crops in particular could be ignored, much in the way that the pharmaceutical industry ignores diseases for which treatments are not sufficiently profitable, for example, third-world malaria.
- Small companies, as well as individual, NGO, and community breeders used to be able to depend on publicly available seed to work with; they are now dependent on the multinationals. There

20. Biotechnology Commission 2005.
21. Palladino 1996.
22. Biotechnology Commission 2005.

is abundant evidence that "privatization" has greatly reduced the ability of all breeders to access genetic material.[23]

- Conflicts of interest for scientists can result if they are involved with both business and universities. Commercial secrecy impedes science and technological advance, publication of research results can be delayed, and academic partners can suffer forfeiture of royalties if discoveries are not maintained as trade secrets.
- Who will train the next generation of plant breeders? Universities and public bodies have traditionally performed this role.
- Publicly funded bodies are often better able to think long term.

ARGUMENTS FOR PRIVATE BREEDING

- The profit motive is one of the best stimuli to progress.
- Competition between companies encourages efficiency and innovation.
- A more diverse gene pool results from a market being divided between several companies.
- Small, independently run R&D companies are among the best at developing and introducing new products.

Market Domination by Multinationals

ARGUMENTS AGAINST

- Fewer companies mean less research; one USDA study shows that there is an inverse relationship between the concentration of companies and R&D spending.[24]
- The needs of growers may be marginalized if breeding companies are increasingly controlled from overseas.
- The needs of the poor in developing countries are unlikely to feature very high in private research programs.

ARGUMENTS FOR

- Multinational corporations ensure that research skills and knowledge are globalized, ensuring effective communication between experts.
- Multinationals are highly effective at technology transfer—a boon for developing countries.
- Funding for capital-intensive projects is often facilitated.

23. Fowler and Lower 2005.
24. Schimmelpfennig et al. 2004.

KEEP OFF! THOSE SEEDS ARE MINE! THE RISE OF PATENT PROTECTION

As private breeding has overtaken public, the issue of protection for plant varieties has grown and has become increasingly contested. The two issues are as intertwined as a vine and the tree on which it grows. No private investor is going to put capital into a project that competitors will benefit from, but on the other hand, too strong an assertion of intellectual property rights can prevent the continual and cumulative improvement of varieties that is central to breeding.

With the wisdom of hindsight it can be see that the advent of F_1 hybridization would change everything. For the first time, seeds were produced from a crop that the grower could not resow and expect to get something similar to the parent. It was the next best thing to having a patent. It can be argued that the seed has a dual nature, which makes it difficult for seed production to be treated like any other commodity apart perhaps from animal breeding—it is not only a product in itself but carries within itself the ability to produce more of the same. F_1 hybrid seeds ended this. Traditional seed could also be described as having a triple character, as it is possible to use seed to breed new and distinct varieties.

Until F_1 hybridization created patents by default, plant breeders had no control over what happened to their varieties; they could spend years perfecting a new plant, but by selling a packet of seed, they enabled someone else to profit from the work that they had done. No wonder that, historically, so many breeders explained what they were doing as being for the public good, contributing to the global commons and improving nature for the benefit of humanity. Absolutely sincere they may have been, but their protestations of disinterested generosity sometimes have a slightly hollow ring—as there was no way they could protect their intellectual property. The dual or treble nature of the seed subverted attempts to commercialize and commodify it, but also undermined attempts at improving it.

Patent law was developed in the late nineteenth century in order to protect the inventors and developers of new products and to stimulate the process of invention. Until then, anyone could copy a good product, manufacture it, and rake in the proceeds. Patent law and its acceptance internationally were a vital part of the process of globalization which accompanied the rapid growth of the smokestacks, railway lines, and telegraph cables of nineteenth-century industrialization. Attempts were made to include plant varieties in early treaties such as the Paris Convention of 1883, but national legislators and lawyers generally balked at the difficulties of patenting living material. This irked many in the seed and nursery trade. Luther Burbank was one:

> A man can patent a mousetrap or copyright a nasty song, but if he gives the world a new fruit that will add millions to the value of the Earth's annual harvest he will be fortunate if he is rewarded by so much as having his name connected with the result. Though the surface of plant experimentation has thus far only been scratched and there is so much immeasurable important work waiting to be done, in this line, I would hesitate to advise a young man to adopt plant breeding as a life work until [Congress] takes some action to protect his unquestioned right to some benefit from his achievements.[25]

After his death, the campaign was carried on by his widow, to be partially rewarded in 1930 by the U.S. Plant Patent Act, which extended protection to vegetatively propagated plant material, excluding tubers. Thomas Edison, in support of the act, is reported to have said that "this [bill] will, I feel sure, give us many Burbanks." It was his hope, and that of many others, that patent protection would stimulate enterprise and plant breeders, who now knew that they would be rewarded.

The 1930 act made little real difference, for it only really affected fruit and some ornamentals. Attempts were made to draft laws that would ensure breeders had a return on their investment in France and Germany during the 1930s, but little farther progress was made until 1961 and the International Convention for the Protection of New Plant Varieties (Union Internationale pour la Protection des Obtentions Végétales; UPOV), which aimed to establish uniformity in plant protection law through Plant Breeders' Rights, devised as an alternative to a patent system—more like a copyright, designed to protect the property rights of breeders. Unauthorized commercial propagation of plant varieties was forbidden, but there were two notable exceptions: farmers had a right to sow seed saved from their own crops, and the "breeders exemption," whereby breeders were allowed to use Plant Breeders' Rights varieties to breed new varieties themselves. This latter exemption recognized the third aspect of the seed discussed above—its capacity to be used to generate farther variety and therefore farther both human progress and business opportunities for the breeder. The exemption was recognition that although breeders had a right to pro-

25. Quoted in Juma 1989, 159.

tect their innovation, the genetic material itself was still part of the global commons. It recognized a fundamental fact about breeding—that it is cumulative. The UPOV convention was followed by most of the industrialized countries slowly tightening up on variety protection.

A plant variety protection (PVP) system was instituted in the United States in 1969 for sexually reproducing plants, but with little support from the seed industry, which was suspicious of any government attempts at regulating their industry, already well protected by the technology of hybridization. Ten years later, the situation was different—biotechnology loomed, which greatly increasing the amount of investment needed for a new variety and so also increased the concern of corporate breeders to protect that investment. Consequently there was a revival of debate about patents. The growing desire of corporate interests in patenting their projects went hand in hand with a desire to discourage public breeding. Thomas Roberts, CEO of DeKalb Agricultural Research, Inc., wrote in the *Seedsman's Digest* in 1979:

> The PVPA [Plant Variety Protection Act] has provided incentives for the improvement of self-pollinated species by private plant breeders. This law is effective because it encourages the private sector to invest research funds on crops they could not otherwise afford to breed. Further encouragement can be provided by minimizing the use of public funds for the improvement of these crops. With the need for agricultural research so great, public institutions should avoid duplicating research efforts being carried on in the private sector; they should limit their applied research to those crops where experience has demonstrated private effort to be inadequate. It is wasteful and counter-productive for public research funds to be used to compete with private research.[26]

It was DeKalb's company that launched the first hybrid wheat in the United States in 1974. However, it had not been a great success, owing to a relative lack of substantial improvement in yields over nonhybrids and difficulties in actually producing the seed.[27] Suggestions have often been made that one of the incentives for corporate production of hybrid wheat is to prevent farmers saving their own seed. As widespread and effective self-certification schemes and PVP protection have spread, it is indeed curious to see how interest in hybrid wheat has waned.

A 1991 revision of UPOV shifted the balance of PVPs away from being like

26. Quoted in Kloppenburg 1988, 146–47.
27. Edwards 2001.

a copyright towards being more like a patent, with farm-saved seed now allowed only as an optional exception, which led to "honesty box"–type systems of royalty payment run by breeders' associations. Restrictions were also placed on further breeding. The seed industry, now with the immense power of major multinationals behind it, was at the time of this writing lobbying for further revisions to create even tighter patent-type protection over plant and seed varieties. Farm-saved seed represents a major percentage of the seed sown, even in industrialized countries; this is seen by the seed industry as a huge loss of potential revenue. While evidence supports the obvious, that patents do stimulate research, there is a great danger that a further restriction of the breeders' exemption will stifle innovation. There are even accusations that the companies which own the seed and breeding industry are so set on bolstering their now very powerful positions that they are trying to stifle all competition and intentionally obstruct progress in breeding. It is now increasingly difficult for developing countries to access advanced varieties or breeding lines, hampering their ability to develop indigenous and regionally appropriate breeding programs.[28] Surveys of breeders in the public sector in America have also indicated that increasing private "ownership" of genetic material is hampering research and development.[29]

Distinct from the patenting of new varieties is the patenting of genes discovered by companies whose researchers have located and established the function of those genes. Since the discovery of the function of one gene may be part of a whole pathway that could play a part in a variety of plant processes, and therefore impact on the development of *any* new variety which used *any one* of these processes, the implications are enormous. This threatens to give yet vastly more power to the corporations which make such discoveries. It is almost possible to imagine a situation in which plant breeding begins to seize up, because key genetic links are "owned" by corporations unwilling to allow others to have access to them without paying exorbitant fees.

"Biopiracy" is a term that is used to describe the patenting of genes by corporations, sparking campaigns by activists in many countries. Patenting, and the disputes it causes, is an issue that is changing constantly, and it can be very difficult to assess what the dangers really are.[30] Patenting genes has

28. See Alston and Pardey 2006 for a comprehensive discussion of the difficulties many developing countries face.

29. Fowler and Lower 2005.

30. See http://www.grain.org for updated information on the patenting issue from a campaigners" perspective.

also raised considerable opposition among breeders, at least in academia; one veteran university breeder recalled that "when the first patent appeared, from Monsanto, for a male-sterile corn line we passed a resolution against it at Cornell . . . . [T]he benefit went to Harvard who had applied for the patent, but the work had been done by graduate students in Texas."[31]

HELPING HANDS OR THIEVING HANDS? PATENTS, DEVELOPING COUNTRIES, AND THE INTERNATIONAL AGENCIES

The issue of patents on plant genetic material has been most widely contested in relation to north-south issues. The 1970s and 1980s saw the beginning of "seed wars," with spokesmen for developing countries claiming that the industrialized world was benefiting from their genetic resources, which they had freely given but were then being sold back to them. Their claims were bolstered by academics in the industrialized world (mostly in the humanities) who were sympathetic to the prevailing trend of Left-nationalist politics in many newly independent countries. There is no doubt that the economies of the rich world have benefited enormously from genes collected elsewhere: a Turkish land race contributed genes against stripe rust in wheat worth $50 million per year (in 1979), while an Ethiopian gene that protected U.S. barley from yellow dwarf disease was worth $150 million a year (in 1983).[32] In many cases these genes are present in crops because of the work and selection skills of countless traditional farmers over many generations.

However, in exchange, the developing world has benefited from the industrialized world's breeding expertise, much of it contributed through aid. It might appear, on balance, that the exchanges of genetic and technical information between north and south have been at least approximately reciprocal. This reciprocity however has occurred under a regime of the free exchange of genetic material—changing this to a regime of restrictive intellectual property rights over wild genes or over genes which are present because of the selection practices of traditional farmers, or indeed over the free and unfettered farther development of modern varieties, clearly carries enormous risks for plant breeding in the future.

That some legal protection for newly developed varieties is necessary is

31. R. Murphy 2006.
32. Kloppenburg 1988.

widely accepted in the plant breeding profession, and there is little doubt that its absence in developing countries has tended to hamper both the development of plant breeding in those countries, and the willingness of other countries to export their plant varieties to countries without protection—Monsanto's experience with "guerrilla plant breeding" in India is an almost comic opera example of what can happen (see chapter 15). The development of protection regimes is now however almost hopelessly mired in a highly emotive debate polarized between multinational corporations and anticorporate activists. Arguments that lack of a PVP regime inhibits plant breeding progress cuts no ice with many in the latter camp, as for them, traditional landraces are good enough, and the idea of plant breeding is implicitly rejected by many of them. "Sovereignty" is the key word, and the replacement of the old and local by the new and the scientifically improved is seen only as making "farmers vassals to American corporations," in the words of one article in *The Ecologist* magazine about Texas A&M University's International Agriculture Office setting up demonstration plots across Iraq in 2004. "Iraqis," the piece went on, "were baking bread for 9,500 years before America existed, have no weight when it comes to deciding who owns Iraq's wheat."[33] It can be surmised with some confidence that 99.9 percent of the readers of the magazine had eaten bread made from scientifically improved grain within the previous 24 hours.

Farmers in developing countries will be the ones who will lose out in this tug of war, particularly the middle-sized farmers with "informal" channels of obtaining seed, who are most at risk of prosecution in an unduly restrictive PVP regime, especially one which forbids farm-saved seed—and it is these who are often the most innovative and most efficient. The wealthy will simply bribe their way around any legal inconveniences, while the poorest, who tend to be outside the system, may not be too badly affected—lurid expectations of "seed police" knocking on the doors of millions of peasant huts in the hope of recuperating a few rupees' worth of "stolen" seed at each one simply do not hold up.

Struggling to cope with the new world of patenting are the international agricultural agencies. The early efforts of the Green Revolution resulted in the setting up of IRRI in 1960 and CIMMYT in 1966 (see chapter 12). These were followed by the International Institute of Tropical Agriculture (IITA), established in Nigeria in 1967 to cover a wide range of African crops, including bananas and plantains, cassavas, sweet potatoes, yams, grains, legumes,

33. J. Smith 2005, 44.

and tree crops; and CIAT, set up in 1967 in Colombia. All of these were to work on a wide range of problems, with a central role in each case for plant breeding. Funding depended largely on the aid budgets of industrialized countries. They have since been joined by others, including the African Rice Centre (WARDA; established in 1971 in Benin), the International Center for Agricultural Research in the Dry Areas (ICARDA; 1977 in Syria), and ICRISAT (1977 in India). Harmonizing and coordinating the work of this portfolio of agencies became an issue in itself, and at the urging of a number of those involved, most notably Robert McNamara, a new body was set up to oversee their work—the CGIAR, which started work in 1971.

The valuable work undertaken by these agencies would be exhausting even to begin to enumerate. It is generally agreed that they have had a major impact on food supplies for the world's poor. They have their critics though. Noting that the CGIAR centers are all in Vavilov centers of diversity, one commentator suggested that the CGIAR system was as much as to collect plant genes from the third world as much as develop new crops for them, making them into the equivalent of the nineteenth-century botanic gardens which played such a crucial role in gathering economically useful plants together for the benefit of the imperial powers.[34] Some of the material that originated in the agency breeding programs has indeed ended up in the fields of the north, 21 percent of the wheat acreage of the United States in 1984 was of CIMMYT or other Mexican ancestry for example—but plenty of wheat of European and North American origin made the journey the other way.[35]

CGIAR was set up following a decade of the "seed wars." The centralized American-dominated system has been undeniably successful: efficient, pragmatic, with clear goals, closely linking seed banks to crop development projects, and a system that could be rapidly replicated in developing countries. In a world where division and difference often seem more important than what we share, an ability to coordinate people working across national and cultural boundaries has been a great achievement. Plant varieties of truly global application have often been the result.

As an example, CIMMYT's breeding methodology has been geared to production of widely adapted, disease-resistant varieties with high yield in a wide range of different environments. Between 1966 and 1990, 1,317 new bread wheat cultivars were released by developing countries, of which

34. This is suggested by Kloppenburg 1988.
35. Perkins 1997.

around 70 percent had at least one CIMMYT-produced parent. In 1988, CIMMYT introduced the concept of the mega-environment, which defined broad, but not geographically contiguous, areas with similar environments, cropping systems, and consumer preferences—varieties bred for a particular mega-environment should be able to have a good chance of producing well throughout the area.[36]

Particularly effective have been the working methods promoted by the CGIAR system. R&D in developing countries is often held back not just by a lack of personnel and facilities but also a simple lack of colleagues or workers in areas, technically nothing to do with breeding, but which impact on breeding work, for example, pathology and soil sciences. The CGIAR system has helped provide links between research personnel and context for their work. The systems approach of multidisciplinary teams which characterizes much CGIAR agency research is in marked distinction to the often hierarchical and authoritarian systems often found in developing countries or in countries where tradition, and the importance it grants to seniority or membership of particular ethnic or caste groups, impedes effective collegiate working. The work of women has been enhanced through gender-awareness research—it should be apparent by now that plant breeding has been a pretty male-dominated world! Closeness to existing university facilities, such as IRRI's being physically close to University of the Philippines College of Agriculture can also help, by linking researchers into wider networks.[37]

That the industrialized nations put money into the CGIAR agencies is not simply a choice to "donate," but also an investment. Breeding work the agencies have carried out has greatly benefited them: it has been calculated that from 1943 to 1984, Australia gained US$747 million in cost savings from using CIMMYT-bred wheats and that from 1970 to 1993, the U.S. economy benefited by somewhere between US$3 and US$14 billion dollars from CIMMYT wheat and IRRI rice.[38] Investment in plant breeding is an incredibly cost-effective way for industrialized countries to give aid to the developing world—while also reaping some of the benefits; the cost of running IRRI from its start-up in 1962, to 1968 was $15 million, the additional value

36. Rajaram and Van Ginkel 2001.

37. Ruttan and Hayami 1973.

38. Alston, Dehmer, and Pardey 2006. The wide disparity in figures represents lack of agreement among economists on what can be represented as cost savings.

of the Asian rice harvest of 1968–1969 was around a billion dollars (although additional production costs have to be factored in).[39]

CGIAR and its agencies went through a difficult time in the 1990s, with accumulating evidence that some breeding programs were beginning to suffer from restrictions caused by tightening PVP measures which restricted access to genetic material or restricted its use. At the very least, growing restrictions on the freedom of exchange of genetic material began to result in higher costs for breeding.[40] More crucially, the early 1990s saw a serious funding crisis. Staff at several centers had to be made redundant and some programs cut.[41] From the late 1990s, the crises seemed to have past and funding levels improved once again. Problems, however, remain: funding is much more tied to particular projects by donor governments or other donors, the rise of intellectual property rights over plant genetic material is slowly starving the CGIAR system, and the role of the agencies has become steadily wider, and less concerned with crop breeding. This latter development in part reflects the interests of governments under pressure from consumers in the rich world to concentrate on what they perceive of as important environmental issues, but which might not be seen as priorities by the global poor.[42]

As time has gone on, the CGIAR system has proved increasingly "inclusive," in terms of being more open to questioning the way research is done, in particular the relationship between the educated researcher and uneducated farmer. Participatory research has become more common, and there has been a major focus on the role of women in agriculture and in plant breeding research. Sustainability has also been prioritized.

CHALLENGING THE BOUNDARIES— PARTICIPATORY BREEDING

It looks like a day out. A group of women in their best clothes talking and laughing, meeting relatives and friends from distant villages, making cautious conversation with others from communities beyond their normal range of social contact. The festive atmosphere emphasizes the event's

39. L. R. Brown 1970.
40. Rajaram and Van Ginkel 2001.
41. CGIAR 1996.
42. Alston, Dehmer, and Pardey 2006.

importance, but beneath it is a serious sense of purpose. This is a visit to a breeding station, where farming women have come to report on their assessments of some new bean varieties. Organizing this event has not been easy, a number of taboos have had to be delicately negotiated, and some of the participants have come having faced disapproval from older and more traditional relatives. The event has also incurred disapproval from more conventionally minded local breeders. Welcome to the brave new world of participatory breeding.

A constant criticism directed at interventions by the industrialized countries into the farming systems of poorer countries, whether the relationship was that of colonialism or aid, has been that of a failure to respect the traditions and the methods of farmers. All too often, growers in the poor world have been regarded as backward, their methods inadequate and mired in superstition. Over the years, odd voices have been raised to point out how traditional farmers are in fact building on centuries, even millennia, of accumulated experience, particularly of local conditions. From the 1960s on, these voices grew in strength and number, boosted by the confidence and assertiveness of those from newly independent nations and a growing criticism of "neocolonialism" within the industrialized world. Quieter voices within the scientific and development communities have drawn attention to the intelligence of particular techniques used by traditional farmers.[43]

A greater respect for the third-world farmer can be seen as part of the increasing awareness of racism and a rejection of older ideas of unquestioned western superiority. In breeding the fruit of this transformation can be seen in the field of participatory breeding, which promises not only more effective breeding, but also the radical alteration and democratization of the relationship between breeder and grower. It has developed over the last two decades of the twentieth century in the world of the international agency and publicly funded bodies. Issues of patents and corporate control of knowledge will almost certainly reduce its impact in the world of private breeding.

Involving farmers— especially small traditional farmers—in breeding challenges a whole series of boundaries, each one of which represents a potent division of status and power, especially in very hierarchical traditional societies: science over technology, education over lack of education, theory over practice, men over women. "Science for the people" and "let yourself

43. A good early example of such work on mixed cropping systems in Africa is Leakey 1970.

by guided by the masses," may sound like slogans, but they very clearly describe the spirit of participatory breeding and indicate its enormous potential as an effective tool.

At its simplest, participatory breeding involves getting farmers to trial a limited number of varieties already produced by breeders—participatory varietal selection. Breeders may involve farmers earlier in the process, with experimental lines or crosses produced by breeders—client-orientated breeding. Such an approach has been used in developing new varieties for the marginal areas of India, Bangladesh, and Nepal, which saw little benefit from the advances of the 1960s and 1970s. A deeper level of participation involves multiplying the number of stages, such as distributing elite lines to farmers, getting feedback and then working on crosses, and then distributing these for farther selection. Farmers may also be involved working alongside breeders in carrying out crosses—empowering for farmers and with potential to reduce costs for breeders.[44]

One of the most interesting and radical examples of participatory breeding was the Rwandan bean program, run during the late 1980s by CIAT in collaboration with the Institut des Sciences Agronomiques du Rwanda (ISAR), completed tragically just before the civil war and genocide of 1994. Its work, however, has been carried on by PABRA (see chapter 12).

The Rwanda bean project built on a history of teamwork in CIAT, where agronomists, breeders, anthropologists, pathologists, and nutritionists regularly work together. Experiments were designed by an anthropologist attached to the team based on observations of how local farmers carried out their own experiments with new varieties and techniques. Farmers, nearly all women, were brought into the breeding station to discuss how trials should be set up; they were interviewed about what they saw as desirable characteristics and gave their opinions on the lines being trialed. Breeding lines were then distributed and the women were questioned at a variety of stages.

In this program, the element of participation led to a confrontation with one of the established Green Revolution breeding strategies, that of creating varieties which yield well in a wide variety of locations. The sheer complexity of hilly Rwanda with its huge range of soils, microclimates, and farming systems that rely heavily on intercropping, meant that outcomes would inevitably involve a wider range of variety types than would have been conventionally regarded as desirable. Having a wide range of varieties to some

44. Gypmanteseri 2003, Witcombe and Joshi no date.

extent mimics the range of genotypes to be found in landraces, with the advantages of limiting pest and disease impact.

Women do most of the farming in Rwanda; involving women in breeding work inevitably challenges traditional local mores which can create problems, and in this case led to initial social stigmatization for some of the women. Later on, as the fruits of their work began to be seen, and more importantly, eaten, by the whole community, opinions changed and they grew in status. As the program became established, communities were asked to select representatives to become involved. Some researchers involved in the project did find it difficult to accept the involvement of the women—they felt that their expertise was threatened, that they were being pushed into extension work (lower status), and that their sense of authorship over varieties was being reduced. However the program's successful outcomes apparently helped to overcome many of these feelings, as researchers began to appreciate that the new varieties were not only better but would also be more widely distributed.

Production increases were as high as 38 percent over existing varieties, with lines selected by farmer participation yielding more than those selected by breeders alone—a massive vindication of the program's approach. Varieties selected through participation also had a much greater chance of being grown in later years than varieties previously released by ISAR.

Participatory breeding can involve some imaginative approaches. During the 1990s, one breeder who worked for a Thai government-funded planting and breeding station had the idea of planting out new fruit tree crosses or selections along roads, giving responsibility for looking after them to the families whose property ran along side. Once they started fruiting he would organize a competition, with prizes for those who brought along the best fruit. Receiving the prize was conditional on a government officer coming along to take cutting or grafting material from the tree in question.[45]

Participatory breeding can also integrate scientific breeding more deeply with traditional knowledge if based on making selections from landraces. Rather than bringing new genetic material into a region from outside, locally adapted material is taken as the basis from which to start, with breeders working with local people on selecting the best lines and evaluating crosses made from them.[46] Indeed, the earliest published reports of participatory breeding, during early 1990s, grew out of biodiversity conservation work.

45. Leakey 2007.

46. Danial et al. 2007.

It is clear that participatory breeding is a powerful new methodology; it can be seen as the culmination of many years of researchers and activists raising concerns over the "colonial" relationship between scientist-breeders and practitioner-farmers. Its future depends on a willingness to discuss ways in which it can be made genuinely participatory and rooting its methods in assisting rural communities with real empowerment.

RESPECTING NATURE, BUT WHAT IS NATURAL? THE DILEMMAS OF BREEDING FOR ORGANIC AGRICULTURE

More and more food products on the shelves of the shops describe themselves as "organic" or "natural," even nonfood goods like shampoo and tobacco are doing so. Capturing public distrust of science and worries induced by a succession of crises over the content of food, the organic industry has expanded hugely since the last decade of the twentieth century. Organic agriculture as a philosophy and a practice has been around since the 1920s, but for most of its history was regarded as a fringe movement.

The arrival of the organic industry at the "goods in" gate of the supermarket has made growers look at how their crops are bred; commercial success depends upon yields being produced from crops that were bred for conventional agriculture and that do not always succeed under organic regimes. This involves a different set of practices—notably, no synthetic fertilizers and no synthetic pesticides. In addition, the wholesale opposition of the industry to GM technology is pushing questions of breeding higher up the agenda. Breeding for organic agriculture is a very new field, concentrated in the Netherlands, and set to grow and change fast.

Pest and disease resistance is clearly of major importance for organic growers, with a preference for complex horizontal resistance over vertical (see chapter 12). The variety of resistance shown by landraces is regarded favorably, and some early breeding work on wheat is attempting to create genetically diverse populations through composite crosses, created by randomly crossing wheat populations. The rejection of transgenic techniques for introducing resistance genes may strike the nonbeliever as strange, yet this reflects a fundamental fact about the organic movement—that it is not based in science or a pragmatic approach to realizing certain objectives; rather, it is a fundamentally ideological approach to growing plants. It is very difficult, in fact, to find any consistency across the variety of positions held by the movement. This difficulty is reflected in the variety of views held about what are, and are not, acceptable breeding techniques. The biodynamic movement, rooted in the mystical philosophy of German Rudolf Steiner developed during the 1920s, is the most limiting—preferring to use open-pollinated plants, because of a desire to respect "natural" species boundaries.[47] The mainstream movement does accept F_1 hybridization but the use of CMS and other hybridization technology which prevents plants reproducing "naturally" is generally rejected. Many organic growers do seem to have an inbuilt resistance to F_1 hybrids, however. The use of colchicine to induce polyploidy and induced mutation are also rejected in principle. Despite this, practice seems to be different, as some varieties produced through induced mutation are used—the malting barley 'Golden Promise' produced through gamma-radiation is very popular with organic brewers.[48]

COUNTING THE COST OF MONOPOLIES—SOME CONCLUSIONS

"The dissolution of the [British] Plant Breeding Institute was a disaster . . . . [T]he chemical companies who bought up plant breeding had a total lack of understanding of the complexity and the long-term nature of the business," is a view shared by many in the agricultural world.[49] That plant genetic material has become increasingly "owned" or made inaccessible for reasons of commercial confidentiality has undoubtedly greatly hampered research and development. Plant breeding was once one of the most open and collegiate of worlds, with the doctrine of the free exchange of plant varieties for breeding work quite exceptional in comparison to the rest of the commercial world—a result of plant breeding's rather odd position, balanced between the scientific (and therefore noncommercial) world and the profit-driven business world. Breeders could not only exchange material, they could exchange ideas; there was an international camaraderie, the free networking of scientists and breeder had allowed a synergy of brainpower to develop. Wholesale privatization has begun to destroy this. On top of which there are many suggestions that now that they control the market, the agroindustrial companies which own plant breeding resources have become lazy—"the companies have proved that they cannot innovate" in the words of a biotechnology academic.[50] If the breeders of the future are going

47. See Brinch 2005 for a detailed exposition of the biodynamic view. Interestingly, the movement appears to question the intrinsic value of F_1 hybrids.

48. Maplestone 2007.

49. Leakey 2007.

50. D. Murphy 2007a.

to have to work in separate little boxes, all jealously guarding their commercial secrets from each other, then one of the great advantages of globalization is lost. The consequences could be bleak indeed.

It is not just privatization that is leading to breeders and others engaged in research to working in increasing isolation. The growth of "bioprospecting" by companies looking for new products or genetic material in the developing world has led to moves to clamp down on what was once a very open system of exchange of biological material between scientists. In the past, European and American scientists and others were able to tap the entire world for genetic material—the view was that nature was a public good, a global commons. Much good came of this—the development of crops, medicines, and products that have benefited everybody. But there has also been shameless exploitation, as the plant knowledge of poor communities has been used by commercial interests seeking new sources of genes—shareholders and workers in the developed world benefit, but not the people of the country from whence the genetic material came. Even if traditional knowledge is not being exploited, many in the developing world feel that they have a right to "their" plants, so if anyone is to make money out of them, they have a right to a share. However, given the extreme inequality typical of developing countries, beneficiaries of any "profit sharing" are likely to be ambitious local elites who often use progressive-sounding postcolonial rhetoric in order to establish a moral case for attempts at fencing off the genetic resources of their countries from outsiders.

The Convention on Biological Diversity adopted at the Earth Summit in Rio de Janeiro in 1992 was a milestone in attempts to protect the earth's genetic diversity. However, in giving nations rights to protect their biodiversity from commercial exploitation, it has arguably contributed to a situation where scientists and other researchers and innovators are unable to gain access to genetic material, as a series of bureaucratic obstacles are now increasingly being placed in the way of exporting any live plant material. Needless to say, those most able to overcome these obstacles (which often need bribes to smooth the way) are large multinational corporations. Breeding for the public good, or innovative, speculative, or noncommercial breeding will inevitably suffer. The danger is that an idealist attempt at conserving resources will end up hampering our ability to make progress for the whole human race.

Ornamental plant breeding is also in danger of suffering as nations attempt to limit "export" of seed or plants from their national floras. Much of the most innovatory ornamental plant business is conducted on a small scale with relatively little financial reward for seed collectors or for the nurseries which grow the plants on back home. On the other hand, some larger concerns are able to make considerable sums from prospecting for new genes from the wild. An example has been the 'New Guinea' *Impatiens*, resulting from expeditions organized by Longwood Gardens in the United States and the USDA in 1970. The beneficiaries have been commercial seed producers, breeders, and nurseries in the developed world, but can the inhabitants of New Guinea expect anything in return? Arguably, since no traditional knowledge has been used, and wild plants are part of a theoretical global commons, no one in New Guinea can expect any reward. But the Rio treaty aims at encouraging nations to think about conserving their natural resources as potential sources of income. However, hardly any countries have mechanisms in place to encourage plant collectors to engage with making agreements for paying a proportion of profits into their revenues. One possible way out, being pioneered in Lebanon, is for the country itself to organize ornamental plant prospecting and development.[51]

SUMMARY

Growth in the recognition of the importance of the diversity of crops has grown during the twentieth century, especially as the Green Revolution began to cause a steep increase in the rate of extinction of landraces from the 1950s onward. The preservation of diversity is vital for future breeding, as landraces and older varieties may have genes that contemporary varieties do not. The methods and mechanisms needed to preserve diversity have been highly contested, primarily between developing countries and the industrialized world. Nevertheless, the 1960s and 1970s saw the establishment of a number of international bodies to promote breeding work for poorer countries and difficult agricultural environments.

Plant breeding during the twentieth century involved both publicly funded bodies and private industry. F_1 hybrids, by creating a "virtual patent," enabled corporations to begin a slow march towards taking over plant breeding from public bodies. From the 1970s onward, a growing interest in biotechnology was one of several factors which led to agrochemical and other companies, not traditionally involved in breeding, to take over seed and breeding companies in the industrialized world. This has been paral-

51. R. Brown 2007.

leled by a growth in patent-type legislation aimed at restricting access to genetic material. Many professionals and commentators argue that corporate interests threaten the future of plant breeding and its ability to provide benefits for all humanity. However, various other developments in plant breeding, such as participatory breeding, have sought to challenge traditional hierarchies and divisions in actively seeking farmer involvement in breeding.

FIFTEEN ❦ CONCLUSIONS

At the break of dawn, a band of young people, dressed in the hooded head-to-foot white overalls used by farmers working with pesticides, dash across a field, jump over a gate, and head into the rows of mustard plants. Slashing wildly with an assortment of sickles, garden shears, spades, and even one machete, the band are clearly not involved in harvesting a crop. The mustard plants are cast aside and trampled underfoot as the group moves on. Another test plot of a genetically engineered crop has been destroyed.

This scene has happened in Britain, France, Germany, America, India, and elsewhere. The issue of GM crops has clearly touched a nerve, engendering opposition on a considerable scale, much to the shock and surprise of the scientific and agricultural communities and the food industry. Opposition has varied considerably from country to country, and markedly in tone: some scientists have expressed disquiet about possible negative effects on human health or the environment, but some commentators and members of the public have expressed their opposition in terms which can only be described as near-hysterical.

There is no point in going over the well-rehearsed arguments about GM crops. Instead it might be more useful to consider the whole story of plant breeding, and then look at how the new technology fits into the big historical narrative, in particular considering to what extent it is true that, as supporters of GM technology maintain, there is no real, qualitative, difference between it and previous plant-breeding methods. It is also worth looking at where our story so far has left us, at what the motivations were for the move to GM crops. The opposition to GM also needs examination, particularly since previous new plant-breeding technologies caused remarkably little public interest, let alone disquiet—and even less sickle-wielding vandals attacking mustard plants. Is opposition to GM part of a much wider attack on plant breeding, on modern agriculture, or indeed on science and modern society?

In 1978, Daniel Nathans and Hamilton Smith received the Nobel Prize in Physiology or Medicine for their isolation of restriction endonucleases, compounds that can be used to cut the DNA molecule at specific places. The very basis of genetic information could now be utilized with precision;

alongside a variety of other techniques, it enabled researchers to extract a particular piece of DNA from one organism and place it into the genetic code of another. GM technology was simply the most dramatic and the most powerful of a whole new toolbox of laboratory-based technologies generally referred to as "biotechnology." For the breeder, biotechnology's various techniques enabled them to perform crosses that had never been possible before and to propagate plants on a scale that could previously have only been considered possible in fairy tales. The joy of GM technology in particular was twofold. One was its precision—a breeder could locate the gene for the trait they wanted to transfer and transfer it, and it alone, without having to import the rest of the genome, as well. So much work in plant breeding was taken up with complex breeding programs to ensure that a wanted trait was maintained while unwanted ones were bred out. The other joy was being able to import traits from *any* living thing—breeders were no longer restricted to working with close relatives. Inheritance always used to be thought of in vertical terms, with genes being passed *down* breeding lines; now it could be thought of a horizontal, with genes being transferred from one species to another, directly going from genome to genome. This, however, was the technology's philosophical Achilles heel. That one gene could be transferred between, for example, a fish, and a plant, set off alarm bells. For many it was man playing God. Evoking Mary Shelley's fictional monster, a journalist who almost certainly knew nothing about genetics invented the term "Frankenfood," a headline writer's dream but a nightmare for researchers and for those who had invested in the technology.[1]

THE GRAND NARRATIVE—
A SUMMARY OF PLANT BREEDING HISTORY

To begin to assess how GM technology fits into the plant breeding story is one reason to get a broad perspective over time. It is also useful to look at what has driven progress in plant breeding, specifically, at why particular developments happened at the time they did—the perspective of the past can be a useful mirror for the present.

Broadly speaking, as plant breeding has progressed from unconscious selection through to lab-based biotechnology, the following trends can be noticed:

1. The first use of the term "Frankenfood" was in a letter to *The New York Times*, June 16, 1992.

YIELD

- Traditional farmers value reliability over yield.
- Pioneers in new agricultural areas value fitness for conditions over yield.
- Yield becomes important when demand threatens to outstrip supply, as when increasing populations have to be fed from a finite supply of arable land.
- Consumers with levels of income above basic security value a variety of factors above yield, notably flavor and method of production.

GENETIC "REACH"

- Traditional farmers did not carry out hybridization but took notice of good accidental hybrids.
- Deliberate hybridization made the combining of desired traits possible, and the introduction of new traits from related species possible.
- Mendelian genetics made hybridization easier to plan.
- Knowledge of plant reproduction is used to drive ever-widening crosses between plants previously thought incompatible.
- Laboratory techniques make more wide crosses possible.
- GM technology brings all of life into the plant breeder's grasp.

SOURCES OF VARIATION

- Introduction of new species, varieties, and sports always important.
- Imperial expansion and pioneering agriculture drive widespread systematic plant introduction.
- Mendelian genetics drives a search for desirable traits.
- Induced mutation used to increase the rate at which random change can introduce variation.
- The need to have access to as wide a range of genetic material as possible drives the systematic collection of landraces.
- GM technology reduces the importance of landraces as sources of genes, as genes can now be sought anywhere. Key genes, however, become targets of bioprospecting, leading to accusations of biopiracy.

SKILLS OF PLANT BREEDERS

- In traditional societies, all active farmers (i.e., most of the population) are involved in conscious or unconscious selection.

- "Gentleman farmers" or astute peasants become the main agents of change as agriculture becomes larger scale.
- Seed merchants, wealthy amateurs, and university-based research pioneers drive early hybridization and more sophisticated selection systems.
- Professional breeders trained in genetics take over. Farmers no longer play any part in breeding.
- Multidisciplinary teams of geneticists, pathologists, statisticians, agronomists.
- As plant breeding becomes more laboratory based, breeders need to be less and less directly familiar with plant material or realities of growing—which may not necessarily make them better breeders.

GENETIC DIVERSITY WITHIN CROPS

- Traditional agriculture brought into being vast numbers of landraces, each one genetically mixed and constantly evolving.
- Conscious selection and deliberate hybridization result in vastly less diversity in the industrialized world, both within the now genetically pure varieties, and diversity across the number of species being grown.
- F_1 hybridization resulted in a further reduction of diversity, to the extent that it began to imperil crop health—as with the southern corn blight crisis.
- Modern crop varieties, however, are far more diverse than older ones, in that they contain within them genetic material from a wide range of sources; as an example, an average of less than ten different landraces had been used in breeding a rice variety in 1950 but more than sixty in IRRI-bred rice in 1990.[2]

Despite these broad tendencies, the whole question of expanding or contracting levels of diversity both within crops and between them is a complex one, which will be looked at below.

GLOBALIZATION

- Crops have been spreading along trade routes ever since humanity began to farm and to trade.

2. Evans 1997.

- The general tendency has been for certain crops to spread way beyond their homelands. In some cases this can be linked to their productivity and adaptability, or, breedability—corn being the supreme example. In other cases, good crops spread and are bred for new environments for other, more political reasons; one cannot help feeling that if the Chinese had invaded Britain in 1600, we would long ago have had rice fields in Norfolk.
- As some crops extend their global reach, others contract, as so many African grains have before the march of corn. However, there are also pressures in the other direction.

INTELLECTUAL PROGRESSION

- Ancient and traditional peoples saw crop varieties as gifts of the gods and treated seed with a religious reverence. Empirical evidence about crops was generally expressed in relationship to the supernatural.
- As human understanding of the natural world grew, crop selection became a technology—a trial and error process of experimentation based on empirical observation, presented as evidence. All supernatural elements were lost, but a large element of intuition remains.
- Mendelism allowed plant breeding to become an applied science, driven by genetics but relying heavily on other sciences too. Prediction of performance was based on knowledge of genetic outcomes. Intuition still had some role.
- Unraveling of plant genomes allowed breeders to see plants as a series of molecular interactions with precise linkages between genes and chemical components of plants. The role of intuition was almost eliminated.

OF EMPIRES AND HARVESTS—WHAT DRIVES PLANT BREEDING?

Looking back over the whole history of plant breeding, one question stands out: what has driven humanity to select and breed crops at the times and in the places that they did? For thousands of years people grew crops, which for the most part changed slowly, usually imperceptibly so. Change in the last two centuries has been extraordinarily rapid—is this simply because all social, economic, and technological change has been rapid over this time, and plant breeding got swept along too? What are the forces that drive progress in plant breeding? This is not just a question for the historian, for its answer should help us to cast light on our present situation, and on the progress we need to make in plant breeding. Understanding motives and driving forces better enables us to control and direct the end result of those motives and forces.

"What drives plant breeding?" is a question about technology—does technology drive history, or does history drive technology? The first suggests that machines make history, that particular technologies bring about changes in economic organization, with social, cultural, and political changes in train, the latter that we have greater control over technology, that particular technologies arise because economic and social forces drive technological creativity. One way to look at technological progress is simply to say that necessity is the mother of invention; applied to plant breeding, this would imply that progress was made when more mouths needed to be fed, or newly conquered lands tilled. A deeper analysis would bring into play the role of how plant breeding as a technology (and genetics as a science) has been driven by social and economic forces, and, using a terminology borrowed from Marxism, how ruling classes maximize their opportunities for wealth creation and accumulation—technology for the last three centuries has been driven by the owners of capital.

Technological change before the European Enlightenment was slow and patchy, with some technologies clearly having an impact on agricultural progress, such as the metal plow with a moldboard, which arrived in Europe only after the Roman period, enabling the fertile but heavy soils of northern Europe to be tilled and so allowing far more people to live off the land. Agricultural changes allowed more people to be fed, and contributed to wealth creation. However, there comes a point at which existing agricultural technology can no longer feed a larger population—the Malthusian crisis. Historians have traced a number of such boom-and-bust "Malthusian cycles." One in particular, Graeme Snooks, suggests that a civilization can escape from the Malthusian crisis and seek to continue to maintain economic growth (very low for most of history) by one of three routes: trade, conquest, or technology. Historically, few civilizations have been in a good enough geographical position to benefit greatly enough from trade; the level of technology change is immensely slow, so conquest has been the way forward—at least until the civilization's resource base is unable to support any farther conquest—and it collapses.

Conquest is a good analogy for how in preindustrial civilizations, agriculture was able to feed more people. Farming can produce more food through a number of means: improved crops, improved plant and animal nutrition, improved technology (e.g., plows and hillside paddy fields), more efficient use of labor, and taking more land into cultivation. As we have seen, in most societies agriculture has been a low-status occupation, so few resources or minds were devoted to making technical progress, with the result that a more efficient use of labor and developing new lands were the main means used (see chapter 3). There comes a point, however, when the producers cannot be made to produce any more (or will suffer or rebel if they are taxed too heavily); the usual answer is to expand the area under cultivation, which often results in the development of marginal areas. Such areas tend to be very vulnerable to changes in weather patterns or to degradation through the farming methods used, which makes the whole civilization then vulnerable to ecological collapse. That farming activities damage their environment (e.g., through deforestation) and contribute to the collapse of historical civilizations is well established.[3] Plant breeding could not offer a way forward—crop plants were regarded as more or less immutable "gifts of the gods"—the idea that they could be markedly improved was simply beyond either the experience or the philosophical conception of the vast majority of pre-Enlightenment peoples.

That crop plants could be molded and improved for human benefit needed a conceptual breakthrough—that humans could do this, and had a right to, *and* a breakthrough in experimental methodology. *Both* were provided by the post-Renaissance movement in Europe, which gave us the Enlightenment and the Scientific Revolution. Arguably the only other civilizations that could have achieved this were China and Japan. Japan, as we have seen, made considerable progress in agricultural improvement, including plant breeding, during the Tokugawa era. It is interesting to look at where the world centers of pre-industrial–era agricultural improvement were: the Netherlands and Flanders, Japan, England, and, to a lesser extent, France and Italy. All were densely populated, with the first two having some of the highest population to available land ratios of anywhere on the globe. The Netherlands and England were notable in having innovative societies with a relatively decentralized political structure, which gave landowners and better-off small farmers an incentive to involve themselves in agricultural improvement and crop innovation. They were also notably free of the worst effects of the religious and intellectual conformity enforced by either churches or rulers in many other societies. Necessity provided the pressure,

3. Eisenberg 1998; Diamond 2005.

and relatively open societies the opportunity, for active plant breeding to be born.

Improved crop varieties (as distinct to new crops, like fodder legumes, turnips, or New-World introductions like corn) did not make much impact until the nineteenth century. By this time, the interlinked series of revolutions—scientific, technological, commercial, and philosophical—that Europe had experienced was well on its way to impacting on the rest of the world. Improved crop yields fuelled population and economic growth, which in turn helped fuel the era of imperial conquest. Conquest brought about an immense interchange of crop plants between Old and New Worlds. The need for varieties which would thrive in their new homes drove plant breeding forward, on all fronts, from imperial botanic gardens making selections for plantation agriculture to peasant farmers selecting corn seeds to plant the next year on tiny plots. Economic and social forces were clearly driving plant breeding—but at the same time, the knowledge that plants *could* be changed through selection, and as time went on, through hybridization, made certain economic and political decisions possible. If the knowledge and the ability to improve crops had not been there, much of the colonial occupation of the world would have been severely circumscribed—the railway lines would not have been driven across the prairie and the pampas without the yields and the security of new wheat and corn varieties. Plant breeding may have been driven by necessity, but it also acted as one of the handmaidens of European conquest and settlement.

With Mendelism, plant breeding changed from being a technology to an applied science. Discoveries in genetics allowed an increasing number of options to be presented to those investing in plant breeding. This provides an interesting lesson in the relationship between technology and history. During the twentieth century, breeders and those working in areas related to breeding have had a natural tendency to become enthusiastic about new technologies. But as so often with other technologies, so boosters for it have appeared, loudly proclaiming that it will solve a whole string of problems, and revolutionize breeding. Inevitably, reality is different. Norman Simmonds, reviewing a fifty-year career in breeding, was able to write an article titled "Bandwagons I Have Known" and describe how "if the bandwagon is a good one . . . it becomes a gravy-train, a seat on it nearly guarantees funds, grants and other goodies."[4]—an example of technology driving history, perhaps. "Colchiploidy" did not feed the world, and mutation breeding took

4. Simmonds 1991, 7.

so many decades to become useful that by the time it did, more advanced biotechnologies were set to make it obsolete. GM technology, too, has attracted the starry-eyed as well as the appalled; one commenter, writing in *Venture* magazine in 1986, predicted that "in 5 or 10 years, Saudi Arabia may look like the wheat fields of Kansas."[5]

HENRY FORD AND THE GENETIC BOTTLENECK

"Any customer can have a car any color they like, so long that it is black," Henry Ford once snarled. His statement has become almost a definition of what has become to be known as "Fordism"—the system of mass production and mass consumption that dominated the industrialized countries for most of the twentieth century. Capitalism's productivity and economies of scale brought about an era in which a decent standard of living became *possible* for almost everyone in the industrialized countries. It also brought a uniformity—the range of products was limited, although they changed frequently in small details and occasionally in big ones as technology moved on. The degree to which everyone benefited was influenced largely to the extent to which the state became involved in evening out the effects of economic crises and economic exploitation. After World War II, governments effectively formed a social compact with capital and trade union movements in the name of economic and social stability—part of which was the considerable state investment in plant breeding, either directly or through support of universities and other research institutions.

One technological advance that worked well within a world defined economically by Fordism was F_1 hybridization. As we have seen (chapter 10), F_1 hybridization was not the only route to greatly increasing corn yield or quality; if the dice of history had been rolled differently, F_1 hybridization may not have seemed such an attractive option, or it would have had to compete with other techniques. For business, its ability to ensnare farmers into coming back for seed every year was clearly very attractive. Since the technique worked so well for corn and brought investors in seed companies impressive profits, it is easy to see how it became the obvious route for much plant breeding activity.

The southern corn blight episode highlighted the dangerously narrow genetic base caused by the use of CMS in corn breeding (see chapter 14). Yet this lack of diversity was also part and parcel of Fordism. In every field of

5. Quoted in Kloppenburg 1988, 203.

manufacturing, the era of mass production offered few real choices; brand names and cosmetic changes disguised the fact that consumers were offered little choice in a wide range of products. In agriculture, the growth of capital-intensive industrial farming made a wide range of plant varieties unnecessary and often undesirable. The era of Fordism was about large-scale production, employing economies of scale that made products cheap enough for everyone to buy—too much variation in either raw materials or end products, risked adding to costs.

Scientific plant breeding contributed to making the twentieth century unprecedented for the quality of life it enabled the masses to have. It was not the case in the developing world—here the specter of famine loomed, brought about not just by the operation of a Malthusian cycle but also by the disruptions brought about by imperial rule. Famine was averted through the Green Revolution, which can be seen as an attempt by the United States, to export the Fordist model of agriculture, particularly in the stress on varieties which would perform across continents, in a wide range of climates and habitats. The Green Revolution can also be seen as parallel to America effectively becoming the world's banker under the Bretton Woods agreement of 1944.

The 1970s saw a change, a transformation from mass-production manufacturing to a more diffuse system, dubbed "post-Fordism." In post-Fordist economies, the advantages of economies of scale are reduced, largely because computer technology allows for much greater fine detailing of production. An increasingly knowledge-based economy also promotes a networking model of business relationships rather than a strictly hierarchical one. In addition, consumers have reacted against what came to be seen as the blandness of Fordist-era consumerism. The post-Fordist era is the era of niche manufacture and marketing. During the 1980s, the word "choice" became the keyword—in food products as much as in anything else. Informed and increasingly wealthy consumers did not just want pasta in twenty different shapes. They expected to be able to have the option of buying twenty different shapes as: wholemeal, organically grown, wheat-free, organic wholemeal, organic wheat-free, hand-made, traditionally made, to say nothing of all the multi-colored, and occasionally obscenely shaped, variants sold as souvenirs in Italy. In the post-Fordist era, one is expected to be able to express one's individuality by choosing from a bewildering range of niche products.

The implications for plant breeding are that diversity is again becoming important. Consumer preferences for choice, for different production methods, and for heritage products, are beginning to add to other pressures, which together make the twentieth century look like rather a bottleneck in terms of plant diversity. Diversity reduces the ability of disease to spread through a particular crop. Diversity also allows for cropping seasons to be extended—vital in fruit and vegetables. A major study would be required to trace the number of varieties of a given crop which were of commercial importance during the twentieth century, but the general indications are that variation has been increasing through the last few decades of the century. It is known, for example, that in California the number of plum, peach, and nectarine varieties that produced more than a thousand cases of fruit doubled between 1983 and 1995 and that the number of wheat varieties in the United States has increased from 126 varieties in 1919 to 469 in 1984.[6]

The number of crops that have attracted the attention of growers and breeders is also increasing. Again, the early twentieth century appears to have been a bottleneck. Some traditional societies often cultivated a larger range of crops and food plants than we do, in addition to eating or somehow using a great many species which grew as "weeds" in their fields or gardens—up to several hundred in parts of Mexico or Indonesia.[7] The diet of modern humanity sounds almost restricted in comparison. The following illustrates just how dependent we have become on a few crops: there are two hundred and fifty thousand described higher plant species, of which thirty thousand are known to be edible and of which seven thousand have been used for food on a regular basis. Of these a hundred and fifty have become important crops, and fifteen supply more than 90 percent of all food. Three leading crops, wheat, rice, and corn, supply almost two-thirds of all food energy and more than half of all protein derived from plants.[8] A major trend established at the end of the twentieth century is the greater interest being shown in "minor crops," usually plants that have fed traditional cultures but were swept aside by corn or wheat. Amaranthus (species of *Amaranthus*) was once a major grain crop for the Aztecs, but Spanish disgust at its ritual association with human sacrifice had led to its suppression; it had then been disregarded until the 1970s when it was discovered that its grain contains high levels of lysine, the amino acid usually deficient in grains. Consequently there was a rush of interest in breeding it; in 1986, the U.S. Agency for International Development started grant aid for its study in a number of

6. Brush 2004.
7. Brush 2004; Smil 2000.
8. Smil 2000.

developing countries in South America, Africa, and Asia.[9] Quinoa, with seed containing 19 percent protein, and an amino acid balance superior to any other grain crop, is also now attracting attention from agronomists.

FUNDAMENTAL RUPTURE OR LOGICAL PROGRESSION?—GM CROPS AND HISTORY

Opponents of GM crops like to stress how they are qualitatively and utterly different to anything that has gone before; their supporters, that they are not that new or distinct, indeed, that "humanity has always practiced genetic modification." Looking back on plant breeding history, we can appreciate that both standpoints involve a sleight of hand. Gene manipulation is clearly of a different order to tribal farmers picking over their crops for next year's seeds, and yet there is a logic about the way that plant breeding history has progressed. Breeders have gradually been able to rein in more genes to produce the plants they want and to better understand how genes control the final product. GM technology finally gives breeders the precision that they have always wanted. It is salutary to remember mutation breeding. The use of such a random and imprecise procedure in an attempt to increase variation puts many worries about GM into perspective.

The precision of GM technology is not what has caused the problems. It is the ability to cast the breeder's net over all of life that has brought plant breeders blinking into the headlines. Before GM technology, breeders were using an inventive variety of laboratory techniques to get related plants to cross; now, suddenly, they could identify a gene in any organism and incorporate it. This is clearly a huge leap. The fact that a gene simply codes for a limited set of chemical reactions, and that only very limited numbers of genes can be used, a tiny proportion of an entire genome, failed to register with many opponents, let alone the general public. The simple fact that a fish gene could be incorporated into a plant set alarm bells ringing. This was clearly a major qualitative change in what plant breeding could do.

Mary Shelley's *Frankenstein* is one of the defining novels of the whole post-Enlightenment period. It encompasses so many of our fears about science and technology running out of control, a more up-to-date version of Goethe's story *The Sorcerer's Apprentice*, itself based on a tale by the Roman writer Lucian (ca. 125 CE–180 CE). For much of the popular press, the plant breeder has become the modern sorcerer, someone with power to do good

9. Juma 1989.

or ill—but also someone who simply has too much power. That such concerns should appear so suddenly, after a previous lack of interest in breeding technologies, says more about the opponents of GM than the technology itself, as does the relative lack of opposition in the United States or China and the strength of opposition in India and Europe. Clearly there are political and cultural factors involved.

GM technology arrived at the time it did primarily because the science behind it had developed sufficiently. However, there has also been a growing rumble of concern about world food security since the high point of the Green Revolution in the 1960s. Half the advances in plant productivity had been brought about by breeding—the other half through irrigation, crop protection, and nutrition, but in the world's most important croplands there was no room for improvement in the latter three factors, so it was clearly breeding that was going to have to provide future improvements. Breeding was beginning to show diminishing returns; one study published in 1999 showed that the efficiency of translating investment into breeding research into yields was down by 75 percent and that there was an ever-increasing effort to produce a constant linear rise in yields, when many predictions made by bodies such as the FAO and the World Bank suggested an exponential increase was needed to keep pace with rising populations.[10] The great worry during the 1970s was that the Green Revolution was a unique coming together of many favorable factors, which would not be repeated. By the end of the twentieth century, the global supply of arable land was clearly peaking, and pressures from a growing environmental movement made farther forest clearance for agriculture politically less acceptable. It was clear that breeders needed more tools in the toolbox. GM technology offered the hope of making improvements, possibly even something as dramatic as cereals which could work with bacteria to make their own fertilizer by fixing nitrogen (like legumes) or making photosynthesis more efficient. Saudi Arabia may not turn into Kansas, but the possibility of making really major improvements in crop physiology was one of the main areas of hope.

Agriculture since the 1960s has also been under increasing pressure to reduce its reliance on chemicals for crop protection. Again, breeding offered a way forward. Pest and disease resistance is highly plant-specific, so the option of being able to transfer genes for resistance across large taxonomic gaps appeared very attractive—GM technology appeared to be ideally suited to do this.

10. Mann 1999; Evans 1997; Smil 2000.

Before moving on to look at the wider context of the opposition to GM crops, and its relationship to a wider agenda critical of scientific plant breeding it is worth looking at a few points salient to wider breeding issues:

- There is nothing to suggest the GM technology inherently involves higher risks, to plant health, human health, or the environment. Each crop variety needs evaluation on its own terms. There is a strong tendency for evidence against one crop, usually in the form of misquoted press coverage of a scientific study, to be used as evidence against the entire technology; on this basis, the first railway crash should have been sufficient to terminate all farther development of wheeled transport.
- The consensus amongst scientists working in biotechnology is that a strict regulatory environment and thorough testing of new GM crops is necessary, or at least advisable, even though nothing of the sort has ever been proposed for the less predictable mutation breeding. Regulation helps allay public fears.
- GM technology is one more, albeit powerful, tool. It does not, and cannot, replace conventional plant breeding. A GM variety will still need farther breeding to produce varieties suitable for different environments and particular purposes.
- Given the above, the testing of every single new variety based on the same gene insertion into a crop species seems unnecessary and expensive, as does any testing of gene insertion carried out with species which are sufficiently closely related to be able to be bred conventionally. The cost of regulation is an issue, as greater costs can severely limit the availability of new crops for poorer farmers.
- GM crop acreage has been growing at around 15 percent per year from 1996, with around one-fifth of the world's population regularly eating food of GM origin. There have been no medically attested health problems associated with them. Environmental problems have been limited to some very localized crossover of herbicide resistance to wild grasses on field margins.

ATTACK OF THE LITERARY LOCUSTS—THE OPPONENTS OF PLANT BREEDING

That opposition to GM crops is part of something wider is indicated by the following invective:

> The United States of America declared a war on Indian rice back in the early 1960s when India's No 1 scientist mole, Dr M.S. Swaminathan, stole the gene bank of rice, evolved over decades . . . and passed it over to the Americans. India had 120,000 varieties of rice seeds; today, no more than 50 are available. This is United States' war on genetic diversity, ably supported by scientist-criminals like MS Swaminathan. And the worst crime of the Indian Government is to have these criminals head Agriculture Policy of India. At age 75, Swaminathan lives to lord it over many Expert Committees, including crucial ones in the Ministry of Agriculture. The deadly tentacles of this hydra-headed monster extend in many ministries of the Government of India.[11]

That this should be directed at Dr. Swaminathan, the most socially progressive of all the world's leading plant breeders, is chilling.[12] Clearly, this is not just a dispute about plant breeding . That much of the radical critique of the Green Revolution has stated or implied that plant breeding was unnecessary, and in fact was a cover for "gene theft," is made explicit by the above quotation.

That plant breeding, a "benign activity," in the words of Norman Simmonds, should attract opposition, may seem shocking. That there is opposition is a subject that needs addressing in some depth.

Opposition to GM crops is beginning to spill over to affect plant breeding and genetics as a whole. Fewer young people are choosing to work in plant breeding—a reflection of the damaging effect the GM controversy is having on recruitment to the profession.[13] Genetics as a whole is beginning to become a dirty word in some quarters.

Opposition to scientific plant breeding comes from several directions:

- Marxist/leftist critique of capitalism. Anything carried out by capitalists is in the name of profit, and therefore against the interests of everyone else.
- Neo-romanticism. There has always been a current of opposition to industry and science. Originally an artistic movement, romanticism's

11. Shrivastava 2006.

12. The Web site of the M. S. Swaminathan Research Foundation (http://www.mssrf.org) gives an idea of the social remit of Dr. Swaminathan's work. Unlike many breeders, he has taken great care to actively address issues of poverty, caste exclusion, and gender bias.

13. Fowler and Lower 2005.

modern version has tended to become more political and far-reaching in its critique of modernity. The influential counterculture writer Theodore Roszak, writing in the 1960s, saw scientific thinking as being the root cause of what is wrong with modernity.[14] Romanticism is a great basis for poetry and art, but a very poor one for science, technology, or public policy.

- Postmodernism. An intellectual current that rejects all single sources of authority and explanatory theory in all realms of human life. It has had a particular influence on science studies. Some science studies theorists have suggested, largely building on the work of philosophers Jean-François Lyotard and Thomas Kuhn, that science is just another "narrative" and therefore that a scientist's explanation of something is no different from, and no more valid than, that of a tribal shaman's.
- Antiglobalization. One of the strongest manifestations of a global consciousness has been paradoxically, a distinctly global movement against globalization. The movement addresses a range of issues of social, economic, and political injustice, but tends to link them all to a simplistic analysis, with a strong anti-Western—particularly anti-American, bias. It is characterized by ideological elements drawn from all the above.

Opposition to scientific plant breeding, like any political issue, can be seen as spreading across a spectrum. On the one hand are those who share many of the concerns of plant breeders themselves over plant patenting, but who extend their interest to a wider-ranging critique. The leading NGO in this respect is GRAIN, who claims to represent the interests of third-world farmers, and in a manifesto-type statement declares:

> Research was pushed off the farm and moved into far away institutes and labs run by Western scientists. In developing countries, the "green revolution" is probably the most dramatic example of this push to industrialize agriculture and do away with a broad mosaic of local and diversity-based production systems.[15]

Many African and Asian plant breeders would object to be called "Western," but the use of the word illustrates a basic fact about much of the anti-

14. Roszak's *Where the Wasteland Ends* (1972) was particularly influential.
15. GRAIN n.d.

scientific rhetoric of the last few decades. What has tended to link the various ideological strands mentioned above is a declared opposition to the Enlightenment project—that science and rationality, linked to the universal recognition of certain basic human rights, can lead us on a path of progress. Antienlightenment thinking holds that tradition is superior to modernity, tribal knowledge to scientific, third world and eastern superior to "Western." The supposedly harmonious relationship between tribal peoples and their environments is idealized despite much evidence to the contrary. Lesser and more lifestyle-orientated trends include the "rediscovery" of traditional gardening techniques, endlessly compiled in books on "natural gardening; the organic food industry; and the enormous interest in "heirloom" flowers and vegetables, "as grown by grandmother." Consumers in developed countries may accept and welcome modernity in most areas of their lives, but in the field of agriculture, about which the vast majority know very little, another standard reigns. Consumers seem to want their meat, eggs, and vegetables to come from a farm that looks like something out of a child's picture book. Indeed, industrial agriculture has been responsible for many iniquities—battery hens, nitrogen pollution, and habitat destruction—but it is simplistic in the extreme to claim that all modern farming inevitably leads down this road. It is also simplistic to think that we can continue to feed the human race in the manner to which it is accustomed without science-based intensive farming.

Mendelian genetics and F_1 hybridization arrived at a time when technology and progress were regarded as both unstoppable and desirable, especially in the United States. But over the twentieth century, the link that the Enlightenment made between scientific and technical progress, cultural improvement, and political liberty has been gradually weakened. The Holocaust, the atom bomb, and the failures of nuclear power ensured that, for many, the link was well and truly severed. The optimism of post-Enlightenment progress gave way to the bland consumerism of the Fordist era, which threatened to reduce us all to "cogs in the machine." This was supplanted by "postmodern pessimism," with many wishing to turn their backs on modernity. In fact—very few actually do want to reject modernity—except when they go food shopping.

Skepticism about the Enlightenment is unsurprisingly not shared by scientists. Such attitudes are generally those of nonscientists; in the words of one figure who was both scientist and novelist, C. P. Snow, "If we forget the scientific culture, then the rest of Western intellectuals have never tried, wanted, or been able to understand the industrial revolution, much less accept it. Intellectuals, in particular literary intellectuals, are natural Luddites."[16]

Of all the critics of scientific plant breeding, indeed of modern agriculture generally, none is more vociferous or influential, than Vandana Shiva.[17] There is no doubting the injustices or the environmental damage she recounts—her solution, though, is to return to the past, to an India of village communities farming with traditional methods and landrace crops. Her politics are very much within the Gandhian tradition of village-based development, but the original Gandhians, as we have seen with regard to cotton, were not afraid of engaging with science (see chapter 5). Shiva's voluminous writings and public lectures (critics describe her as jetting around the world telling us all to be peasants) encapsulate better than anyone else a position that has become disturbingly influential with many educated liberal-minded people. Science is accused of being reductionist, of dividing humanity from nature, intrinsically linked to the exploitation and oppression of third-world people and of women, part of a brutal Western attempt at world domination. Criticism of the Green Revolution, which has segued into an outright and total opposition to GM technology, is a central plank of this particular brand of political philosophy.

There is a dark side to this philosophy. In attacking globalization and scientific rationality, its proponents have a tendency to idealize and romanticize the traditional and local; indeed the extreme cultural relativism of postmodernism has merged with left-wing postcolonial politics to support assertions of nativist and identity politics. However, as Meera Nanda, a leading Indian critic of what she calls "reactionary postmodernism," points out, "the populist left opposition to the Green Revolution, GM crops, and other science intensive initiatives, is routinely co-opted by the ultra-nationalist, autarkic, elements of the Hindu right." Shiva has been interviewed and favorably quoted by *The Organiser*, the journal of the Rashtriya Swayamsavak Sangh (RSS), a Hindu nationalist organization, the sight of whose members marching in formation wearing khaki shorts, is a powerful and frightening reminder of its original inspiration—Hitler's brownshirts.[18] Identity politics is the natural playground of the political Far Right. In rejecting the universality of Enlightenment values, antiscience critics on the Left have found

16. Snow and Collini 1993, 22.

17. An example of her influence is her giving a BBC Reith lecture, one of the most highly regarded annual lecturing slots in Britain (Shiva 2000).

18. Nanda 2003b, 30.

themselves sharing a bed with those on the opposite end of the political spectrum.

The traditional agricultural systems favored by Shiva would never be able to provide enough food for a country or a world that is rapidly urbanizing. In addition, there is another powerful reason why high-yielding crops and the science-based industries behind them are needed—they enable more to be grown per unit of land. This reduces the amount of forest, savannah, and other land, currently occupied by pastoralists, tribal peoples, or wild nature, that needs to be converted to arable. It should not forgotten that throughout history, agricultural output was increased primarily through felling forests, draining wetlands, and plowing wild landscapes. Lovers of nature should not forget that the greatest destroyer of what they love is agriculture—the sight of well-tended fields of scientifically created and scientifically managed crops should be welcomed, for it is they that allow the forests to remain standing and the wetlands stay undrained. There is a powerful environmental argument for sustaining the momentum of the Green Revolution and its high-productivity crops.

Failure to invest in plant breeding for the future, which inevitably includes a big dose of GM technology, risks the human race hitting its Malthusian limits. Graeme Snooks, the historian discussed earlier with regard to technological progress, ventures to suggest that a civilization that fails to invest in technology is doomed to collapse.[19] Rwanda provides a possible—and terrible—warning. Although its genocide was ethnic in character, extreme population pressure was undoubtedly the major factor in setting off this country's slide into civil war. Densely populated, with farmers tilling ever-smaller plots, often on land only marginally suitable for crops, Rwanda had only received very belated assistance with crop breeding and agricultural development. There is nothing inefficient about small farms, indeed, quite the contrary; many studies have shown that small farms are often more efficient than larger ones. A 1977 International Labour Organisation study, for example, showed that in east and southeast Asia, yield, employment, and added value per hectare increases as average farm size decreases.[20] But Rwanda's traditional agriculture, while undoubtedly rich in knowledge and techniques, lacked the modern varieties and scientific back up needed to support the higher productivities needed by its increasing population. What

19. Snooks 1996.
20. Sneep and Hendriksen 1979.

happened there serves as a terrible warning to those who would wish to turn their back on science.

THE FUTURE IS CREOLE

Before summing up and looking toward the future, there is one final story to tell. It concerns the marrying of the latest technology and the oldest processes in plant breeding. It is a salutary tale, as much as for the urban romantics who think they know what is best for farmers as much as for powerful multinational corporations.

In March 2002, following its approval by the Indian government, Mahyco-Monsanto released a number of cotton hybrids containing a gene for the production of a compound lethal to caterpillars, which had been derived from a bacterium. 'Bt cotton,' as it was known, was already in cultivation in India—effectively illegally. Since 1998, however, anti-GM activists had been campaigning against the cotton, with Vandana Shiva denouncing them as "seeds of suicide, seeds of slavery, seeds of despair."[21] Farmers, however, were desperate to obtain cotton that would not fall victim to bollworm and to avoid the costs and the dangers of using pesticides. In a situation familiar to producers of software and fashion goods, whereby Asian markets are flooded with fake goods, seeds of the Bt cotton "escaped" from Mahyco-Monsanto's test plots and were used to breed new "unofficial" Bt cotton varieties. "Disappearance" of seeds from test plots is the bane of plant breeders the world over—farmers know that among them are potentially much better plants than the ones they grow. So much for rural conservatism, or indeed the love of traditional landraces.

By 2005, it was estimated that 2.5million hectares were under "unofficial" Bt cotton, twice the acreage as under the ones that had been sown from Monsanto's packets. The unofficial Bt cotton varieties had been bred, either by companies operating in an ambiguous legal position or by farmers themselves. A veritable cottage industry had sprung up, a state described as "anarcho-capitalism," whereby small-scale breeders were crossing reliable local varieties with the caterpillar-proof Bt plant. Hundreds of Bt cotton varieties were the result. The world's first GM landraces had arrived, a blend of tradition and science and a powerful illustration that old and new technologies can not only coexist but should both be valued.

21. Herring 2006, 471.

Shiva's "Operation Cremate Monsanto" had spectacularly failed, its anti-GM stance borrowed from Western intellectuals had made no headway with Indian farmers, who showed that they were not passive recipients of either technology or propaganda, but could take an active role in shaping their lives. What they did is also perhaps more genuinely subversive of multinational capitalism than anything GM's opponents have ever managed.

What the Indian breeders who were combining landrace and other local cottons with Bt varieties were doing was an active version of what we saw happening in Chiapas, Mexico (see chapter 2), whereby by HYV corns became *acriollado*—creolized. This was in striking contrast to the political activists in Chiapas who were busy doing the opposite—stuffing kernels of their traditional corns into packets in a fridge to try to prevent cross-fertilization with the modern world. The two approaches can be seen as a metaphor for the responses of traditional societies to globalization: one takes the best of the new and blends it with their own culture, the other, like the proverbial ostrich, tries to deny it is happening by seeking refuge in identity politics and keeping the outside world at bay. On an international level, it is the countries that have actively picked and chosen from the offerings of Western and capitalist culture, ethics, and science that are making strides: China and southeast Asia. Those that have not adapted, but put up barriers to the West, have shown a tendency to stagnate intellectually, politically, and economically. The concept of the creole crop is also a good metaphor for the bearers of the future: people of mixed heritage, with inventive new tongues, techniques that marry the past and the future, and a confidence that they are in control.

To turn to plant breeders themselves. Traditionally, good breeders have not just been geneticists, but all-round crop scientists. There is a danger, however, that as valuable biotechnology and GM techniques are, that they will draw resources away from traditional plant breeding methods or even from other high technology breeding methods. Some commentators point out that new plant breeders today are much less well-versed in agricultural and plant sciences than they use to be, a result of an approach to breeding where nobody ventures beyond the laboratory walls. It is not just a sad day when the only entire piece of plant life a breeder comes into contact with is the rubber plant in reception, but a very worrying day—as effective breeding relies entirely on understanding the context for which plants are being bred. The great breeders and promoters of breeding have historically been those like Norman Borlaug and Henry A. Wallace, who constantly got out into the fields, felt rust-infected wheat between their fingers, and talked endlessly to farmers, both the rich and the poor. Breeding has never just been applied science either, but has always had a bit of "art" about it—a good breeder had a level of intuition which guided him or her in particular directions. Intuition has played a smaller and smaller role, but it is unlikely that a need for it will ever go away.

Plant breeders have tended to be shy, not natural seekers of publicity. That so few have told their story is one reason why there have been so few histories of plant breeding. A profession more prepared to discuss its purpose and to be more assertive might be in a stronger position to overcome the obstacles that increasing privatization and corporate control of genetic resources is placing in its way. It might also be stronger in its ability to resist ill-informed criticism.

Plant breeding has provided us with the ability to live in increasing numbers on an increasingly crowded planet. It has also helped furnish us with plants to beautify and improve our surroundings. Like so much of what else helps us to enjoy life and live it to the full—scientifically based medicine, political and religious liberty, the freedom to travel, and modern communications—it is a product of the Enlightenment. Our ability to survive as a species and to coexist with the myriad other species with which we share the planet depend on our recognizing the fruits of the Enlightenment and defending its spirit: its rationality, its empiricism, and its call to everyone's right to life, liberty, and the pursuit of happiness.

TECHNICAL NOTES

An index to terms included here follows on page 435.

1. BINOMIAL NAMES, SELECTIONS, AND HYBRIDS

Since the Swedish botanist Linnaeus (1707–1778, also known as Carl von Linné) developed the so-called binomial system in the eighteenth century, all plants (and animals) have names, usually derived from Greek or Latin, composed of two parts and written in italics. An example is *Rudbeckia fulgida*, the black-eyed Susan, a North American wildflower popular as a garden plant for its deep yellow flowers with their distinctive dark center. The first name is the *genus*, which may have only a single species or hundreds (as in *Rhododendron* or *Euphorbia*); the second is the *specific name*. In combination with the genus name, this gives the plant its unique appellation as a distinct *species*.

Labels in garden centers are often a first introduction to the binomial naming system, these often have a third name, usually in single quotation marks, for example, *Rudbeckia fulgida* 'Goldsturm,' which indicates that this is a garden selection, or *cultivar* (see below). Many hybrids between two species (*interspecific hybrids*) are indicated with an "x," such as *Viola x wittrockiana*—the common pansy. Sometimes the second name in a binomial is in quotation marks: *Rudbeckia* 'Juligold.' This should indicate that it is a hybrid between at least two different species of *Rudbeckia*; however, common usage often drops specific names for well-known garden varieties.

So, there are a lot of *selections* and *hybrids* in the garden center, the vegetable patch, the orchard, and the field. But what do these two words mean?

The difference between selections and hybrids goes to the heart of the subject of this book, as it is a fundamental distinction and one of enormous importance to the development of cereal crops, vegetables, fruit, industrial crops, and ornamental plants. Selections are simply plants judged superior and *selected* out from the fellows by human agency, from whom they may be only slightly different or on the other hand, quite dramatically different. *Hybrids* are the result of a cross—sometimes natural, sometimes deliberate—between two distinctly genetically different plants: either different species or two distinct races or identities within the species. Strictly speaking it is more correct to speak of "breeding" in connection with hybrids than with selections; "plant breeding," however, is generally understood to cover both.

There are different conventions in use for the way plant names are used. Here, a horticultural convention is followed, with genetically defined plant variety names put between quotation marks. The word *variety* is used to cover both selections and hybrids, when the distinction is not important.

2. CLONES AND CULTIVARS

Plants may be propagated from seed or very often, vegetatively, through cuttings, division, grafting, and such. Seed is produced as part of sexual reproduction, potentially involving the blending of genes from both parents—as a result, the progeny will have a different combination of genes to the parents, which will generally result in a different physical appearance (see technical note 23: Phenotype and Genotype). Vegetative propagation, such as division, the taking of cuttings, or grafting, does not involve genetic blending, so the new plants will be genetically identical to the (single) parent. For example, all potato (*Solanum tuberosum*) 'King Edward,' daffodil 'King Alfred' (a *Narcissus* cultivar), apple (*Malus domestica*) 'Herefordshire Russet,' and rose 'Iceberg' (a *Rosa* cultivar) are genetically identical—they are *clonal*. In horticulture, clonally propagated plants are known as *cultivars*; the term can also mean that plants are also grown from seed, but seedlings must show uniformity and be able to produce uniform seedlings in turn. This latter meaning is a use of the term more common in agriculture—individuals of a cultivar do not have to be genetically identical but must have an acceptable level of conformity.

3. MUTATIONS

A mutation is a sudden change in a plant's genetic material due to a small change in the coding of a gene. The results are usually deleterious but occasionally enhance its ability to survive; it is through such mutations that evolution advances. Some mutations enhance a plant's utility or attractiveness to humanity, for example, the early wheats that did not shed their seed at maturity and double flowers that end up as garden plants.

4. CHROMOSOME SETS AND POLYPLOIDY

Most plants are *diploid*, with one set of chromosomes from the male parent and one from the female: "AA." In a flower, the male pollen and the female ovule each contain only half the parental set of chromosomes—the *haploid* condition: "A." *Polyploidy* occurs when chromosome numbers double (or triple, etc.). It can occur spontaneously within a species (*autopolyploidy*)—"AAAA"—or when two species hybridize (*allopolyploidy*). Hybridization between "A" and "B" can result in diploid progeny—"AB"—or the chromosomes can double—"AABB" (the *tetraploid* condition). Complex crosses result in combinations such as *hexaploid* bread wheat, with two sets of chromosomes from three different parents: "AABBDD." A plant with four sets of chromosomes is a *tetraploid*, one with six sets a *hexaploid*; a *polyploid* is any plant with more than two sets. The rutabaga is a good example of a tetraploid, thought to be a cross between the cabbage (*Brassica oleracea*) and the turnip (*Brassica rapa*) that arose spontaneously in seventeenth-century Bohemia; it is actually one of several *Brassica* allopolyploid crosses. Polyploids are often larger and more vigorous than diploids.

Tetraploids cannot interbreed with individuals from their parental stock, because the chromosomes in a diploid gamete cannot pair up with those in a haploid gamete (there is now a two to one ratio), so effectively a new species is formed; this can happen in nature as well as in the garden, the field, or the laboratory.

A great many crops are polyploid; apart from tetraploid durum wheat and hexaploid bread wheat, bananas are *triploid* (having three chromosome sets) and strawberries *octoploid*. The banana's triploidy means that it has a set of genetic data inadequate for reproduction—and so seeds do not form. The lack of seeds (large and teeth-cracking enough to have been used by anticolonial rebels as bullets more than once in the nineteenth century) makes bananas edible (see below). Many citrus are triploid and are consequently seedless.

5. HYBRIDS

A hybrid (sometimes popularly called a *cross*) is the progeny of two plants that are distinct in some way; they could be separate species or other discrete naturally occurring entities, such as geographical races or subspecies, or man-made entities, such as cultivars, or simply just two genetically different plants. Often, however, *hybrid* has been used more specifically to describe a cross between two separate species; these may be sterile (as with the proverbial mule), or they may be fertile. Hybridization can only occur (at least, outside the laboratory) between closely related species, so botanists have tended to use it to prove close relationships.

The word is also used another way—to describe a first generation (F_1) cross between two distinct selections of a plant—as in "hybrid maize" or "hybrid rice." Owing to the confusion that this much more specific meaning has, these will be referred to throughout as F_1 hybrids (see chapter 10).

6. INBREEDING AND OUTBREEDING (SELF-POLLINATION AND CROSS-POLLINATION)

Sexual reproduction mixes genes and therefore achieves constant variation, which ensures that a species can adapt to changing circumstances. However not all plants mix genes every generation; some, often from stressful environments (including many annuals) play safe by *self-pollinating* or *inbreeding* so that they fertilize or pollinate themselves, although in practice there is also usually a low level of pollination between plants. Others, however, *cross-pollinate* or *outbreed*, with a variety of mechanisms to reduce the possibility of self-pollination. As might be expected, self-pollinators tend to produce progeny very similar to their parents, outbreeders, since they mix genes far more, produce more varied offspring. Wheat and barley are examples of inbreeders, maize and potatoes of outbreeders.

Many domestic plants show a strong tendency to self-pollinate, whereas their ancestors were outbreeding cross-pollinators. This is because early farmers had a tendency to unconsciously select for self-pollinators. Selection for particular characteristics over many generations resulted in plants whose offspring reproduced with consistency and predictability the desired character—which self-pollinators do and cross-pollinators do not. In many cases this "went against nature," as many plants have mechanisms (physical or genetic) which limit self-pollination (and therefore self-pollinating) and so maximize genetic variation. Mutations which made these mechanisms (called *self-incompatibility* mechanisms) fail would have been selected for—at first unconsciously, but once understood, they were consciously sought out by breeders. Tomatoes are a good example; the ancestors of the domestic, self-pollinating, tomato are strongly cross-pollinating.

7. GENETIC DIVERSITY

Each individual plant that is the result of sexual reproduction will have a different genetic makeup. *Genetic diversity* refers to how wide a level of variation there is in a given population—which may be a species or a geographical race, or some other population. This can often be appreciated by looking at a number of individuals: wild flowering currents (*Ribes sanguineum*) in the Columbia River Gorge of Oregon, for example, which vary between deep red and palest pink, indicating a high level of diversity (at least assuming that this one characteristic reflects diversity across the plant's full set of genetic material). Wild daffodils (*Narcissus pseudonarcissus*) in Herefordshire, England, however, look remarkably alike—it is almost impossible to find one which is different in any way from its fellows. This indicates a low level of diversity, probably because the species is not a native and was initiated from a very small number of individuals.

8. GRAFTING

Grafting refers to the joining together of two different cultivars of a species, or very closely related species. The most familiar examples are apples, where one cultivar provides the *rootstock* and the other the *budstock* or *scion*. The budstock is the variety we are familiar with: 'Cox's Orange Pippin,' 'Fuji,' and so forth. The rootstock controls the vigor and longevity of the tree. The budstock is always a clone of the plant from which it was cut.

Grafting is often linked with hybridization in the public imagination; in fact, they have nothing to do with each other!

9. PHOTOPERIODISM

Plants respond to light in a variety of different ways. Many have adapted to particular climate zones by initiating certain types of growth at different times of year—the trigger being day length. Onions in northern latitudes, for example, often only begin to produce bulbs when days are getting shorter, each cultivar being triggered on a particular day. Clearly, when plants are moved from their region of origin, their cycle may not fit in with the growing season in their new home.

10. GENE POOL

A *gene pool* refers to all the genes available to take part in reproduction within a particular population. A rare and geographically isolated plant species might have a very small gene pool in the wild, but move it to a botanic garden and plant it in an order bed alongside some relatives from other regions, and the new gene pool will consist of all the genes of all the plants it can cross with.

11. GENETIC DRIFT

Genetic drift is often observed in small isolated populations. Small populations are, on a simple basis of probability, more likely to have atypical mixtures of genes than large ones. As a consequence, recessive genes (see technical note 12: Recessive Gene) are more likely to come together and be expressed; other gene combinations unlikely in "normal" populations are also likely to occur. The result is that the population will, through successive generations, drift away from the norm.

12. RECESSIVE GENE

Mendelian genetics describes genes as existing in two forms: *alleles*, which can be either dominant or recessive. When these are recombined in reproduction, the dominant one is expressed—so that when "A" (dominant) and "a" (recessive) are combined as "Aa," the progeny will have the character defined by "A," not "a." Recessive alleles are only expressed when they come together: "aa." Mendel's explanation was radically different to the theory of *blended inheritance*, widely believed in the nineteenth century, in which it was held that characteristics in the parents blended in reproduction. Indeed, reproduction often does result in blending of parental characters, but this was explained within a Mendelian framework by the idea of *polygenes*, developed in the early twentieth century (see chapter 11).

13. GERMPLASM

Genetic material available for a plant to breed with, or—as the term is more generally used—available for a breeder to work with, is called *germplasm*. Sometimes the term is used as an alternative to "seed" or "propagating material," emphasizing that it is the genes in the plant material rather than the material itself that is important to the breeder.

14. NITROGEN AND LODGING

Plants need nutrients—chemical elements in a form which they can absorb and utilize. The two most fundamental ones are nitrogen and phosphorus. Nitrogen is needed by plant foliage, and the more nitrogen a plant receives the more its foliage will grow—except that different species and, within species, different varieties, vary enormously in their response to nitrogen, some growing faster and more luxuriantly with higher levels and therefore having more resources to put into flowers and seed. Others respond more slowly. Lush growth, however, is more prone to attack by fungal pathogens: rusts, moulds, mildews, and so forth, so there is a trade-off. In addition, lush foliage growth tends to be top-heavy and physically weak, making plants that are fed high levels of nitrogen more likely to fall over, or *lodge*.

15. MONOECIOUS AND DIOECIOUS

Although the majority of plants have male and female parts on the same plant, some species have separate male and female plants. In either case, the male pollen fertilizes the female ovules via a structure called the style. The ovules, after fertilizing, turn into seed.

Most of the "flowers" with which we are familiar are *monoecious* and *hermaphroditic*. They are monoecious because both male and female organs are found on the same plant and hermaphroditic because they are both found together in the same floral structure. Maize is monoecious but not hermaphroditic, the male flowers being the pollen-bearing tassels atop the plant and the female flowers being the silks to be found in leaf axils lower down. It is this physical separation between the two that drew early botanists to be able to approach the very fact of gender in plants and that enabled early hybridization experiments to be a relatively straightforward business.

Some plants bear male and female flowers on completely separate plants—examples are date palms and willows. These are *dioecious.*

16. HYBRID VIGOR OR HETEROSIS

It has long been noticed that hybrid plants frequently show increased vigor—this is known as *hybrid vigor*, or, *heterosis*. The reason for hybrid vigor has been explained by a number of theories, but the exact cause has yet to be definitely established.

17. SEGREGATION

A diploid plant will have one gene, or, more correctly, one allele (see technical note 12: Recessive Gene), inherited from one parent and another from the other. At *meiosis*, when chromosomes split to form haploid pollen or ovule cells, the genes will separate—or *segregate*—leading to a separation of characters. For early plant breeders, the key breakthrough, even before Mendel worked out the details, was to realize that a trait may disappear and reappear from one generation to another. For more on segregation and the expression of characters see the following.

18. F_1 AND F_2 GENERATIONS

One of the reasons for Mendel's making the breakthrough in genetics that had eluded others before him was his rigorous and comprehensive following through of all the individual plants in each generation. Generations are given the designation "F" for filial: F_1 is the first generation following a deliberate cross, F_2, the one after, and so forth. The segregation of characteristics which follows from an F_1 to an F_2 generation is one that confused many before Mendel, and is key to understanding the importance of the first generation in creating hybrids.

As an example, if we suppose that a plant may have either rough stems or smooth stems, the rough stem gene—"R," is dominant, so that plants with "RR," "Rr," or "rR" will all have rough stems; only those with "rr" will be smooth-stemmed.

If an "RR" plant is crossed with an "rr," then the alleles of each parent separate (segregate) into different pollen or egg cells (more correctly called ovules). An "RR" plant only produces pollen or ovules with allele "R," an "rr" only with allele "r"; thus all offspring are found with a combination of "R" with "r," and the progeny are all uniformly "Rr" = rough-stemmed.

	R	R
r	rR	Rr
r	rR	Rr

If two of the F_1 generation are crossed, or selfed, to produce an F_2 generation, the pollen (or ovules) of each parent will segregate and there will be equal quantities of "R" and "r" pollen will be formed. When the pollen fertilizes the two different ovules at random, three kinds of progeny result:

	R	r
R	RR	Rr
r	rR	rr

The segregation of the pollen or ovule is reflected in the segregation of smooth stem plants (¼ "rr") from the rough-stemmed parental type (¼ "RR" and ½ "Rr"). This is why sowing the seed of F_1 hybrids does not result in progeny which are either consistent with or identical to the parents.

19. MASS SELECTION, PEDIGREE, AND PURE-LINE BREEDING

Mass selection is the age-old method of simply collecting seed from the plants that have been chosen as seed parents for the next generation and sowing it, then saving seed from that generation, and so on. In essence there is no difference between a prehistoric farmer and a modern professional doing this—the difference lies in the selection criteria used for selecting the parents of the seed to be kept for sowing next year. As selection criteria became more demanding over time and more informed by science, particularly in identifying what crop outcomes were wanted (yield, reliability, disease-resistance, etc.), the process became more exact and sophisticated.

Pedigree breeding is much more precise but covers a variety of approaches, all of which involve developing a *lineage*, so that one parent plant is used to produce the next generation of plants, from which again parent plants are chosen. Although hybridization is not necessarily the start of the process, it often does begin with the creation of a cross and then selecting the plants with the desired combination of characteristics from the first (F_1) generation, then sowing their seed, and again making a selection in the second (F_2) generation, and so on for as many generations as is needed until the desired combination is seen in enough plants as to make the sowing of their seed produce a result that can be regarded as reasonably predictable. In self-pollinating crops this repeat sowing process tends to narrow the number of variants, and in particular to eliminate *heterozygous* combinations, that is, pairs of dissimilar alleles ("Rr"), and to concentrate *homozygous* combinations, that is, pairs of identical alleles ("RR" or "rr"). Homozygous combinations will breed true; heterozygous ones would segregate to give homozygous plants ("RR" or "rr") as well as heterozygous ("Rr").

Pure-line selection assumes that a variety will self-pollinate; if one individual with the desired characteristics is chosen, and then its seed sown, the progeny will have the same genotype as the parent. Pure lines are useful in that they are predictable, and if they can be made to cross (e.g., by hand-pollination) then a hybrid between two distinct lines can be made. One disadvantage of pure-line selection is that the genetically based variation in ability to withstand stresses and pests of landraces is lost; another is that high-yielding plants (or others selected for) may be weak growers or have other undesirable characteristics.

20. GLUTEN

Wheat grains contain—as do all seeds—an embryo plant. They also contain *endosperm*, a mass of starchy material designed to feed the emerging seedling on germination. As well as starch, the endosperm contains *gluten*, which is a combination of proteins. Gluten has an elastic quality, which, when bread flour is being digested by yeast, traps the carbon dioxide given off by the yeast, creating the bubbles which allow bread dough to rise. It is also responsible for the chewy texture of bread and other wheat-based products. Rye and barley have lower levels of gluten; oats have none.

High-gluten flour makes good bread, and so achieving high gluten levels has been a goal of plant breeders ever since the late nineteenth century. Gluten-rich wheat is known as "hard" or "strong." Wheats from severe climates tend to have a higher gluten content—hence the popularity of Canadian wheat with bakers. Breeding wheat with a high enough gluten content has been an important goal in milder climates.

Some older wheat types, such as spelt, contain a different type of gluten—one less likely to cause problems in that proportion of the population who suffer from an allergy to some of the gluten proteins. 'Kamut' wheat is similar, and partly as a consequence is marketed as a health product. 'Kamut' is an interesting example of a landrace (probably originally from Egypt) that has enjoyed a new lease on life after being registered as a new variety in 1990 by the Quinn family of Montana. Early marketing literature made out that it had been discovered in an ancient Egyptian tomb, but its origins, though obscure, are now known to have been more prosaic (Quinn 1999).

21. SPORTS

Sports is common horticultural parlance for mutants.

More accurately described as *somatic mutations*, bud sports have long been a means of improving existing varieties, taking advantage of the fact that fruit trees (apples in particular) sometimes produce shoots with slightly different fruit characteristics; these can then be used to provide material for grafting and so be propagated as a new variety. Some varieties have produced whole families of offspring in this way—the 'Delicious' apple alone has been a source of over a hundred, including many that are sold simply as 'Red Delicious.' The great advantage of bud sports is that they offer the reliability and familiarity of the old variety, but are sufficiently different to include an element of novelty.

Generally, bud sports are somatic mutations, that is, they do not involve sexual reproduction. They start as a single cell and eventually form a bud that will grow into a new shoot or branch.

22. LAMARCK

Jean-Baptiste de Lamarck (1744–1829) was a French naturalist—and one of those unfortunate characters who is chiefly remembered today not for what he got right but for what he got wrong. He advanced an early form of the theory of evolution and the idea that evolution progressed in accordance with natural laws. This included a theory of the inheritance of acquired traits, which suggested that morphological and physiological traits acquired during life could be passed onto progeny so that, for example, an apple tree that grew into a tight bent shape because of strong prevailing winds would produce seedlings that would have a similar shape. It was only during the twentieth century that the theory was completely disproven; however it was revived by Lysenko in the USSR (see chapter 9).

23. PHENOTYPE AND GENOTYPE

It was Wilhelm Johannsen who first proposed the use of these two terms. The *phenotype* of a plant is the result of genetic material of the plant (*genotype*) interacting with the environment; it is the expression of the genotype—the plant you see before you. At the very simplest level, it is easy to appreciate that since some alleles are dominant and others recessive, the presence of genes for recessive characters can be masked. A plant with genotype "AA" will have its appearance, or phenotype, dictated by the dominant gene "A"; one with "aa" will have its phenotype dictated by the recessive gene "a"; those with genotype "Aa" will be dictated by "A," so the "a" is not expressed. In the latter case the presence of the recessive gene in the genotype will only be known after the plant has been bred with either a pure recessive "aa" or another "Aa," and a quarter of the progeny are observed to have the phenotype dictated by "a."

24. FEED CONVERSION RATIO

The *feed conversion ratio* is an indication of how efficient it is to convert plant food to animal flesh. This is also an indication of the relative sustainability of various kinds of animal husbandry, as clearly, the more inefficient the ratio, the more energy is used to produce the fertilizer to grow the crops, and the more land is used up.

Plant breeding focused on efficient conversion of resources (nutrients such as nitrogen and phosphorus and water) into energy and protein is clearly vitally important for improving sustainability. It is interesting to consider how much of the earth's surface various diets use: with high-intensity cropping, someone with a largely vegetarian diet needs only 700–800 square meters of the planet's surface to supply their needs; a diet made up of around 30 percent animal products (dairy, poultry, and pork) needs around twice as much land—1,500 square meters; and a diet rich in beef requires as much as 3,000 square meters.

Feed conversion ratios compare units by weight fed to animals and units of usable flesh or other products for human consumption thus obtained. They can be measured in terms of energy or protein. The following gives the ratio as percentages, energy followed by protein (Smil 2000):

salmon	35%–40% / 40%–45%
dairy produce	30%–40% / 30%–40%
eggs	20%–25% / 30%–40%
chicken	15%–20% / 20%–30%
pork	20%–25% / 10%–15%
beef	6%–7% / 5%–8%

From this it can be seen that those who regularly tuck into a steak are using up a lot of the world's resources!

25. INBREDS

Whereas most plants are cross-pollinating (outbreeding)—which has the advantage of shuffling genes every generation and thereby improving the chances of a species remaining flexible in the face of environmental change—many cultivated plants are self-pollinating (inbreeding). If cross-pollinators are forced to self-pollinate, then it is more likely that recessive alleles will end up together. If these result in unfavorable consequences, then the next generation will suffer from a variety of problems. Such is the case with maize, which produces puny plants if female silks are repeatedly pollinated with pollen from their own tassels: *inbreeding depression*. The advantage, however, of self-pollinating is that it only takes a few generations for the alleles of nearly all the genes to become *homozygous*— "RR" or "rr"—not *heterozygous*—"Rr." Producing inbreds is thus a way in which plants with very predictable outcomes can be realized.

It was Mendelian genetics that opened breeders' eyes, and most specifically Edward Murray East's, to the possibilities of *inbreds*—no longer were they poor-quality specimens, but they were carriers of genetic predictability and a means of isolating and fixing particular desired characteristics.

26. OPEN-POLLINATED

This is the opposite of controlled hybridization: allowing the wind to blow pollen from one plant to another in a field, or, in the case of insect pollinated crops, letting the bees do the job. *Open-pollinated* crops are usually a mixture of genotypes, although narrowed down by generations of mass selection. They have the advantage of cheapness and are more adaptable, as they, like landraces, have some level of genetic variation. If plants of the seed crop are kept separate from plants of other varieties of the same crop, seed collected from an open-pollinated crop should *breed true*, that is, produce a population essentially the same as the parent generation.

27. COMBINING ABILITY

This is a measure of how effectively different characteristics can be combined. For example, a breeder may want to bring together in one plant high yield and drought tolerance and may have potential parents—one of which has genes which enhance the former and the other, the latter. It may be difficult to combine them, so the F_1 generation will have few plants with both, or it may be easy, in which case there will be plenty of plants with both characteristics. The development of inbred lines for F_1 maize production during the 1930s required a major study of *combining ability*. Statistical analysis to separate out what may be due to combining ability and what due to other factors was a feature of this work.

28. CYTOPLASMIC MALE STERILITY

Occasionally found as a mutation in a variety of plants, *cytoplasmic male sterility* (CMS) is a boon to the breeder of F_1 hybrids, as the male pollen-bearing parts of the flower fail to develop, so preventing it from being fertilized by itself. It clearly makes mechanical emasculation (detasseling or pulling out stamens with tweezers) unnecessary. If the hybrid line is wanted for further breeding beyond the F_1 generation, however, fertility will need to be restored. During the 1940s, dominant *restorer* genes were discovered, which restored male fertility to plants carrying the sterility gene; the implications of this was that breeders had what was effectively an on-off switch.

29. MULTIPLE-FACTOR HYPOTHESIS AND POLYGENES

Johannsen had shown (see chapter 6) that within pure lines, any variation among plants is the result of environment, not heredity. However, there were still many cases where plant traits were not inherited in qualitative fashion as might be expected from Mendel's work. There was a black and white aspect to the factors that Mendel looked at: his peas were wrinkled *or* they were smooth; there were no *slightly* wrinkled peas. In fact, Mendel deliberately chose such simple qualitative examples in order to understand basic principles. In many cases inheritance involves shades of grey: if a pink flower crosses with a white one, the progeny may show a variety of shades of pink. There is *continuous variation*—hence the pre-Mendelian belief in "blending inheritance."

Nilsson-Ehle and East showed independently of each other that when environment can be ruled out as a cause of continuous variation, then multiple genes are involved, each of which contributes to the same trait, but whose action is cumulative.

These are now known as *polygenes*. The branch of genetics that deals with such patterns of inheritance is called *quantitative* or *biometrical* genetics.

30. INTROGRESSION AND BACKCROSSING

Backcrossing refers to crossing a hybrid variety with one of its parents, then possibly doing likewise with the progeny. The aim is to introduce to the desired breeding line, a trait that the parent has but which has been lost through hybridization. This is often done when a cultivated variety lacks a trait regarded as desirable that a more primitive or wilder form has. It is, however, a slow process; up to ten backcrosses are often necessary.

Introgression refers to the transferring of a limited number of genes into a breeding line from a wild species during backcrossing. This may be deliberate, as part of a breeding program, or accidental, as in the evolution of crops (see chapter 1).

31. PARTITION AND HARVEST INDEX

A major breakthrough was in *partitioning*, the division of the plant's production between its different parts: leaves, stems, roots, and flower/seed heads. Short-straw wheats and rices shifted resources from stems to seed heads, resulting in greater yields. A way of discussing partition in economic terms is by measuring a variety's *harvest index*; this compares the total weight of the plant with the part of the plant that has economic value—in this case, the seed head. Clearly the larger the proportion that goes into the seed head, the more grain can be produced from the same number of plants. For most of the twentieth century the harvest index stayed the same for wheat. Varieties released in the 1970s however had made gains as much as 40 percent. These were the so-called *semidwarfs*. Semidwarfs were amongst the best of the new *high-yielding varieties* (HYVs).

Changing partitioning to the benefit of the head can result in problems—lodging, or falling over—which is why when breeders started to select for increased harvest index, stiff straw was also sought. The new grains also responded well to feeding, particularly with nitrogen, whereas old varieties simply grew more leaves, the semidwarfs grew more seeds.

32. SYNTHETICS

Synthetics are hybrids that are best understood as being a halfway house on the road to the full-blown F_1 hybrid assembled from inbreds. A cross-pollinating crop, such as maize, is inbred to three or four generations to produce partial inbreds with partial homozygosity. Combining ability is tested, and then those F_3 or F_4 plants that combine well are then grown together so that they cross naturally, that is, there is no emasculation of one variety. Synthetics tend to yield better than nonhybrids and have the advantage that if farmers save seed from their own plants, they will be able to keep most of the characteristics of the variety—and will be able to do this for several years. Eventually the variety would segregate and new seed would need to be bought. Being cheaper and simpler to produce, they are a good "appropriate" technology for developing countries: they need less skill to produce and reduce reliance on corporate sources of seed.

33. HORIZONTAL AND VERTICAL RESISTANCE

Resistance to pests and diseases can be described as either *vertical* or *horizontal.* In vertical resistance, one or more genes grant a very precise and absolute resistance to one particular race of an infectious organism, generally a fungus. Horizontal resistance involves several genes and gives a wider and less absolute resistance. Vertical resistance is an attractive option commercially, as it is generally easier to work with (only one gene is involved) and gives total protection. However, it is far more likely that the pathogen will evolve and develop the ability to overcome that resistance. Horizontal resistance is much less vulnerable to the pest or disease overcoming it, but breeding for it may involve "going backwards"—backcrossing with more primitive strains, thus reviving undesirable characteristics such as the presence of toxins. It is these toxins that kill pests, but they can also make crops unpalatable, even dangerous. If there are no such problems, horizontal resistance is much more of an "appropriate technology" for poorer farmers because of its greater stability.

34. THE IMPORTANCE OF THE CULTIVAR

The growers of ornamental plants are very insistent that a distinction be made between a *cultivar* and any individuals grown from its seed. Plants that are propagated vegetatively maintain a unique genome, thereby allowing the propagation of very genetically complex plants, the result of chance crossing or crossings with a low probability of fertility. There is absolutely no guarantee that anything grown from their seed will have the same characteristics. The selling of seedlings from cultivars by unscrupulous nurserymen has been a problem through the ages—but very tempting if a desirable plant is slow to propagate. Modern laboratory techniques, such as meristem culture and micropropagation, however, have allowed some rarities to be mass-produced (such as orchid hybrids, from as early as the 1960s).

35. GENE BANKS

Most seeds can be preserved for very long periods if kept cold and dry, making them ideal for storage in *gene banks.* Since the 1960s, gene bank creation and management has become a science in itself. Here are some of the key issues:

- Long-term storage of vegetative material is also possible: cuttings, divisions, and so forth., in cold storage or as living plants in *field gene banks.*
- The phenotype is not necessarily a useful guide to the genotype; therefore, it is important that material is collected and preserved which may not appear immediately useful.
- Field gene banks are possible for some crops, but they take up space and are vulnerable.

- To be of use, gene banks need functioning and accessible databases and directories to enable researchers to access material.
- For security reasons, duplicates are vital.
- The creation of *base* collections and *active* collections. The base collection is stored under optimum storage conditions, the active at another location, and is occasionally sown, in order to refresh the collection; this fresh seed can then be used to refresh the base collection if and when this is deemed necessary. However over time this can result in an accumulation of genetic changes.
- Landraces are dynamic, so they are constantly adapting to changes in local conditions, pathogens etc, so seedbanks can arrest evolution.

INDEX TO TERMS USED IN TECHNICAL NOTES

BIBLIOGRAPHIC ESSAY

GENERAL

As a general text on plant breeding, G. S. Chahal and S. S. Gosal's *Principles and Procedures of Plant Breeding* is impossible to beat, especially since chapters often start with a brief summary of the history of the technique in question—historical material can then be followed up with their references. N. W. Simmonds's *Principles of Crop Improvement* has long been a standard text but tends to be less accessible to the nonspecialist. R. W. Allard's *Principles of Plant Breeding* is also very useful, covering domestication and changes in plants under traditional selection.

N. W. Simmonds's *Evolution of Crop Plants* (Harlow, UK: Longman, 1976) is invaluable as a reference to all major and many minor crops, and particularly useful for unraveling the genetics involved. It is, however, now somewhat dated.

One of the most useful studies—and the nearest to a general history of plant breeding there is—is J. H. Perkins's *Geopolitics and the Green Revolution: Wheat, Genes and the Cold War* It covers far more than the title suggests! Coming out too late to be included in this study, but looking very promising, as a general history is D. Murphy's *People, Plants and Genes, The Story of Crops and Humanity.*

ONE. ORIGINS

The origins of crops is an area that has been thoroughly researched, and is one that continues to generate new theories. A number of articles on origins are gathered together in the following books, which draw mainly on archaeological and anthropological evidence: J. B. Hutchinson, ed., *Evolutionary Studies in World Crops* (Cambridge, Cambridge University Press, 1974); D. R. Harris and G. C. Hillman, ed., *Foraging and Farming: The Evolution of Plant Exploitation* (London, Unwin Hyman, 1989); P. J. Ucko and G. W. Dimbleby, ed., *The Domestication and Exploitation of Plants and Animals* (London, Duckworth: 1969); and C. W. Cowan and P. J. Watson, ed., *The Origins of Agriculture: An International Perspective* (Washington, DC: Smithsonian Institution Press, 1992).

J. R. Harlan's *Crops and Man* is one of the best books on the subject of origins by one of the leading authorities—not just on origins from an archaeological standpoint (which dominates most accounts), but also from the perspective of a geneticist and plant breeder. J. Holden, J. Peacock, and T. Williams's *Genes, Crops and the Environment* provides a good general account and continues on to look at a variety of other historical plant breeding issues, up to and including gene banks and GM technology.

The issue of pre-Columbian management of the landscape of the Americas is a fascinating one, covered in C. C. Mann's essay "1491," published in *The Atlantic*; it is a wide-ranging summary of the current thinking about early land management in the New World. S. Budiansky, in *Nature's Keepers* (New York, Free Press: 1995), also discusses this in some detail. The topic is also raised in E. Eisenberg's magisterial *The Ecology of Eden*; he also covers shifting agriculture—the book is an immensely interesting and thought-provoking account of our relationship with nature and the way we change it.

The best account of the disputes over the origin of corn is J. Doebley's "George Beadle's Other Hypothesis: One-Gene, One-Trait," in *Genetics*. Beadle outlined his theory in "Teosinte and the Origin of Maize," in a *Journal of Heredity* article in 1939. Mangelsdorf produced a great deal of material on the subject. Two sources are particularly notable, as they indicate how his ideas developed over time: his original collaboration with P. G. Reeves, also titled "The Origin of Maize," published in a paper in the *Proceedings of the National Academy of Science U S A* in 1938, and a 1958 paper in *Science*, "Ancestor of Corn" (1958).

TWO. LANDRACES

Comparatively little is known about landrace selection. One of the best sources is *Farmers' Bounty* by S. B. Brush, a fascinating account that concentrates on exploring the incredible range of diversity to be found among traditional crops. The book has the advantage that it never sinks to romanticizing the crops and the people discussed; Brush is particularly informative on the modern-day status of landraces, and his bibliography is extremely comprehensive. The same author cowrote, with M. R. Bellon, "Keepers of Maize in Chiapas, Mexico," in *Economic Botany*, a very interesting account of the interplay between old and new in a notoriously restive province. Information on Navdanya, the Indian landrace lobby group, can be found at http://www.navdanya.org, and there are links through to other similar Indian organizations.

D. Christian's *Maps of Time: An Introduction to Big History* discusses the issue of slow rates of innovation in premodern agrarian societies. Ideas about conquest and growth rates in preindustrial societies are derived from G. D. Snooks's *The Dynamic Society: Exploring the Sources of Global Change*; again, a "big history" book but one that develops a more powerful argument—the endogenous nature of historical change.

THREE. "IMPROVEMENT"

The Agricultural Revolution involved little progress in plant breeding as such, but an understanding of it remains essential for an appreciation of the background to the great progress made in breeding in the century following. These sources offer an overview of the period in Britain, and to a lesser extent in western Europe: M. Overton's *Agricultural Revolution in England*, J. D. Chambers and G. E. Mingay's *Agricultural Revolution, 1750–1880*, D. Grigg's *The Dynamics of Agricultural Change*, and Slicher van Bath's *The Agrarian History of Western Europe AD.500–1850*. Grass as forage is looked at specifically in G. Harvey's very readable *The Forgiveness of Nature—The Story of Grass*.

The history of plant breeding is intimately bound up with a history of the seed trade; unfortunately this has been little researched, but useful sources include a two-part article by M. Thick, "Garden Seeds in England before the Late 18th Century," in the *Agricultural History Review*, and B. W. Sarudy's "Nurserymen and Seed Dealers in the 18th Century Chesapeake" in the *Journal of Garden History.*

Developments in fruit and vegetable gardening under Henri IV in France are dealt with briefly and tantalizingly in C. Mukerji's *Territorial Ambitions and the Gardens of Versailles*—a fascinating book, but nothing to do with plant breeding! Information on fruit in Britain is covered by F. A. Roach's *Cultivated Fruits of Britain*.

Song China's own agricultural revolution is discussed by J. Needham and F. Bray in *Science and Civilisation in China*, and is put into a wider context in Bray's *The Rice Economies: Technology and Development in Asian Societies*. Japan's progress in the Tokugawa era is discussed by R. T. Shand in *Technical Change in Asian Agriculture* and by T. Sato in an essay "Tokugawa Villages and Agriculture" in C. Nakane and S. Oishi's *Tokugawa Japan: The Social and Economic Antecedents of Modern Japan.*

As background and context, histories of technology offer insights into the historical processes behind early plant breeding, particularly enlightening has been J. Mokyr's idea of "industrial enlightenment" as developed in *The Gifts of Athena: Historical Origins of the Knowledge Economy.*

FOUR. VEGETABLE MULES

There are two exhaustive studies of pre-Mendelian hybridization, although both miss increasing amounts of material as the nineteenth century advances: C. Zirkle's *The Beginnings of Plant Hybridization*, and H. F. Roberts's *Plant Hybridization before Mendel*. Zirkle has written elsewhere on the subject, notably a paper in *Agricultural History*, "Plant Hybridization and Plant Breeding in Eighteenth-Century American Agriculture." More recent, and digestible, is a study that focuses on 'Fairchild's pink,' *The Ingenious Mr. Fairchild*, by M. Leapman. Corn is the focus for another study of this early period: Wallace and Brown's *Corn and Its Early Fathers*.

That the churches (Catholic and Protestant) found evolution difficult to accept is widely known, but they were not wholly opposed to evolutionary ideas, and a remarkable insight into the role of some British clergymen in early genetics can be appreciated by attempting to read The Honorable and Very Reverend Dean of Manchester William Herbert's "On Hybridisation amongst Vegetables," published in the *Journal of the Horticultural Society of London* in 1846; the prose is extraordinarily wordy by today's standards but worth persevering with.

Flower shows are discussed by B. Elliott in "Flower Shows in Nineteenth-Century England" in *Garden History* and his *The Royal Horticultural Society, A History 1804–2004*.

For some reason, the strawberry has attracted more detailed historical study than practically anything else. S. Wilhelm and J. E. Sagan's *A History of the Strawberry: From Ancient Gardens to Modern Markets* and G. M. Darrow's *The Strawberry: History, Breeding, and Physiology* provide a huge amount of detail; the latter is particularly useful on Duchesne. Sugar beet deserves a more in-depth study; the main sources found were J. W. Robertson-Scott's *Sugar Beet, Some Facts and Some Illusions* and an essay by G. H. Coons in the *USDA Yearbook, 1936*—"Improvement of the Sugar Beet." Material on sugar cane, discussed as part of a wider study of technology transfer in colonial and postcolonial economies is W. W. Ruttan and Y. Hayami's "Technology Transfer and Agricultural Development," in *Technology and Culture*.

We are indebted to V. Orel for raising the profile of the Czech contribution to agricultural and genetic history with his article "The Influence of T. A. Knight (1759–1838) on Early Plant Breeding in Moravia," in *Folia Mendeliana in Acta Musei Moraviae*, and with a much larger-scope book-length study, *Gregor Mendel: The First Geneticist*. Mendel's own role as a plant breeder has been discussed in a number of papers by Orel and M. Vávra, including "Mendel's Program for the Hybridization of Apple Trees" in the *Journal of the History of Biology*.

A study of the criticism of hybridization as ungodly would make an interesting study but there is little recorded; Leapman and Zirkle both discuss it. Lindley 1844 and Anon 1881 in the *Gardeners' Chronicle* and Wilks in the *Journal of the Royal Horticultural Society*, Vol. 24, are the only references I have found to what appears to what could have been widespread anti-hybridisation feeling; they could well be the tip of an iceberg.

FIVE. EMPIRE

Central to the use of plants in the imperial era were the networks of botanical gardens—L. H. Brockway gives an account of the role of the British Royal Botanical Gardens in *Science and Colonial Expansion: The Role of the British Royal Botanical Gardens*.

There is frustratingly little on the great Vilmorin family—and corporate takeovers have apparently dispersed the company records. The best source seems to be "The Role of the Vilmorin Company in the Promotion and Diffusion of the Experimental Science of Heredity in France, 1840–1920" by J. Gayon and D. T. Zallen in the *Journal of the History of Biology*. There is also some material in C. E. Mahan's *Classic Irises and the Men and Women Who Created Them*, which has a lot of information about early French plant breeders—much which has nothing to do with irises!

The development of cereals, notably wheat, is well covered in a variety of sources. Roberts's *Plant Hybridization before Mendel* covers early ground, as does C. Darwin's *The Variation of Animals and Plants under Domestication* and H. de Vries' *Plant Breeding: Comments on the Experiments of Nilsson and Burbank*. "Wheat" in *Biologist*, by F. G. H. Lupton is a good summary. Cereals in Britain are covered by *Wheat in Great Britain* by J. Percival and a paper in *Agricultural History Review* by J. R. Walton, "Varietal Innovation and the Competitiveness of the British Cereals Sector, 1760–1930." There is a good source on Rimpau, "An Early Scientific Approach to Heredity by the Plant Breeder Wilhelm Rimpau (1842–1903)," by A. Meinel, in *Plant Breeding*.

Over the Atlantic, A. L. Olmstead and P. W. Rhode give a good overview in "Biological Innovation in American Wheat Production: Science, Policy and Environmental Adaptation," in *Industrializing Organisms: Introducing Evolutionary History*. The same authors take a wider look at "wheat imperialism" in "Biological Globalization: The Other Grain Invasion" in a paper for *The Historical Society*. A general account of early U.S. plant breeders and other key innovators in agriculture can be found in a popular text by P. De Kruif, *Hunger Fighters*; no sources are given and it is something of a breathless hagiography, but it is very informative. Saunders is covered in J. W. Morrison's "Marquis Wheat—A Triumph of Scientific Endeavor"; early hard wheats in "Turkey Wheat: The Cornerstone of an Empire," by K. S. Quisenberry and L. P. Reitz; and Carleton's battle with the millers in "The Durum Wheat Controversy" by M. W. M. Hargreaves—all in *Agricultural History*. A fascinating portrait of Carleton can be found in "Durum Wheat" by T. Isern for North Dakota Public Radio.

The Potato by L. Zuckerman is a general history; there is a little more on breeding in "Selecting and Breeding for Better Potato Cultivars," by G. R. MacKay, in *Improving Vegetatively Propagated Crops*, as well as a more popular account in A. Romans's *The Potato Book*. J. Dunnett's *A Scottish Breeder's Harvest* has material on Scottish breeders.

Taiwan's rice episode is covered in C. Juma's *The Gene Hunters: Biotechnology and the Scramble for Seeds*, Ruttan and Hayami's "Technology Transfer and Agricultural Development," and *Technical Change in Asian Agriculture* by R. T. Shand. The Mauritius revolt is covered by W. K. Storey in an article in *Agricultural History*, "Small-Scale Sugar Cane Farmers and Biotechnology in Mauritius: The 'Uba' Riots of 1937."

Cotton's complex origins are discussed in "Cotton," by V. Santhanum and J. B. Hutchinson, in *Evolutionary Studies in World Crops*; cotton in Egypt in E. R. J. Owen's *Cotton and the Egyptian Economy 1820–1914*, cotton in the American South by J. H. Moore's "Cotton Breeding in the Old South" in *Agricultural History*. The very interesting story of cotton in India is covered in considerable depth in an article in the Indian *Economic and Political Weekly*, "Suicide Deaths and the Quality of Indian Cotton," by C. Shambu Prasad.

The Johnny Appleseed story is told by M. Pollan in an intriguing sideways look at plant breeding in *The Botany of Desire*; he is very eloquent about American apple history in general. During the 1980s and 1990s, the *Fruit Varieties Journal* ran a series of articles on the origins of leading varieties; two on apples were "The 'Delicious' Apple" by C. D. Fear and P. A. Domotom, and "Golden Delicious Apple—Famous West Virginian Known around the World" by T. A. Baugher and S. Blizzard. Strawberry sources are listed above under "Vegetable Mules."

SIX. BREAKTHROUGH

A key figure of the late nineteenth and early twentieth centuries is Liberty Hyde Bailey; the biography *Liberty Hyde Bailey* by A. D. Rodgers is not only about him, but also gives a thorough and lively portrait of these most exciting times for the plant sciences and for agriculture in the United States. From an academic point of view, W. E. Huffman and R. E. Evenson's *Science for Agriculture: A Long-Term Perspective* is a comprehensive source for information on U.S. state funding of agricultural research during this time.

Perkins's *Geopolitics and the Green Revolution* includes material on Bailey and this era, while Juma's *The Gene Hunters: Biotechnology and the Scramble for Seeds* has some interesting material on the role of introductions. Further material on the interest in Mendel in the United States can be found in: D. B. Paul and B. A. Kimmelman's "Mendel in America: Theory and Practice, 1900–1919," in *The American Development of Biology*, and B. A. Kimmelman's "The American Breeders' Association: Genetics and Eugenics in an Agricultural Context, 1903–13" published in *Social Studies of Science*.

Information on the great hybridization conferences is obtainable from the reports: W. Wilks's "Report of the First International Conference on Hybridisation and Cross-Breeding" in *The Journal of The Royal Horticultural Society* and the *Third International Conference on Genetics; Hybridisation (The Cross-Breeding of Genera or Species)* published as a separate report by the RHS. *Proceedings of the International Conference on Plant-Breeding and Hybridization* in New York was published by the Horticultural Society of New York. There was also a fourth conference, held in Paris, organized by Henri and Philippe de Vilmorin, the report of which I unfortunately was unable to track down; the menus in particular should have made good reading.

An article in *Heredity*, "The Early Days of Genetics," by R. C. Punnett gives a flavor of the times when Bateson was introducing Mendel in Cambridge. Rümker was clearly a major figure in Germany; there is a short online biography, "Kurt von Rümker, 1859–1940," by V. Klemm.

N. Roll-Hansen discusses Svalöf and Johannsen in a book and several papers: "Svalöf and the Origins of Classical Genetics" in *Svalöf 1886–1986 Research and Results in Plant Breeding*, "The Role of Genetic Theory in the Success of the Svalöf Plant Breeding Programme" in *Sveriges Utsädesförenings Tidskrift*, and "The Genotype Theory of Wilhelm Johannsen and Its Relation to Plant Breeding and the Story of Evolution" in *Centaurus*. Perkins discusses this period, with particular attention to Johannsen, in *Geopolitics and the Green Revolution*. Another source is E. Åkerberg's "A Historical Survey of the Breeding Research" in *Svalöf 1886–1986 Research and Results in Plant Breeding*. De Vries' contemporary account makes interesting reading (*Plant Breeding*).

SEVEN. GERMINATION

Acceptance and rejection of Mendelian genetics is discussed by Perkins in *Geopolitics and the Green Revolution*. Differing approaches to plant breeding, their political contexts, and their relationships to the social positions of those who promoted them are discussed in a number of papers: O. Roberts's "Social Imperialism and State Support for Agricultural Research in Edwardian Britain" in *Annals of Science*, and P. Palladino's "Between Craft and Science: Plant Breeding, Mendelian Genetics, and British Universities, 1900–1920" in *Technology and Culture* and "Wizards and Devotees: On the Mendelian Theory of Inheritance and the Professionalization of Agricultural Science in Great Britain and the US, 1880–1930" in *History of Science*. Beaven is a particular focus in Palladino's "Science, Technology, and the Economy: Plant Breeding in Great Britain, 1920–1970" in *Economic History Review*.

Bateson and the origins of Mendelian-led breeding in Britain are discussed in E. J. Russell's *A History of Agricultural Science in Great Britain, 1620–1954* and R. Olby's "Scientists and Bureaucrats in the Establishment of the John Innes Horticultural Institution under William Bateson" in *Annals of Science*. F. Engledow's obituary of Biffen, "Rowland Harry Biffen," in *Obituary Notices of Fellows of The Royal Society*, gives a good portrait of the man. Sir George Stapledon is discussed in "Sir George Stapledon (1882–1960) and the Landscape of Britain," in *Environment and History*, by R. J. Moore-Colyer.

The controversy over Mendel in Germany is covered in "The Reception of Genetic Theory among Academic Plant-Breeders in Germany, 1900–1930," by J. Harwood, and also by N. Roll-Hansen in "The Role of Genetic Theory in the Success of the Svalöf Plant Breeding Programme," both in *Sveriges Utsädesförenings Tidskrift*. The importance of German plant collecting expeditions is raised by R. Pistorius and J. Van Wijk in *The Exploitation of Plant Genetic Information: Political Strategies in Crop Development*. Good sources on France during this period are Gayon and Zallen's "The Role of the Vilmorin Company in the Promotion and Diffusion of the Experimental Science of Heredity in France, 1840–1920" and P.Gouyon's "L'introduction de la génétique en France," a conference paper available online.

Histories of Burpees can be found on the company Web site and in J. Lowe's "Burpees Celebrates its Centennial" in *Flower and Garden Magazine*.

EIGHT. LUTHER BURBANK

W. S. Harwood's *New Creations in Plant Life* is a thorough but utterly—and sometimes even ludicrously—starry-eyed contemporary portrait of Burbank; very much a period piece, it gives quite an insight into the mentality of the times. W. L. Howard's "Luther Burbank: A Victim of Hero Worship," in *Chronica Botanica*, is sufficiently far-removed from Burbank's life and the reaction against him to get some objectivity; it is self-consciously fair and was written in an attempt to try to reassess his reputation. P. Palladino's "Wizards and Devotees" locates his work in the context of the development of genetics in the United States and makes comparisons with the very different social context in Britain.

NINE. "LET HISTORY JUDGE"

The N. I. Vavilov Research Institute of Plant Industry Web site has some basic historical material on the early days of plant breeding and collection in Russia and the USSR—see VIR's "Organization of VIR's Work for Plant Genetic Resources" and Loskutov's *Vavilov and His Institute : A History of the World Collection of Plant Genetic Resources in Russia*. There are three very thorough accounts of the Lysenko affair. Z. Medvedev's *The Rise and Fall of T. D. Lysenko* is all the more interesting for having been written "on the inside"; D. Joravsky's *The Lysenko Affair* is a comprehensive study of the whole business and firmly nails the myth that Lysenko's approach to plant breeding was inspired by Marxism; and W. G. Hahn's *The Politics of Soviet Agriculture* is useful for wider context.

For those who want to appreciate the extraordinary Michurin story from original sources, I. V. Michurin's *Selected Works* makes fascinating reading. Gluttons for punishment may find a copy of T. Lysenko's *Agrobiology: Essays on Problems of Genetics, Plant Breeding, and Seed Growing* (Moscow: Foreign Language Publishing House, 1954) on the library shelf, too.

The story of the involvement of the SS in the theft of genetic material comes from *Biologists under Hitler* by U. Deichmann. There is also a rather tendentious account by C.-G. Thornstrom and U. Hossfeld, "Instant Appropriation-Heinz Brücher and the SS Botanical Collecting Commando to Russia 1943" on the Web site of Bioversity International.

TEN. HYBRID

Two essential sources on corn are *Corn and Its Early Fathers* by H. A. Wallace (the man himself) and W. L. Brown, which tells the story from the very early days; and *The Hybrid-Corn Makers: Prophets of Plenty* by A. R. Crabb, which goes into much more detail on the late-nineteenth- and early-twentieth-century origins of hybrid corn. M. Goodman's account of the crop, "Maize," in N. Simmonds's *Evolution of Crop Plants* is a useful summary. A paper in *Agricultural History* by E. Anderson and W. L. Brown, "The History of the Common Maize Varieties of the United States Corn Belt," is a discussion of early farmer hybridization—revealed partly through oral history research. Another paper, "Background to US hybrid corn," by A. F. Troyer, in *Crop Science* is also informative on the nineteenth century. Not surprisingly, given his status as corn king, P. Mangelsdorf has plenty to say on all aspects of the crop in his *Corn, its Origin, Evolution and Improvement*.

The Illinois corn test is discussed in detail by I. L. Goldman in "The Intellectual Legacy of the Illinois Long-Term Selection Experiment," in *Plant Breeding Reviews*.

There is little specifically on Henry A. Wallace's career in corn, but there is at least one very detailed biography: J. C. Culver and J. Hyde's *American Dreamer: A Life of Henry A. Wallace*. A very sympathetic account is available online; —see the Henry A. Wallace program script, produced by Iowa Public Radio; for a more skeptical account from one of the United States' greatest political historians, see A. Schlesinger's "Who Was Henry A. Wallace? The Story of a Perplexing and Indomitably Naive Public Servant," published by *The Los Angeles Times* and also available online.

Barbara McClintock continues to be an endless source of fascination. A basic biography of her is available online, The Barbara McClintock Papers, published by *Profiles in Science*. There is also a book—*The Tangled Field—Barbara McClintock's Search for the Patterns of Genetic Control*, by N. C. Comfort, but there are questions over the accuracy of some of the technical material included.

D. Fitzgerald has written a lot on the corporate takeover of corn breeding, but from a relatively dispassionate viewpoint, in *The Business of Breeding, Hybrid Corn in Illinois, 1890–1940*, and she discusses the question of whether or not farmers were "de-skilled" in "Farmers Deskilled: Hybrid Corn and Farmers' Work," in *Technology and Culture*. P. Kloppenburg, in *First the Seed: The Political Economy of Plant Biotechnology, 1492–2000,* develops the argument that hybrid corn was driven by business as part of a wider argument about the need for capitalism to turn the seed into a commodity; he offers a scholarly, readable, and incisive Marxist analysis, but like many such analyses, there is the definite sound of square pegs being hammered into round holes. He is essential reading for anyone interested in looking at deeper into the hybrid story—and one of the most stimulating of all the books listed here. Pistorius and Van Wijk, in *The Exploitation of Plant Genetic Information*, take the argument further but do not have such a political axe to grind. Paul and Kimmelman's "Mendel in America" gives a rounded account of the debate over hybrid corn and its background, including a discussion of the genetics of what could have been possible alternatives. The classic Marxist account of hybrid corn though is by J. D. Berlan and R. C. Lewontin: "The Political Economy of Hybrid Corn" in *Monthly Review* follows on from a wider discussion of how capitalism has taken over US agriculture, previously published in the journal; essentially, they claim to show how hybrid corn is one big fraud.

A fascinating study of corn in Africa has been written by J. C. McCann, *Amazing Grace, Africa's Encounter with a New World Crop 1500–2000*, from which comes the Rhodesian/Zimbabwean corn story.

ELEVEN. CORNUCOPIA

So extensive and disparate a field plant breeding becomes during the twentieth century that there are few sources to turn to for an overview. All the sources discussed under "General" above are useful, however, at least for summarizing developments and giving references for more detailed research. Two journals are vital for getting an overview, both of the research and for the occasional historical article: *Euphytica* and *Plant Breeding Reviews.* The latter also often has biographical material on plant breeders of importance. Inevitably, in a book aimed at the general reader, much material only comprehensible to the professional has been left out; two sources would be of interest to those in the field: "Plant Breeding, Development and Success" by D.C. Smith in *Plant Breeding, A Symposium held at Iowa State University*, and *Historical Perspectives in Plant Science,* edited by R. H. Burriss and K. J. Frey.

Mutation breeding is covered very extensively in *Mutation Breeding, Theory and Practical Applications*, by A. M. Van Harten, which includes a lot of fascinating historical and biographical material. All that is missing is a DIY guide to building a gamma-field in the garden.

Yuan Longping may well be remembered as one of the "greats" of the last quarter of the twentieth century; his story is told by J. Li and Y. Xin in "Longping Yuan: Rice Breeder and World Hunger Fighter" in *Plant Breeding Reviews.*

The story of the Californian tomato harvesting machine has become a classic tale of mechanization; my source, W. H. Friedland and A. Barton's "Tomato Technology" was published in *Society*, although this episode has been written up by many others; L. Winner's study of the social and philosophical implications of technological advance, *The Whale and the Reactor*, is one that contextualizes it as part of a wider set of issues.

TWELVE. GREEN REVOLUTION

There is a vast literature on the Green Revolution. It tends to be divided into "insider" accounts—although there are frustratingly few of these—academic-type studies, and critical narratives. Perkins's *Geopolitics and the Green Revolution* is a good general account that contextualizes the events in wider plant breeding history.

The Man Who Fed the World, by L. Hesser, is a biography of Norman Borlaug; *Campaigns against Hunger*, by E. C. Stakman, R. Bradfield, and P. C. Mangelsdorf, is an invaluable and detailed memoir of work in Mexico by three key participants. The story behind IR-8 and other earlier International Rice Research Institute–bred rices is covered in "Breeding History, " by T. Hargrove and W. R. Coffman, in *Rice Today*. *Beyond IR8, IRRI's Second Decade*, published by IRRI, is a follow-up.

Rice Science and Development Politics by R. S. Anderson, E. Levy, and B. M. Morrison, is a good and objective account of rice issues, including the story of IRRI. HYV rice and its social and economic effects in a range of countries are covered in *Modern Rice Technology and Income Distribution in Asia*, edited by C. C. David and K. Otsuka, and in I. Palmer's *The New Rice in Asia: Conclusions from Four Country Studies*.

The Green Revolution Reconsidered, edited by P. B. R. Hazell, and C. Ramasamy, is a collection of in-depth studies in part of Tamil Nadu state, India; their introduction is a good roundup of the various studies undertaken on the Green Revolution in the two decades since its inception. *The Green Revolution in India: A Perspective*, by B. Sen, gives a perspective from India. *New Seeds and Poor People*, by M. Lipton, is one of the best accounts of the economics of why so many stayed poor, despite the Green Revolution.

G. Conway's *The Doubly Green Revolution* provides one of the most readable and fairest overviews, written in the context of looking ahead to the need for future production increases and sustainability. Two papers by J. Harriss in *Green Revolution? Technology and Change in Rice-growing Areas of Tamil Nadu and Sri Lanka* ("Implications of Changes in Agriculture for Social Relationships at the Village Level: The Case of Randam" and "Social Implications of Changes in Agriculture in Hambantota District") are informative about social changes and the rural to urban shift in social networks.

Issues of social unrest are raised in F. R. Frankel's *India's Green Revolution: Economic Gains and Political Costs*. Nanda takes this case further with her arguments that agricultural change helped stimulate the rise of Hindu nationalism in *Prophets Facing Backward, Postmodern Critiques of Science and Hindu Nationalism in India*; she makes some very interesting observations on the mind-set of rural India, and on social commentators in the country.

Critics of the Green Revolution who argue their case include: A. Pearse, in *Seeds of Plenty, Seeds of Want*; K. A. Dahlberg, in *Beyond the Green Revolution, The Ecology and Politics of Global Agricultural Development*; and K. Griffin, in *The Political Economy of Agrarian Change*

C. Hewitt de Alcantara's *Modernizing Mexican Agriculture: Socioeconomic Implications of Technological Change, 1940–1970* is the most powerful critique in many ways, as it is thoroughly evidence based and is published by the United Nations Research Institute for Social Development.

F. M. Lappé and J. Collins make the best-argued radical critique in *Food First: the Myth of Scarcity*. V. Shiva's *The Violence of the Green Revolution* is the author's main text on the subject; her 2000 Reith lecture "Poverty and Globalisation" (available online) gives a good flavor of her voluminous—and repetitive—writings.

THIRTEEN. ORNAMENT

Information on the history of ornamental plant breeding is frustratingly thin and scattered. Two books by R. Gorer, *The Development of Garden Flowers* and *The Flower Garden in England*, give an overview of both ornamental plants and planting design, but only for Britain. A. Coats's *Flowers and their Histories* is an A–Z but has little on breeding.

The florists are discussed by R. E. Duthie in two papers in *Garden History*: "Florists' Feasts and Feasts after 1750" and "English Florist Societies and Feasts in the Seventeenth and the First Half of the Eighteenth Centuries." *Tulipomania: The Story of the World's Most Coveted Flower and the Extraordinary Passions It Aroused* by M. Dash is a history of tulipomania; A. Pavord's very detailed and readable account in *The Tulip* gives much more of the history of the flower. However, a definitive scholarly account of tulip madness has yet to be written. Japanese "floristry" is covered in what was certainly the most opulent and beautiful book consulted during this research—*History and Principle of the Traditional Floriculture in Japan*, by S. Kashioka and M. Ogisu.

The Early Horticulturalists, by R. Webber, is one of the few books on the early nursery trade in Britain, covering the period until the early Victorian era. *The Genetics of Garden Plants*, by M. B. Crane and W. J. C. Lawrence, covers many commonly grown genera—but is now very dated. *History of the Rose*, by R. E. Shepherd, is one of the few even remotely comprehensive histories of our favorite flower. About the only source on rhododendron hybridizing is J. Brown's *Tales of the Rose Tree*, in a lively account of the genus in cultivation. Histories of other genera may also be tracked down in the *Journal of the Royal Horticultural Society*.

Appreciating changes in nineteenth-century planting design and fashion in plants is an essential background for understanding the context in which breeders work. B. Elliott in *Victorian Gardens* and V. Scourse in *The Victorians and their Flowers* provide much of this background. P. C. Newcomb's *Popular Annuals of Eastern North America, 1865–1914* discusses the development of annuals in North America and includes details and facsimile material from original catalogs. A fascinating take on many different aspects of horticulture in Victorian England is provided by M. Waters's *The Garden in Victorian Literature*.

For those interested in plant breeding as an art form, the work of George Gessert is an invaluable source. A number of his essays have been published in the journal *Leonardo.*

FOURTEEN. OWNERSHIP AND DIVERSITY

An area of rapid change and controversy generates a lot of material. *The Exploitation of Plant Genetic Information: Political Strategies in Crop Development*, by R. Pistorius and J. Van Wijk, gives a very thorough and fair account of the various disputes over the "ownership" of plant genes, the establishment of seed banks, the early history of the international breeding bodies, and the story of plant patent laws in the context of the relations between agriculture and capital since the mid-nineteenth century. C. Fowler and R. Lower's paper in *Plant Breeding Reviews*, "The Politics of Plant Breeding." is also a valuable account of a variety of issues and controversies in this area, with a major review of the literature and sources of information. W. E. Huffman and R. E. Evenson, in their *Science for Agriculture: A Long-Term Perspective*, provide a succinct academic account of public/private issues. From the anticorporate activists, one of the best-written and most thoroughly researched sources is *Hungry Corporations, Transnational Biotech Companies Colonise the Food Chain*, by H. Paul and R. Steinbrecher.

C. Juma, a Kenyan author and a major figure in the world food security scene, in *The Gene Hunters: Biotechnology and the Scramble for Seeds*, looks at a wide range of issues connected with biodiversity and crop breeding, with many interesting plant breeding stories. Kloppenburg (*First the Seed*) also provides an invaluable account. It is he who makes the point about the dual character of seed resisting its commodification; the idea of the triple character of seed is made by Pistorius and Van Wijk. "The End of Farm-saved Seed?" on the GRAIN Web site (2006) is a useful and detailed discussion of the history of plant patenting with very extensive citations.

The 1970 Southern corn blight crisis is discussed in detail in *Altered Harvest: Agriculture, Genetics, and the Fate of the World's Food Supply* by J. Doyle, who uses it to build a case against GM crops and corporate farming. A short account is also given by P. Mangelsdorf in "Donald Forsha Jones" in *Biographical Memoirs.* The affair of Dr. Richharia's rice is possibly one of the most interesting in the whole field of variety "ownership"—I suspect it deserves a book. Juma discusses it with a nonpartisan voice; C. Alvares's feature in the *Illustrated Weekly of India*, "Crushed, but Not Defeated : An Interview with Dr. Richharia," is based on an interview by a definite sympathizer, while G. N. Reddi's essay, "Rebuilding the Genetic Resource Base through Farmer-Scientist-Activist Alliance," on the Web site of the *International Development Research Centre*, also gives an account in the context of a radical "grassroots" approach to agricultural research.

Information on ownership issues in the Netherlands can be found in Kloppenburg, and a number of articles in the Dutch plant breeding journal *Euphytica*: J. C. Dorst's "A Quarter of a Century of Plant Breeding in the Netherlands" and H. De Haan's "Factors Affecting the Breeding of Agricultural Crops in The Netherlands."

A study that suggests that increased concentration of ownership of plant breeding resources limits innovation is by D. Schimmelpfennig, C. E. Pray, and M. Brennan's "The Impact of Seed Industry Concentration on Innovation: A Study of U.S. Biotech Market Leaders," published by the *USDA Economic Research Service*. The GRAIN Web site has a considerable amount of material on food sovereignty issues—some well supported, some partisan.

CGIAR's Web site provides plenty of information on their work, and links through to all the other agencies. Other agency websites vary; that of IRRI is particularly good on its history and the accessibility of historical material. A paper by J. M. Alston and P. G. Pardey, "Developing-Country Perspectives on Agricultural R&D: New Pressures for Self-Reliance?" published by the *International Food Policy Research Institute*, gives a concise account of the history of the CGIAR and the other agencies in the context of changes in agricultural R&D in the developing world.

"Client-Orientated Breeding (COB): Bridging Yield Gaps and Addressing Food and Income Security," by K. Joshi in the online version of *New Agriculturalist*, is an extremely useful account of the gradient of participatory breeding approaches. "The Promise of Participation: Democratizing the Management of Biodiversity," in GRAIN's *Seedling* by M. Pimbert, looks at these issues within a wider discussion of participatory methods in research and development. Details of the very inspiring Rwanda bean project can be found in "Partners in Selection, Bean Breeders and Women Bean Experts in Rwanda" published by the CGIAR Gender Program.

Issues in the rapidly developing world of organic breeding are covered in "Sustainable Organic Plant Breeding, Subproject 1—Discussion Paper: Defining a Vision and Assessing Breeding Methods," by Lammerts van Bueren et al., published by the *Louis Bolk Institute*; "Organic Plant Breeding," by M. S. Wolfe, included in the *UK Organic Research 2002: Proceedings of the COR Conference*; and "Ecological Cereal Breeding and Genetic Engineering," by C. Karutz, published by the Swiss Research Institute for Organic Agriculture. The organic movement has a very different history from country to country; Britain's is covered in an informative and readable account by P. Cornford—*The Origins of the Organic Movement*.

FIFTEEN. CONCLUSIONS

Two books in particular have shaped my thoughts on history and its Malthusian or economically driven cycles: D. Christian's *Maps of Time*, and G. D. Snooks's *The Dynamic Society*. Any discussion of how agriculture impacts ecology and therefore the resource base of civilizations is inevitably colored by J. Diamond's very wide-ranging and powerfully-argued *Collapse, How Societies Choose to Fail or Survive*; he makes a very strong case for the relevance of Malthus to the Rwandan genocide. The unsustainability of much ancient agriculture is discussed by E. Eisenberg in *The Ecology of Eden*.

M. R. Smith and L. Marx, ed., in *Does Technology Drive History, The Dilemma of Technological Change*, bring together a number of points of view on the question of whether technology does drives history.

Fordism and postmodernism are discussed in *The Condition of Postmodernity* by D. Harvey—this is a readable, refreshingly jargon-free, and wide-ranging account of subjects usually deeply buried in material with precisely the opposite qualities. On the subject of declining trust in science, see Marx's essays "The Idea of 'Technology' and Postmodern Pessimism" in Smith and Marx's *Does Technology Drive History* and "Reflections on the Neo-Romantic Critique of Science" (in *Daedalus*) and O. Bennett's *Cultural Pessimism: Narratives of Decline in the Postmodern World.* J. Turney forms an interesting discussion of the issues around science, genetics and popular culture in *Frankenstein's Footsteps, Science, Genetics and Popular Culture.* The dark sides of postmodernism and left-wing postcolonial thinking are explored in *The Seduction of Unreason, The Intellectual Romance with Fascism, from Nietzsche to Postmodernism*, by R. Wolin, and *Prophets Facing Backward, Postmodern Critiques of Science and Hindu Nationalism in India*, by M. Nanda, who offers by far the most intelligent critique of Vandana Shiva and the Indian anti-science polemicists.

N. Simmonds was not only one of the greatest historians of crop plants, but a noted teacher and a sometimes-controversial breeder, as well. His trenchant thoughts in "Bandwagons I Have Known," published in the *Tropical Agriculture Newsletter*, are salutary and well worth reading. Fowler and Lower's paper "The Politics of Plant Breeding" is an extremely valuable essay, covering much ground—free from any political dogma but arguing powerfully that something is going terribly wrong in modern breeding.

V. Smil's *Feeding the World, A Challenge for the Twenty-First Century* is simply the best all-round introduction to—and overview of—all the problems of this great question, full of basic facts about all the interconnected issues and remarkably readable; his succinct, balanced and dispassionate scholarship is a joy—a shame, though, about his sarcastic remarks about vegetarians! Similarly, not about plant breeding history but a very readable account of crop breeding in the developing world, is R. Manning's *Food's Frontier.*

The story about Bt cotton in India is found and discussed in great detail in R. J. Herring's "Why Did 'Operation Cremate Monsanto' Fail? Science and Class in India's Great Terminator-Technology Hoax," in *Critical Asian Studies.*

WORKS CITED

Please note that *The Garden* has also been known as the *Journal of the Royal Horticultural Society* since 1977. Online addresses are given only for publications which may be accessed by the general public.

Åkerberg, E. 1986. "A Historical Survey of the Breeding Research," in *Svalöf 1886–1986, Research and Results in Plant Breeding*, ed. G. Olsson, 9–34. Stockholm: LTs Förlag.

Alexander, J., and D. G. Coursey. 1969. "The Origins of Yam Cultivation," in *The Domestication and Exploitation of Plants and Animals*, ed. P. J. Ucko and G. W. Dimbleby, 405–26. London: Duckworth.

Aliev, J., P. Gandilian, P. Naskidashvili, and A. Morgounov. 2001. "Caucasian Wheat Pool," in *The World Wheat Book: A History of Wheat Breeding*, ed. A. P. Bonjean and W. J. Angus, 831–49. Paris: Lavoisier.

Allard, R.W. 1999. *Principles of Plant Breeding*. New York: John Wiley.

Alston, J. M., and P. G. Pardey. 2006. "Developing-Country Perspectives on Agricultural R&D: New Pressures for Self-Reliance?" in *Agricultural R&D in the Developing World: Too Little, Too Late?* ed. P. G. Pardey, J. M. Alston, and R. M. Piggott, 11–28.Washington, DC: International Food Policy Research Institute.

Alston, J. M., S. Dehmer, and P. G. Pardey. 2006. "International Initiatives in Agricultural R&D: The Changing Fortunes of the CGIAR," in *Agricultural R&D in the Developing World: Too Little, Too Late?* ed. P. G. Pardey, J. M. Alston, and R. M. Piggott, 313–60. Washington, DC: International Food Policy Research Institute.

Alvares, C. 1986. "Dr. Richharia's Story—Crushed, but not Defeated." http://www.satavic.org/richharia.htm (accessed March 22, 2007).

Ambrosoli, M. 1997. *The Wild and the Sown, Botany and Agriculture in Western Europe, 1350–1850*. Cambridge: Cambridge University Press.

Anderson E., and W. L. Brown. 1952. "The History of the Common Maize Varieties of the United States Corn Belt." *Agricultural History* 26 (1): 2–8.

Anderson, E. 1967. *Plants, Man and Life*. Berkeley: University of California Press.

Anderson, R. S., E. Levy, and B. M. Morrison. 1991. *Rice Science and Development Politics*. Oxford: Clarendon Press.

Anon. 1845. Editorial. *Gardeners' Chronicle and Agricultural Gazette* 10, 155.

Anon. 1846. Editorial. *Gardeners' Chronicle and Agricultural Gazette* 36, 601.

Anon. 1881. Editorial. *The Gardeners' Chronicle* 15, 48.

Anon. 1923. Obituary. *The Rev. William Wilks. Journal of The Royal Horticultural Society* 48:157–160.

Anon. [former FAO Schwartz employee]. 2006. Personal communication. October.

Apps, D. 1986. Introd. to *Daylilies*, by A. B. Stout. New York: Sagapress.

AScribe Newswire. 2002. "Zapatista Seed Saving Project Puts Its First Collection of Traditional Corn Seeds Into Deep Freeze Storage in Highlands of Chiapas, Mexico." *Global Exchange*. http://www.globalexchange.org/countries/americas/mexico/biodiversity/354.html. (accessed March 5, 2007).

Bailey, L. H. 1896. "The Improvement of Our Native Fruits," *Yearbook of the US Dept. of Agriculture 1896*. Washington, DC: Government Printing Office.

———. 1917. "Victor Lemoine, Plant Hybridist." *The Garden Magazine*, May, 234. http://www.earthlypursuits.com/GardenMag/GardenMag0517–234.htm (accessed March 14, 2007).

Baker, C. J. 1984. "Frogs and Farmers, The Green Revolution in India, and Its Murky past," in *Understanding Green Revolutions, Agrarian Change and Development Planning in South Asia*, ed. T. P. Bayliss-Smith and S. Wanmali, 37–53. Cambridge: Cambridge University Press.

Banga, O. 1976. "Carrot," in *Evolution of Crop Plants*, ed. N. W. Simmonds, 291–93. Harlow, UK: Longman.

The Barbara McClintock Papers. See McClintock, Barbara.

Baugher, T. A., and S. Blizzard. 1987. "Golden Delicious Apple—Famous West Virginian Known around the World." *Fruit Varieties Journal* 41 (4): 134–42.

BBC. n.d. "Robert Bakewell (1725–1795)." http://www.bbc.co.uk/history/historic_figures/bakewell_robert.shtml (accessed October 25, 2006).

Beadle, G. W. 1939."Teosinte and the Origin of Maize." *Journal of Heredity* 30:245–47.

Beales, P. 1992. *Roses*. London: Harper Collins.

Beaty, J. Y. 1954. *Plant Breeding for Everyone*. Boston: Charles Branford.

Bellon, M. R., and S. B. Brush. 1994. "Keepers of Maize in Chiapas, Mexico." *Economic Botany* 48 (2): 196–209.

Bennett, O. 2001. *Cultural Pessimism: Narratives of Decline in the Postmodern World*. Edinburgh: Edinburgh University Press.

Berlan, J. D., and R. C. Lewontin. 1986. "The Political Economy of Hybrid Corn." *Monthly Review* 38 (3): 35–47.

Besler, B. 2000. *The Garden at Eichstätt*. Facsimile edition, Cologne: Taschen.

Bhalla, G. S., and G. K. Chadha. 1983. *Green Revolution and the Small Peasant: A Study of Income Distribution among Punjab Cultivators*. New Delhi: Concept Publishing.

Biotechnology Commission. 2005. "Who Shapes the Research Agenda in Agricultural Biotechnology—Plant Breeding Case Study." London: Agriculture and Environment Biotechnology Commission. http://www.aebc.gov.uk/aebc/subgroups/ra_plant_breeding.pdf (accessed March 23, 2007).

Blumer, M. A. 1996. "Ecology, Evolutionary Theory and Agricultural Origins." in *The Origins and Spread of Agriculture and Pastoralism in Eurasia*, ed. D. R. Harris, 25–50. Washington, DC: Smithsonian Institution Press.

Bonjean, A. P., G. Doussinault, and J. Stragliati. 2001. "French Wheat Pool," in *The World Wheat Book: A History of Wheat Breeding*, ed. A. P. Bonjean and W. J. Angus, 127–65. Paris: Lavoisier.

Borlaug, N. 1983. "Contributions of Conventional Plant Breeding to Food Production." *Science* 219:689–93.
Bourne, V. 2007. "Amos Perry, 20th century breeder and nursery owner." *The Garden* 132 (8): 524–27.
Boyce, J. K. 1987. *Agrarian Impasse in Bengal: Institutional Constraints to Technological Change*. Oxford: Oxford University Press.
Boze Down Vineyard. 2006. "The Wine Police—or—What do Belgians Know about Wine Anyway?" http://www.bozedown.com/Thewinep.htm (accessed March 15, 2007).
Braun, H.-J., N. Zencirci, F. Altay, A. Atli, M. Avci, V. Eser, M. Kambertay, and T. S. Payne. 2001. "Turkish Wheat Pool," *The World Wheat Book: A History of Wheat Breeding*, ed. A. P. Bonjean and W. J. Angus, 817–30. Paris: Lavoisier.
Bray, F. 1986. *The Rice Economies: Technology and Development in Asian Societies.* Oxford: Blackwell.
Brinch, P. 2005. "Seeds of Concern." *Bio-Dynamic Agricultural Association Newsletter* December, 2005, 4. http://www.biodynamic.org.uk/farming-gardening/seeds/cms-hybrids-are-prohibited.html (accessed November, 10, 2008).
Brockway, L. H. 1979. *Science and Colonial Expansion: The Role of the British Royal Botanical Gardens*. New York: Academic Press.
Brown, L. R. 1970. *Seeds of Change, The Green Revolution and Development in the 1970s*.New York: Praeger.
Brown, W. L. 1990. "USDA Contributions to Plant Genetics." *Agricultural History* 64 (2): 315–18.
Brown, J. 2004. *Tales of the Rose Tree*. London: Harper Collins.
Brown, R. 2007. Personal communication. April.
Browning, F. 1998. *Apples*. New York: North Point Press.
Brush, S. B. 2004. *Farmers' Bounty*. New Haven, CT: Yale University Press.
Buller, A. H. R. 1919. *Essays on Wheat*. New York: Macmillan.
Burpee. n.d. "The Legacy of W. Atlee Burpee." http://www.burpee.com/contentarticle.do?itemID=574&KickerID=100270&KICKER (accessed August 22, 2008).
Burriss, R. H., and K. J. Frey., eds. 1996. *Historical Perspectives in Plant Science*. Ames: Iowa State University Press.
Cahoon, C. A. 1986. "The Concord Grape." *Fruit Varieties Journal* 40 (4): 106–7.
Cahoon, G. A. 1996. "History of the French Hybrid Grapes in North America." *Fruit Varieties Journal*. 50 (4): 202–16.
Cameron, J. W., and R. K. Soost. 1976. "Citrus." In *Evolution of Crop Plants*, ed. N. W. Simmonds, 261–64. Harlow, UK: Longman.
Campbell, G. K. G. 1976. "Sugar Beet." In *Evolution of Crop Plants*, ed. N. W. Simmonds, 25–28.Harlow, UK: Longman.
Carter and Co. 1900–1920. Seed catalogs.
CGIAR. See Consultative Group on International Agricultural Research.

Chahal, G. S., and S. S. Gosal. 2002. *Principles and Procedures of Plant Breeding, Biotechnological and Conventional Approaches*. New Delhi: Narosa.
Chambers, J. D., and G. E. Mingay. 1966. *The Agricultural Revolution, 1750–1880*. London: Batsford.
Chang, T. T. 1976. "The Rice Cultures." *Philosophical Transactions of the Royal Society* B.275: 901–6.
Chesshire, C. 2007. Personal communication. May 22.
Chidham and Hambrook Parish Council. 2006. "Chidham and Hambrook—Village History." http://www.chidhamandhambrook.info/everyday/your_village/history.htm (accessed Aug 29, 2006).
Christian, D. 2004. *Maps of Time, An Introduction to Big History*. Berkeley: University of California Press.
Coats, A. 1956. *Flowers and their Histories*. London: Hulton Press.
Collins, G. N. 1910. "Increased Yields of Corn from Hybrid Seed." In *USDA Yearbook, 1910*. Washington, DC: Government Printing Office.
Comfort, N. 2001. *The Tangled Field—Barbara McClintock's Search for the Patterns of Genetic Control*. Cambridge, MA: Harvard University Press.
Consultative Group in International Agricultural Research (CGIAR). 1996. Financial Report. Washington, DC: CGIAR.
Consultative Group in International Agricultural Research (CGIAR) Gender Program. 1994. "Partners in Selection, Bean Breeders and Women Bean Experts in Rwanda." *CGIAR Gender Program*, Washington DC: CGIAR. http://www.cgiar.org/publications/gender/index.html (accessed March 26, 2007).
Conway, G. 1997. *The Doubly Green Revolution*. London: Penguin.
Coons. G. H. 1936. "Improvement of the Sugar Beet." In *USDA Yearbook, 1936*. Washington, DC: Government Printing Office.
Cornford, P. 2001. *The Origins of the Organic Movement*. Edinburgh: Floris Books.
Crabb, A. R. 1947. *The Hybrid-Corn Makers: Prophets of Plenty*. New Brunswick, NJ: Rutgers University Press.
Crane, M. B., and W. J. C. Lawrence. 1952. *The Genetics of Garden Plants*. London: Macmillan.
Culver, J. C., and J. Hyde. 2000. *American Dreamer: A Life of Henry A. Wallace*. New York: Norton.
Curran, J. R. 2004. *The Great American Apple Wizard*. Salisbury, MD: Endeavor Publishing.
Czembor, H. J., J. H. Czembor, I. Menke-Milczarek, and J. Zimny. 2001. "Polish Wheat Pool," in *The World Wheat Book: A History of Wheat Breeding*, ed. A. P. Bonjean and W. J. Angus, 219–42. Paris: Lavoisier.
Dahlberg, K. A. 1979. *Beyond the Green Revolution: The Ecology and Politics of Global Agricultural Development*. New York: Plenum Press.
Dalrymple, D. G. 1988. "Changes in Wheat Varieties in the United States, 1919–1984." *Agricultural History* 62 (4): 20–36.

Damiana, A. B. 2003. "The Early History and Spread of Coffee." *Asian Agri-History* 7 (1) 67–74.
Danial, D., J. Parlevliet, C. Almekinders, and G. Thiele.2007. "Farmers' Participation and Breeding for Durable Disease Resistance in the Andean Region." *Euphytica* 153 (3): 385–96.
Darlington, C. D., J. B. Hair, and R. Hurcombe. 1951. "The History of the Garden Hyacinths." *Euphytica* 5 (2): 233–52.
Darrow, G. M. 1966. *The Strawberry: History, Breeding, and Physiology*. New York: Holt, Rinehart & Winston.
Darwin, C. 1875. *The Variation of Animals and Plants under Domestication*. 2nd ed. London: John Murray. http://darwin-online.org.uk (accessed May 20, 2007).
Dash, M. 1999. *Tulipomania : The Story of the World's Most Coveted Flower and the Extraordinary Passions It Aroused*. London: Gollancz.
David, C. C., and K. Otsuka, K. 1994. *Modern Rice Technology and Income Distribution in Asia*. London: Lynne Riener Publications.
Davies, D. R., and C. L. Hedley. 1975. "The Induction by Mutation of All-Year-Round Flowering in Streptocarpus." *Euphytica* 24 (1): 269–75.
Day, B. E. 1978. "The Morality of Agronomy," in *Agronomy in Today's Society*, ed. J. W. Pendleton, 19–28. Special publication no. 33. Madison, WI: American Society of Agronomy.
Day, R. D., and I. Singh. 1977. *Economic Development as an Adaptive Process: The Green Revolution in the Indian Punjab*. Cambridge: Cambridge University Press.
De Baedemaeker, J. 1994. "Adapting Crop Properties for Efficient Mechanisation." *Acta Horticulturae* 355:77–54.
De Beer, G. 1966. "Genetics: The Centre of Science." *Proceedings of the Royal Society of London*. B.164 (995): 154–66.
De Haan, H. 1958. "Geert Veenhuizen (1857–1930), The Pioneer of Potato Breeding in The Netherlands." *Euphytica* 7: 31–37.
———. 1959. "Factors Affecting the Breeding of Agricultural Crops in The Netherlands." *Euphytica* 8 (3): 183–95.
De Kruif, P. 1928. *Hunger Fighters*. New York: Harcourt, Brace, and World.
De Tapia, E. M. 1992. "The Origins of Agriculture in Mesoamerica and Central America," in *The Origins of Agriculture: An International Perspective*, ed. C. W. Cowan and P. J. Watson, 143–71. Washington, DC: Smithsonian Institution Press.
De Vries, H. 1906. *Species and Varieties, Their Origin by Mutation*. London: Kegan Paul, Trench, Trübner & Co.
———. 1907. *Plant Breeding: Comments on the Experiments of Nilsson and Burbank*. London: Kegan Paul, Trench, Trübner & Co.
Deichmann, U. 1996. *Biologists under Hitler*. Cambridge, MA: Harvard University Press.
DePauw, R., and A. Hunt. 2001. "Canadian Wheat Pool," in *The World Wheat Book:*

A History of Wheat Breeding, ed. A. P. Bonjean and W. J. Angus, 479–515. Paris: Lavoisier.
Destler, M. C. 1968. "'Forward Wheat' for New England: The Correspondence of John Taylor of Caroline with Jeremiah Wadsworth, in 1875." *Agricultural History* 42 (3): 201–10
DeWan, G. 2007. "The Blooming of Flushing." *Newsday*. http://www.newsday.com/community/guide/lihistory/ny-history-hs329a,0,6961092.story (accessed November 24, 2007).
Diamond, J. 1997. *Guns, Germs and Steel: A Short History of Everybody for the Last 13,000 Years*. London: Jonathan Cape.
———. 2005. *Collapse: How Societies choose to fail or survive*. London: Penguin.
Doebley, J. 2001. "George Beadle's Other Hypothesis: One-Gene, One-Trait. *Genetics* 158:487–93.
Dorst, J. C. 1958. "A Quarter of a Century of Plant Breeding in the Netherlands." *Euphytica* 7 (1): 9–20.
Downing A. J. 1869. *The Fruits and Fruit-Trees of America*. New York: John Wiley.
Doyle, J. 1985. *Altered Harvest: Agriculture, Genetics, and the Fate of the World's Food Supply*. New York: Viking Press.
Dressendorfer, W. 1998. *Die Pflanzenwelt des Hortus Eystettensis*. Munich: Schirmer/Mosel.
Dudley, J. W., and R. J. Lambert. 2004. "100 Generations of Selection for Oil and Protein in Corn." *Plant Breeding Reviews* 24 part 1: 79–110.
Dunnett, J. 2000. *A Scottish Breeder's Harvest*. Wick: North of Scotland Newspapers.
Duthie, R. E. 1982. "English Florists' Societies and Feasts in the Seventeenth and the First Half of the Eighteenth Centuries." *Garden History* 10 (1): 17–35.
———.1984. "Florists' Societies and Feasts after 1750." *Garden History* 12 (1): 8–38.
Duvick, D. 1996. "Plant Breeding, an Evolutionary Concept." *Crop Science* 36:539–48.
Earley Local History Group. 2006. *Suttons Seeds: A History 1806–2006*. Reading, UK: Earley Local History Group.
East, E. M. 1919. *Inbreeding and Outbreeding: Their Genetic and Sociological Significance*. Philadelphia : J. B. Lippincott.
Edwards, I. A. 2001. "Hybrid Wheat," in *The World Wheat Book: A History of Wheat Breeding*, ed. A. P. Bonjean and W. J. Angus, 1019–43. Paris: Lavoisier.
Eigsti, O. J. 1955. *Colchicine in Agriculture, Medicine, Biology, and Chemistry*. Ames: Iowa State University Press.
Eisenberg, E. 1998. *The Ecology of Eden*. New York: Knopf.
Elliott, B. 1986. *Victorian Gardens*. London: Batsford.
———. 2001. "Flower Shows in Nineteenth-Century England." *Garden History* 29 (1): 171–84.
———. 2004. *The Royal Horticultural Society, A History 1804–2004*. Chichester, UK: Phillimore.

Engledow, F. 1950. "Rowland Harry Biffen." *Obituary Notices of Fellows of The Royal Society* 7:9–26.
Evans, L. T. 1997. "Adapting and Improving Crops: the Endless Task." *Philosophical Transactions of the Royal Society* B.352 (1356): 901–6.
Evison, R. J. 1998. *The Gardener's Guide to Growing Clematis.* Portland, OR: Timber Press.
Eyssen, R. 1994. "Apple Breeding for Quality, Disease Resistance and Growth Habit." *Acta Horticulturae* 355:173–82.
Fear, C. D., and P. A. Domotom. 1986. "The 'Delicious' Apple." *Fruit Varieties Journal* 40 (1): 2–4.
Feldman, M. 2001. "Origin of Cultivated Wheat," in *The World Wheat Book: A History of Wheat Breeding*, ed. A. P. Bonjean and W. J. Angus, 3–58. Paris: Lavoisier.
Ferguson, J. H. A., and F. Garretsen. 1968. "The Plant Breeder and the Computer." *Euphytica* 17: 274–76.
Ferguson, N. 2003. *Empire: How Britain Made the Modern World.* London: Penguin.
Fiala, J. L.1998. *Lilacs, the Genus Syringa*. Portland, OR: Timber Press.
Findlay, A. 1905. *The Potato: Its History and Culture.* Fife, UK: Cupar.
Fitzgerald, D. 1990. *The Business of Breeding, Hybrid Corn in Illinois, 1890–1940.* Ithaca, NY: Cornell University Press.
———. 1993. "Farmers Deskilled: Hybrid Corn and Farmers' Work," *Technology and Culture* 34 (2): 324–43.
Flegr, J. 2002. "Was Lysenko (Partly) Right? Michurinist Biology in the View of Modern Plant Physiology and Genetics." *Rivista di Biologia/Biology Forum* 95: 259–72. http://www.natur.cuni.cz/~flegr/MANUSCRI/lysenko/lysenko5.htm (Accessed May 21, 2007).
"Four Iowans Who Fed The World" 2003. *Iowa Heritage Illustrated* 84:2 (Summer).
Fowler, C., and R. Lower. 2005. "The Politics of Plant Breeding." *Plant Breeding Reviews* 25: 21–55.
Frankel F. R. 1971. *India's Green Revolution: Economic Gains and Political Costs.* Princeton, NJ: Princeton University Press.
Friedland, W. H., and A. Barton. 1976. "Tomato Technology." *Society* 13 (6): 34–42.
Fussell, B. 1992. *The Story of Corn*. New York: North Point Press.
Gale, M. D., and S. Youssefian. 1985. "Dwarfing Genes in Wheat," in *Progress in Plant Breeding*, ed. G. R. Russell. Cambridge: Butterworths.
Gallagher, J. P. 1989. "Agricultural Intensification and Ridged-Field Cultivation in the Prehistoric Upper Midwest of North America," in *Foraging and Farming: The Evolution of Plant Exploitation*. ed. D. R. Harris and G. C. Hillman, 572–84. London: Unwin Hyman.
Gayon J., and D. T. Zallen. 1998. "The Role of the Vilmorin Company in the Promotion and Diffusion of the Experimental Science of Heredity in France, 1840–1920." *Journal of the History of Biology* 31 (2): 241–62.

Gepts, P. 2004. "Crop Domestication as a Long-Term Selection Experiment." *Plant Breeding Reviews* 24 (2): 1–37.
Gessert, G. 1993a. "Flowers of Human Presence: Effects of Esthetic Values on the Evolution of Ornamental Plants." *Leonardo* 26 (1): 37–44.
———. 1993b. "Notes on Genetic Art." *Leonardo* 26 (3): 205–11.
———. 1996. "Bastard Flowers." *Leonardo* 29 (4): 291–98.
———. 1997. "The Rainforests of Domestication: Ornamental Gardens as Sites of Maximum Genetic Diversity among Domesticated Plants." *Leonardo* 30 (2): 129–32.
Glustschenko, I. J. 1950. *Die Vegetative Hybridisation von Pflanzen*. Berlin: Verlag Kultur und Fortschritt.
Goldman, I. L. 2000. "Prediction in Plant Breeding." *Plant Breeding Reviews* 19:15–40.
———. 2004. "The Intellectual Legacy of the Illinois Long-Term Selection Experiment." *Plant Breeding Reviews* 24 (1): 61–78.
Goodman, M. 1976, "Maize," in *Evolution of Crop Plants*, ed. N. W. Simmonds, 128–36. Harlow, UK: Longman.
Gorer, R. 1970. *The Development of Garden Flowers*. London: Eyre & Spottiswoode.
———. 1975. *The Flower Garden in England*. London: Batsford.
Gottschalk, W., and G. Wolff 1983. *Induced Mutations in Plant Breeding*. Berlin: Springer Verlag.
Gourley, J. H. 1922. *Text-book of Pomology*. New York: Macmillan.
Gouyon, P. 2002. "L'introduction de la génétique en France." *Colloque "L'amélioration des plantes, continuités et ruptures,"* Montpellier, October 2002. http://www.inra.fr/gap/vie-scientifique/animation/colloque-AP2002/index.htm (accessed January 10, 2007).
GRAIN. 2003. "Poisoning the Well: The Genetic Pollution of Maize." *Seedling*, January. http://www.grain.org/seedling/seed-03-01-2-en.cfm (accessed March 23, 2007).
———. 2005. "Fiasco in the Field: An Update on Hybrid Rice in Asia." *Seedling*, March. http://www.grain.org/briefings/?id=190 (accessed November 10, 2008).
———. 2006. "The End of Farm-Saved Seed?" http://www.grain.org/briefings/?id=202 (accessed March 23, 2007).
———. n.d. Agricultural Research: For Whom? By Whom? http://www.grain.org/research (accessed Aprii 11, 2007).
Grenfell, D. 1998. *The Gardener's Guide to Growing Daylilies*. Portland, OR: Timber Press.
Griffin, K. 1974 *The Political Economy of Agrarian Change*. London: Macmillan.
Grigg, D. B. 1982. *The Dynamics of Agricultural Change*. London: Hutchinson.
———. 1984. "The Agricultural Revolution in Western Europe," in *Understanding Green Revolutions, Agrarian Change and Development Planning in South Asia*, ed. T. P. Bayliss-Smith and S. Wanmali, 1–17. Cambridge: Cambridge University Press.

Griliches, Z. 1960. "Hybrid Corn and the Economics of Innovation." *Science* 132:275–332.

Gypmanteseri, P. 2003. Personal communication. March.

Hagberg, A, and E. Åkerberg. 1961. *Mutations and Polyploidy in Plant Breeding*. London: Heinemann.

Hadfield, M. 1980. *British Gardeners, A Biographical Dictionary*. London: Zwemmer.

Hahn, W. G. 1972. *The Politics of Soviet Agriculture*. Baltimore, MD: The John Hopkins University Press.

Hargreaves, M. W. M. 1968. "The Durum Wheat Controversy." *Agricultural History* 42 (3): 211–29.

Hargrove, T., and W. R. Coffman. 2005. "Breeding History." *Rice Today* 5(4): 34–38.

Harlan, J. R. 1971. "Agricultural Origins: Centers and Non-Centers." *Science* 174:468–74.

———. 1992. *Crops and Man*. Madison, WI: American Society of Agronomy.

Harland S. C. 1945. "Obituaries: Prof. N. I. Vavilov." *Nature* 156: 621–22.

Harris, D. R. 1989. "An Evolutionary Continuum of People-Plant Interaction," in *Foraging and Farming: The Evolution of Plant Exploitation*, ed. D. R. Harris and G. C. Hillman, 11–26. London: Unwin Hyman.

Harris, F. S. 1919. *The Sugar Beet in America*. New York: Macmillan.

Harriss, J. 1977a. "Implications of Changes in Agriculture for Social Relationships at the Village Level: The Case of Randam," in *Green Revolution? Technology and Change in Rice-Growing Areas of Tamil Nadu and Sri Lanka*, ed. B. H. Farmer, 225–45. London: Macmillan.

———. 1977b. "Social Implications of Changes in Agriculture in Hambantota District," in *Green Revolution? Technology and Change in Rice-Growing Areas of Tamil Nadu and Sri Lanka*, ed. B. H. Farmer, 246–55. London: Macmillan.

Harvey, D. 1989. *The Condition of Postmodernity*. Oxford: Blackwell.

Harvey, G. 2001. *The Forgiveness of Nature—The Story of Grass*. London: Jonathan Cape.

Harwood, J. 1997. "The Reception of Genetic Theory among Academic Plant-Breeders in Germany, 1900–1930." [In Swedish.] *Sveriges Utsädesförenings Tidskrift* 4:187–95.

Harwood, W. S. 1905. *New Creations in Plant Life*. New York: Macmillan.

Hawkes, J. G. 1983. "N. I. Vavilov—The Man and His Work." *Biological Journal of the Linnean Society* 39 (1): 3–6.

Hayes, H. K., and R. J. Garber. 1919. "Synthetic Production of High Protein Corn in Relation to Breeding." *Journal of the American Society of Agronomy* 11:308–18.

Hazell, P. B. R., and C. Ramasamy, ed.1991. *The Green Revolution Reconsidered*. Baltimore, MD: The John Hopkins University Press.

Hedrick, U. P. 1911. *The Plums of New York*. Albany: State of New York Department of Agriculture.

———. 1950. *A History of Horticulture in America to 1860*. New York: Oxford University Press.

Henry, J. 1997. *The Scientific Revolution and the Origins of Modern Science*. Basingstoke, UK: Palgrave.
Herbert, W. 1846. "On Hybridisation amongst Vegetables." *Journal of the Horticultural Society of London* II:1–28.
Herring, R. J.2006. "Why Did 'Operation Cremate Monsanto' Fail? Science and Class in India's Great Terminator-Technology Hoax." *Critical Asian Studies* 38 (4): 467–93.
Hesser, L. 2006. *The Man Who Fed the World*. Dallas, TX: Durban House.
Hewitt de Alcantara, C. 1976. *Modernizing Mexican Agriculture: Socioeconomic Implications of Technological Change, 1940–1970*. Geneva: United Nations Research Institute for Social Development.
Hillman, G. C., and S. Davies. 1983. "Domestication Rates in Wild-Type Wheats and Barley under Primitive Cultivation." *Biological Journal of the Linnean Society* 39 (1): 39–78.
Holden, J., J. Peacock, and T. Williams. 1993. *Genes, Crops and the Environment*. Cambridge: Cambridge University Press.
Hooker, W. 1816. "Account of a New Pear, Called Williams' Bon Chretien." *Transactions of the Horticultural Society of London* 2:250–51.
Hopper, J. E., P. D. Ascher, and S. J. Peloquin. 1967. "Inactivation of Self-Incompatibility following Temperature Pretreatments of Styles in *Lilium longiflorum*." *Euphytica* 16 (2): 215–20.
Horowitz, N. H. 1990. "George Wells Beadle (1903–1989)." *Genetics* 124:1–6.
Horticultural Society of New York. 1902. *Proceedings of the International Conference on Plant-Breeding and Hybridization 1902*. Horticultural Society of New York.
Hossain, M., M. Abul Quasem, M. A. Jabbar, and M. Akash. 1994. "Production Environments, Modern Variety Adoption, and Income Distribution in Bangladesh," *Modern Rice Technology and Income Distribution in Asia*, ed. C. C. David and K. Otsuka, 221–79. London: Lynne Riener Publications.
Howard, L. E. 1953. *Sir Albert Howard in India*. London: Faber and Faber.
Howard, W. L. 1945. "Luther Burbank: A Victim of Hero Worship." *Chronica Botanica* 9:5–6.
Hsu, R. C. 1982. *Food for One Billion, China's Agriculture since 1949*. Boulder, CO: Westview Press.
Huffman, W. E., and R. E. Evenson. 1993. *Science for Agriculture: A Long-Term Perspective*. Ames: Iowa State University Press.
Hunter, H., and H. M. Leake. 1933. *Recent Advances in Agricultural Plant Breeding*. London: Churchill.
Hurst, C. C. 1901. "Mendel's 'Law' applied to Orchid Breeding." *Journal of the Royal Horticultural Society* 26:688–95.
Hutchinson, J. B. 1974a. "The Challenge of the New Agriculture," in *Evolutionary Studies in World Crops*, ed. J. B. Hutchinson, 161–62. Cambridge: Cambridge University Press.
———. 1974b. "Crop Plant Evolution in the Indian Subcontinent," in *Evolutionary*

Studies in World Crops, ed. J. B. Hutchinson, 151–60. Cambridge: Cambridge University Press.
Hyams, E. 1971. *Plants in the Service of Man, 10,000 Years of Domestication*. London: Dent.
ICAR. See Indian Council of Agricultural Research.
Indian Council of Agricultural Research. 2003. "Yellow Revolution." http://www.icar.org.in/icar3.htm#Yellow (accessed March 10, 2006).
International Rice Research Institute. 1980. *Beyond IR8, IRRI's Second Decade*. Los Baños, Philippines: IRRI.
———. 2007. "About Us." http://www.irri.org (accessed March 5, 2007).
Iowa Public Television. 2006. Henry A. Wallace program script. http://www.iptv.org/wallace/docs/Trans_FullProgram.pdf (accessed December 4, 2006).
IRRI. See International Rice Research Institute.
Isern, T. 1997. "Durum Wheat." *Plains Folk*, North Dakota Public Radio. http://www.prairiepublic.org/programs/plainsfolk/transcripts/durumwheat.jsp (accessed May 14, 2006).
Isvilanonda, S., and S. Wattanutchariya. 1994. "Modern Variety Adoption, Factor-Price Differential, and Income Distribution in Thailand," in *Modern Rice Technology and Income Distribution in Asia*, ed. C. C. David and K. Otsuka 173–220. London: Lynne Riener Publications.
James, J. 1964. *Create New Plants and Flowers—Indoors and Out*. New York: Doubleday.
Jatileksono, T. 1994. "Varietal Improvements, Productivity Change, and Income Distribution: The Case of Lampung, Indonesia," in *Modern Rice Technology and Income Distribution in Asia*, ed. C. C. David and K. Otsuka, 129–72. London: Lynne Riener Publications
Jennings, D. L. 1976. "Raspberries and Blackberries," in *Evolution of Crop Plants*, ed. N. W. Simmonds, 251–54. Harlow: UK: Longman.
Jensen, S. 2007. Personal communication. May 8.
Johnstone, P. H. 1938. "Turnips and romanticism." *Agricultural History* 12 (3): 224–55.
Jones, S. E. 2006. *Against Technology, From the Luddites to Neo-Luddism*. New York: Routledge Kegan Paul.
Joravsky, D. 1970. *The Lysenko Affair*. Cambridge, MA: Harvard University Press.
Joshi, K. 2006. "Client-Orientated Breeding (COB): Bridging Yield Gaps and Addressing Food and Income Security." *New Agriculturalist*. http://www.new-agri.co.uk/06-1/perspect.html (accessed March 26, 2007).
Juma, C. 1989. *The Gene Hunters: Biotechnology and the Scramble for Seeds*. Princeton, NJ: Princeton University Press.
Juniper, B. E., and D. J. Mabberley. 2006. *The Story of the Apple*. Portland, OR: Timber Press, Portland.
Karutz, C. 1998. *Ecological Cereal Breeding and Genetic Engineering*. Frick, Switzerland: Research Institute for Organic Agriculture.

Kashioka, S., and M. Ogisu. 1997. *History and Principle of the Traditional Floriculture in Japan*. Privately printed.
Kass, L., and K. Gale. 2006. Personal communication. February.
Khan, M. H. 1975. *The Economics of the Green Revolution in Pakistan*. New York: Praeger.
Kholi, S. P. 1969. "Wheat Varieties in India." ICAR Technical Bulletin, no.18. New Delhi.
Kime, T. [1917?]. *The Great Potato Boom, 1903–1904, Its Rise, Progress and Fall.* Lincolnshire, UK: T. Kime, Mareham-le-Fen.
Kimmelman, B. A. 1983. "The American Breeders' Association : Genetics and Eugenics in an Agricultural Context, 1903–13." *Social Studies of Science* 13(2): 163–204.
King, K. 2001. "Cytoplasm, Inheritance and Mutations." http://www.bulbnrose.org/Heredity/cyto.html (accessed December 20, 2005).
Kingsbury, N. 1991a. "Fashion under Glass." *The Garden* 116:298–302.
———. 1991b. "Lost Treasures of the Glasshouse." *The Garden* 116:655–59.
———. 2003. "An Overview of Current Practice in Ecological Planting Design in Europe and North America." In *Nature Enhanced*, ed. N. Dunnett and J. Hitchmough, 58–96. London: Spon Press.
Kizito E. B., W. Castelblanco, J. Omara, A. Bua, T. G. Egwang, M. Fregune, and U. Gulberg. n.d. "Assessment of Cassava Diversity in Uganda Using SSR Markers." Inter-American Center of Tax Administrations. http://www.ciat-library.ciat.cgiar.org/Articulos_Ciat/Paper%20one-%20Cassava%20diversity%20in%20Uganda.pdf (accessed November, 10, 2008).
Kjærgaard, T. 2003. "A Plant That Changed the World: The Rise and Fall of Clover, 1000–2000." *Landscape Research* 28(1): 41–50.
Klemm, V. 1987. "Kurt von Rümker, 1859–1940." *Von Thaer bis Mitscherlich. Kurzbiographien bedeutender Berliner Agrarwissenschaftler*. Beiträge zur Geschichte der Humboldt-Universität zu Berlin16:22–28. http://141.20.115.193/fakultaet/history/personen/Ruemker_a.htm (Accessed November 4, 2006).
Kloppenburg, J. R. 1988. *First the Seed: The Political Economy of Plant Biotechnology, 1492—2000*. Cambridge: Cambridge University Press.
Knight, T. A. 1806. "Observations on the Method of Producing New and Early Fruits." *Transactions of the Horticultural Society of London* 2:30–39.
———. 1807. "On Raising New and Early Varieties of the Potatoe." *Transactions of the Horticultural Society of London* 2:57–59.
Konstantinova, A. M. 1960. "Michurin Methods in Breeding Forage Crops by Means of Hybridization." In Proceedings of the Eighth International Grassland Congress, 51–53. Reading, UK: University of Reading.
Kral, E. A. 2005. "Orville A. Vogel: Wheat Breeder Helped Found Green Revolution and Invented Scientific Research Equipment Used Worldwide." Nebraska State Education Association. http://www.nsea.org/news/Vogel.htm (accessed June 28, 2006).

Kramer, M. G., and K. Redenbaugh. 1994. "Commercialization of a Tomato with an Antisense Polygalacturonase Gene: The FLAVR SAVR™ Tomato Story." *Euphytica* 79(3): 293–97.
Laidlow, A. 1848. [Untitled contribution.] *Gardeners' Chronicle and Agricultural Gazette* 1, 7.
Lammerts van Bueren, E. T., M. Hulscher, J. Jongerden, M. Haring, J. Hoogendoorn, J. D. van Mansvelt, and G. T. P. Ruivenkamp. 1998. *Sustainable Organic Plant Breeding, Subproject 1—Discussion Paper: Defining a Vision and Assessing Breeding Methods*. Driebergen, the Netherlands: Louis Bolk Institute.
Landes, D. 1998. *The Wealth and Poverty of Nations*. London: Little, Brown & Company.
Lappé, F. M., and J. Collins. 1977. *Food First : The Myth of Scarcity*. Boston: Houghton Mifflin.
Larter, E. N. 1976. "Triticale." In *Evolution of Crop Plants*, ed. N. W. Simmonds, 117–19. Harlow, UK: Longman.
Leakey, C. L. A. 1970. "Heterogeneous Agricultural Populations." *Agricultural Progress* 45:34–42.
———. 2007. Personal communication. March 15.
Leapman, M. 2000. *The Ingenious Mr Fairchild*. London: Headline.
Leighton, A. C. 1967. "The Mule as a Cultural Invention." *Technology and Culture* 8(1): 45–52.
Lesins, K. 1976 "Alfalfa, Lucerne." In *Evolution of Crop Plants*, ed. N. W. Simmonds, 165–68. Harlow, UK: Longman.
Lewontin, R. C., and J. D. Berlan. 1986. "Technology, Research, and the Penetration of Capital: The Case of US Agriculture." *Monthly Review* 38(3): 21–34.
Li, C. C. 1987. "Lysenkoism in China." *Journal of Heredity* 78:339–40.
Li, J., and Y. Xin. 2000. "Longping Yuan: Rice Breeder and World Hunger Fighter," *Plant Breeding Reviews* 17:1–11.
Lin, J. Y. 1994. "The Nature and Impact of Hybrid Rice in China." In *Modern Rice Technology and Income Distribution in Asia*, ed. C. C. David and K. Otsuka, 375–410. London: Lynne Riener Publications.
Lindley, J. 1843. "The Amateur's Garden." *Gardeners' Chronicle and Agricultural Gazette* 26, 444.
———. 1844a. Editorial. *Gardeners' Chronicle and Agricultural Gazette* 27, 443.
———. 1844b. Editorial. *Gardeners' Chronicle and Agricultural Gazette* 28, 459.
Lipton, M. 1989. *New Seeds and Poor People*. London: Unwin Hyman.
Liu, Y.-S., B.-Y. Li, G.-R. Li, and X.-M. Zhou. n.d. "Graft Hybridization and the Specificity of Heredity in Fruit Trees." Henan Vocation-Technical Teachers College, Xinxiang, China. <http://www.chinagene.cn/info_detailed.asp?id=818 (accessed December20, 2005).
Livingston, A. W. 1998. *Livingston and the Tomato*. Repr., Columbus: Ohio State University Press (orig. pub. 1893).

Loskutov, I. G. 1999. *Vavilov and His Institute: A History of the World Collection of Plant Genetic Resources in Russia*. Rome: International Plant Genetic Resources Institute.
Lovelock, Y. 1972. *The Vegetable Book: An Unnatural History*. London: Allen & Unwin, London.
Lowe, J. 1976. "Burpees Celebrates Its Centennial." *Flower and Garden Magazine* 20 (3): 26–29, 45.
Lupton, F. G. H. 1985. "Wheat." *Biologist* 32(2): 97–105.
MacKay, G. R. 1987. "Selecting and Breeding for Better Potato Cultivars." In *Improving Vegetatively Propagated Crops*, ed. A. J. Abbott and R. K. Atkin, 183–93. London: Academic Press.
Mahan, C. E. 2007. *Classic Irises and the Men and Women Who Created Them*. Malabar, FL: Krieger.
Malin, J. C. 1944. *Winter Wheat in the Golden Belt of Kansas: A Study in Adaptation to Subhumid Geographical Environment*. Lawrence: University of Kansas Press.
Mangelsdorf, P. C. 1958. "Ancestor of Corn." *Science* 128:1313–20.
———. 1974. *Corn, Its Origin, Evolution and Improvement*. Cambridge: Belknap Press of Harvard University Press.
———. 1975. "Donald Forsha Jones." In *Biographical Memoirs*, vol. 46. Washington, DC: National Academy of Sciences.
Mandgelsdorf, P. C., and P. G. Reeves. 1938. "The Origin of Maize." *Proceedings of the National Academy of Science of the U S A* 24:303–12.
Mann, C. C. 1997. "Saving Sorghum by Foiling the Wicked Witchweed." *Science* 277:1040.
———. 1999. "Crop Scientists Seek a New Revolution." *Science* 283: 310–16.
———. 2002. "1491." *The Atlantic* 289 (3): 41–53.
Manners, G. 2001. "NERICA- New Rice, Transforming Agriculture for West Africa." *Science in Africa*. http://www.scienceinafrica.co.za/nerica.htm (accessed March 21, 2007).
Manning, R. 2000. *Food's Frontier: The Next Green Revolution*. Berkeley: University of California Press.
Maplestone, P. 2007. Personal communication. Aprii 12.
Marvell, A. 1681. "The Mower, against Gardens." In *Miscellaneous Poems*, ed. G. A. Aitken, 83–84.London: Lawrence & Bullen, 1892. http://www.luminarium.org/sevenlit/marvell/mowagainst.htm (accessed Aprii 20, 2007).
Marx, L. 1978. "Reflections on the Neo-Romantic Critique of Science." *Daedalus* 107 (2): 61–74.
———. 1994. "The Idea of 'Technology' and Postmodern Pessimism" In *Does Technology Drive History, The Dilemma of Technological Determinism*, Smith, ed. M. R. and L. Marx, 101–14. Cambridge, MA: MIT Press.
McCann, J. C. 2005. *Amazing Grace, Africa's Encounter with a New World Crop 1500–2000*. Cambridge, MA: Harvard University Press.
McClintock, Barbara. n.d. The Barbara McClintock Papers. Profiles in Science/

National Library of Medicine. Cornell University, Ithaca. http://profiles.nlm.nih.gov/LL/Views/Exhibit/narrative/cornell.html (accessed Aprii 12, 2006).

McCollum, G. D. 1976. "Onion and Allies." In *Evolution of Crop Plants*, ed. N. W. Simmonds 186–89.Harlow, UK: Longman.

McEachern, G. R. 2003. "A Texas Grape and Wine History." http://aggie-horticulture.tamu.edu/southerngarden/Texaswine.html (accessed March 15, 2007).

McFadden, E. S. 1930. "A Successful Transfer of Emmer Characters to Vulgare Wheat." *Journal of the American Society of Agronomy* 22:1020–34.

McGloughlin, M. 2001. "Designer Foods: Enhancing Nutrition with Biotechnology." American Medical Association Briefing on Food Biotechnology. http://www.whybiotech.com/html/pdf/ama_mcgloughlin.pdf (accessed January 16, 2007).

Medvedev, Z. A. 1969. *The Rise and Fall of T.D. Lysenko*. New York: Columbia University Press.

Meinel, A. 2003. "An Early Scientific Approach to Heredity by the Plant Breeder Wilhelm Rimpau (1842–1903)." *Plant Breeding* 122:195–98.

Mendel, G. 1901. "Experiments in Plant Hybridisation." Introd. By W. Bateson. *Journal of the Royal Horticultural Society* 26:1–32.

Merezhko, A. F. 2001. "Wheat Pool of European Russia," in *The World Wheat Book: A History of Wheat Breeding*, ed. A. P. Bonjean and W. J. Angus, 831–49. Paris: Lavoisier.

Merker, A. 1986. "Triticale—A New Cereal Crop." In *Svalöf 1886–1986 Research and Results in Plant Breeding*, ed. G. Olsson, 132–39. Stockholm: LTs Förlag.

Michurin, I. V. 1949. *Selected Works*. Moscow: Foreign Languages Publishing House.

Minnis, P. E. 1992. "Earliest Plant Cultivation in the Desert Borderlands of North America." In *The Origins of Agriculture, An International Perspective*, ed. C. W. Cowan and P. J. Watson, 121–41. Washington, DC: Smithsonian Institution Press.

Mokyr, J. 2002. *The Gifts of Athena: Historical Origins of the Knowledge Economy*. Princeton, NJ: Princeton University Press.

Moore, E. A. 2005. "I'll Never Forget the Cornfields." *Christian Science Monitor* october 4.

Moore, J. H. 1956. "Cotton Breeding in the Old South." *Agricultural History* 30 (3): 95–104.

Moore-Colyer, R. J. 1999. "Sir George Stapledon (1882–1960) and the Landscape of Britain." *Environment and History* 5:221–36.

Morris, C. E. and D. C. Sands.2006. "The Breeder's Dilemma—Yield or Nutrition?" *Nature* 24:1078–80.

Morrison, J. W. 1960. "Marquis Wheat—A Triumph of Scientific Endeavor." *Agricultural History* 34 (4): 182–88.

Mukerji, C. 1997. *Territorial Ambitions and the Gardens of Versailles*. Cambridge: Cambridge University Press.

Murphy, D. 2007a. *Plant Breeding and Biotechnology: Societal Context and the Future of Agriculture*. Cambridge: Cambridge University Press.

———. 2007b. *People, Plants and Genes, The Story of Crops and Humanity*. Oxford: Oxford University Press.

Murphy, R. 2006. *Plant Breeding and Biotechnology: Societal Context and the Future of Agriculture*. Cambridge: Cambridge University Press.

Nanda, M. 2003a. *Postmodernism and Religious Fundamentalism, A Scientific Rebuttal to Hindu Science*. Pondicherry, India: Navayana.

———. 2003b. *Prophets Facing Backward, Postmodern Critiques of Science and Hindu Nationalism in India*. New Brunswick, NJ: Rutgers University Press.

National Research Centre for Sorghum. 2001. *The Sorghum Project and Dr. N.G.P. Rao*. http://www.nrcsorghum.res.in/heads.asp (accessed March 9, 2007).

Needham, J. and F. Bray. 1984. *Science and Civilisation in China, Vo1.6. Biology and Biological Technology, Part II. Agriculture*. Cambridge: Cambridge University Press.

Nene, Y. L. 2005. "Rice Research in South Asia through the Ages." *Asian Agri-History* 9 (2): 85–106.

Newcomb, P. C.1985. *Popular Annuals of Eastern North America, 1865–1914*. Washington, DC: Dumbarton Oaks.

N. I. Vavilov Research Institute of Plant Industry. 2005. "Organization of VIR's Work for Plant Genetic Resources." http://www.vir.nw.ru/history/vir_act.htm (accessed June 28,2006).

Normile, D. 1999. "Crossing Rice Strains to Keep Asia's Rice Bowls Brimming." *Science* 283:313.

North Carolina State University.2002. "NC State Geneticists Study Origin, Evolution Of 'Sticky' Rice." http://www.sciencedaily.com/releases/2002/10/021023064638.htm (accessed Sept 18, 2006).

Northrup, J. 1904. *Seed Truth*. Minneapolis, MN: Northrup, King & Co.

Norton, J. B. 1902. "Improvement of Oats by Breeding." In Proceedings, International Conference on Plant Breeding and Hybridization. New York: Horticultural Society of New York.

Nulty, L. 1972. *The Green Revolution in West Pakistan: Implications of Technological Change*. New York: Praeger.

Oelke E. A., T. M. Teynor, P. R. Carter, J. A. Percich, D. M. Noetzel, P. R. Bloom, R. A. Porter, C. E. Schertz, J. J. Boedicker, and E. I. Fuller. 1997. "Wild Rice." In *Alternative Field Crops Manual*. University of Wisconsin-Extension. http://www.hort.purdue.edu/newcrop/afcm/wildrice.html (accessed August 16, 2006).

Ohnuki-Tierney. E. 1993. *Rice as Self—Japanese Identities through Time*. Princeton, NJ: Princeton University Press.

Olby, R. 1989. "Scientists and Bureaucrats in the Establishment of the John Innes Horticultural Institution under William Bateson." *Annals of Science* 46:497–510.

Olmo, H. P. 1976 "Grapes." In *Evolution of Crop Plants*, ed. N. W. Simmonds, 294–97. Harlow, UK: Longman.

Olmstead, A. L., and P. W. Rhode. 2004. "Biological Innovation in American Wheat Production: Science, Policy, and Environmental Adaptation." In *Industrializing Organisms: Introducing Evolutionary History*, ed. S. R. Schrepfer and P. Scranton, 43–83. New York: Routledge.

———. 2006. "Biological Globalizaion: The Other Grain Invasion." In Globalization, Empire, and Imperialism in Historical Perspective, 5th Historical Society Conference. Chapel Hill, NC. June 1. http://64.233.169.104/search?q=cache:ZKFTveCrjNMJ:www.bu.edu/historic/06conf_papers/Rhode_Olmstead.pdf+grain+invasion&hl=en&ct=clnk&cd=1 (accessed May 23, 2007).

Orel, V. 1978. "The Influence of T. A. Knight (1759–1838) on Early Plant Breeding in Moravia." *Folia Mendeliana in Acta Musei Moraviae* 13:241–60.

———. 1996. *Gregor Mendel: The First Geneticist*. Oxford: Oxford University Press.

Orel, V., and M. Vávra. 1968. "Mendel's Program for the Hybridization of Apple Trees." *Journal of the History of Biology* 1 (2): 219–24.

Overton, M. 1996. *Agricultural Revolution in England*. Cambridge: Cambridge University Press.

Owen, E. R. J. 1969. *Cotton and the Egyptian Economy 1820–1914*. Oxford: Clarendon Press.

PABRA. See Pan-African Bean Research Alliance.

Palladino, P. 1993. "Between Craft and Science: Plant Breeding, Mendelian Genetics, and British Universities, 1900–1920." *Technology and Cultures* 34 (2): 300–23.

———. 1994. "Wizards and Devotees: on the Mendelian Theory of Inheritance and the Professionalization of Agricultural Science in Great Britain and the US, 1880–1930." *History of Science* 32:409–44.

———. 1996. "Science, Technology, and the Economy: Plant Breeding in Great Britain, 1920–1970." *Economic History Review* 49:116–36.

Palmer, I. 1976a. *The New Rice in Asia: Conclusions from Four Country Studies*. Geneva: United Nations Research Institute for Social Development.

———. 1976b. *The New Rice in Indonesia*. Geneva: United Nations Research Institute for Social Development.

Pan-African Bean Research Alliance. 2006. "Better Beans for Africa." http://www.ciat.cgiar.org/africa/pabra.htm (accessed March 10, 2006).

Pataky, J. K. 2003. "Fifty Years of Supersweet Sweet Corn: It Keeps Getting Better." Lecture, University of Illinois Agronomy Day, August 21.

Paul, D. B., and B. A. Kimmelman. 1988. "Mendel in America: Theory and Practice, 1900–1919." In *The American Development of Biology*, ed. R. Rainger, K. R. Benson, and J. Maienschein, 281–310. Philadelphia: University of Pennsylvania Press. http://www.mendelweb.org/MWpaul.html (accessed October 11, 2006).

Paul, H., and R. R. Steinbrecher. 2003. *Hungry Corporations, Transnational Biotech Companies Colonise the Food Chain*. London: Zed Press.
Paulsen, G. 2003. "Making Wheat a Success in Kansas." http://www.kswheat.com/general.asp?id=325 (accessed April 20, 2005).
Pavord A. 1999. *The Tulip*. London: Bloomsbury.
Pearse, A. 1980. *Seeds of Plenty, Seeds of Want*. Oxford: Clarendon Press.
Percival, J. 1934. *Wheat in Great Britain*. London: Duckworth.
Perkins, J. H. 1997. *Geopolitics and the Green Revolution: Wheat, Genes and the Cold War*. New York: Oxford University Press.
Peterson, T. 1992. "A Celebration of the Life of Dr. Barbara McClintock." http://www.nalusda.gov/pgdic/Probe/v3n1_2/mcclinto.html (accessed Aprii 12, 2006).
Phillips, L. L. 1976. "Cotton." In *Evolution of Crop Plants*, ed. N. W. Simmonds, 443–50. Harlow, UK: Longman.
Pickersgill, B. 1969. "The Domestication of Chilli Peppers." In *The Domestication and Exploitation of Plants and Animals*, ed. P. J. Ucko and G. W. Dimbleby. London: Duckworth.
Pike, C. 2007. Personal communication. January.
Pimbert, M. 2003. "The Promise of Participation: Democratising the Management of Biodiversity." *Seedling*, July. http://www.grain.org/seedling/?id=243 (accessed March 26, 2007).
Pinstrup-Andersen, P., and M. Jaramillo. 1991. "The Impact of Technological Change in Rice Production on Food Consumption and Nutrition." In *The Green Revolution Reconsidered*, ed. P. B. R. Hazell and C. Ramasamy, 85–104. Baltimore, MD: The John Hopkins University Press.
Pistorius, R., and J. Van Wijk. 1999. *The Exploitation of Plant Genetic Information: Political Strategies in Crop Development*. Biotechnology in Agriculture Series. New York: CABI.
Pollan, M. 2001. *The Botany of Desire*. New York: Random House.
Porsche, W., and M. Taylor. 2001. "German Wheat Pool." In *The World Wheat Book: A History of Wheat Breeding*, ed. A. P. Bonjean and W. J. Angus. Paris: Lavoisier.
Price, E. 2001. "East of Eden." *The Garden* 126:456–59.
Provine, W. B. 1971. *The Origins of Theoretical Population Genetics*. Chicago: University of Chicago Press.
Punnett, R. C. 1950. "The Early Days of Genetics." *Heredity* 4(1): 1–10.
Quinn, R. M. 1999. "Kamut®: Ancient Grain, New Cereal." In *Perspectives on new crops and new uses*, ed. J. Janick. Alexandria, VA: ASHS Press. http://www.hort.purdue.edu/newcrop/proceedings1999/v4–182.html (accessed January 10, 2007).
Quisenberry, K. S., and L. P. Reitz. 1974. "Turkey Wheat: The Cornerstone of an Empire. *Agricultural History* 48 (1): 98–110.
Rajaram, S., and M. Van Ginkel. 2001. "Mexico: 50 Years of International Wheat Breeding." In *The World Wheat Book: A History of Wheat Breeding*, ed. A. P. Bonjean and W. J. Angus, 579–608. Paris: Lavoisier.

Rao, N.G.P. 1991. "Advancing Dryland Agriculture: Plant Breeding Accomplishments and Perspectives." *Indian Journal of Genetics*, 51:2, 145–280.
Rao, S., Singh, G. and Chatterjee, M. 2001. "Indian Wheat Pool," in *The World Wheat Book: A History of Wheat Breeding*, ed. A. P. Bonjean and W. J. Angus, 773–816. Paris: Lavoisier.
Reddi, G. N. 2005. "Rebuilding the Genetic Resource Base through Farmer-Scientist-Activist Alliance." http://www.idrc.ca/en/ev-85303-201-1-DO_TOPIC.html (accessed Aprii 16, 2007).
Redfern, R. A. 1972. *Gooseberry Shows of Old*. Ed. Allan Hill. Lancashire and Cheshire Antiquarian Society/Manchester Central Library.
Rentoul, J. N. 1991. *The Hybrid Story*. Melbourne: Lothian Books.
RHS. See Royal Horticultural Society.
Richards, P. 1985. *Indigenous Agricultural Revolution*. London: Hutchinson.
Rick, C. M. 1976. "Tomatoes." In *Evolution of Crop Plants*, ed. N. W. Simmonds, 268–72. Harlow, UK: Longman.
Roach, F. A. 1985. *Cultivated Fruits of Britain*. Oxford: Blackwell.
Roberts, H. F. 1929. *Plant Hybridization before Mendel*. Princeton, NJ: Princeton University Press.
Roberts, O. 1991. "Social Imperialism and State Support for Agricultural Research in Edwardian Britain." *Annals of Science* 48:509–26.
Robertson-Scott, J. W. 1911. *Sugar Beet, Some Facts and Some Illusions*. London: Horace Cox.
Rodgers, A. D. 1949. *Liberty Hyde Bailey*. Princeton, NJ: Princeton University Press.
Roggen, H. P. J. R., and A. J. Van Dijk.1972. "Breaking incompatibility of *Brassica oleracea* by steel-brush pollination." *Euphytica* 21 (3): 424–25.
Roggen, H. P. J. R., A. J. Dijk, and C. Dorsman. 1972. "'Electrically aided' Pollination, A Method of Breaking Incompatibility in *Brassica oleracea*." *Euphytica* 21 (2): 181–84.
Roll-Hansen, N. 1978. The Genotype Theory of Wilhelm Johannsen and Its Relation to Plant Breeding and the Story of Evolution." *Centaurus* 22:201–35.
———. 1986. "Svalöf and the Origins of Classical Genetics." In *Svalöf 1886–1986 Research and Results in Plant Breeding*, ed. G. Olsson, 35–43. Stockholm: LTs förlag.
———. 1997. "The Role of Genetic Theory in the Success of the Svalöf Plant Breeding Programme." *Sveriges Utsädesförenings Tidskrift* 4:196–207.
Rollinson, F. H. S. 1843. "Hybrid Plants." *Gardeners' Chronicle and Agricultural Gazette* 27:461.
Romans, A. 2005. *The Potato Book*. London: Frances Lincoln.
Ronald, W. G., and P. D. Ascher. 1976."Lilium × 'Black Beauty'—A Potential Bridging Hybrid in Lilium." *Euphytica* 25 (1): 285–91.
Rosset, P. 2000. "Do We Need New Technology to End Hunger?" http://www.foodfirst.org/node/230 (accessed March 8, 2007).

Royal Horticultural Society. 1902. "American Hybrid Conference." *Journal of the Royal Horticultural Society* 27:1060–1159.
Russell, E. J. 1966. *A History of Agricultural Science in Great Britain, 1620–1954*. London: Allen & Unwin.
Ruttan, V. W., and Y. Hayami. 1973. "Technology Transfer and Agricultural Development." *Technology and Culture* 14 (2 part 1): 119–51.
Santhanum, V., and J. B. Hutchinson. 1974. "Cotton." In *Evolutionary Studies in World Crops*, ed. J. B. Hutchinson, 196–200. Cambridge: Cambridge University Press.
Sarudy, B. W. 1989. "Nurserymen and Seed Dealers in the 18th Century Chesapeake." *Journal of Garden History* 9 (3): 111–17.
Sato, T. 1990. "Tokugawa Villages and Agriculture." In *Tokugawa Japan: The Social and Economic Antecedents of Modern Japan*, ed. C. Nakane and S. Ōishi, 37–80. Tokyo: University of Tokyo Press.
Schimmelpfennig, D., C. E. Pray, and M. Brennan. 2004. "The Impact of Seed Industry Concentration on Innovation: A Study of U.S. Biotech Market Leaders." http://ssrn.com/abstract=365600 (accessed March 23, 2007).
Schlesinger, A. 2000. "Who Was Henry A. Wallace? The Story of a Perplexing and Indomitably Naive Public Servant." *Los Angeles Times*, March 12. http://www.cooperativeindividualism.org/schlesinger_wallace_bio.html (accessed December 6, 2006).
Schumpeter, J. A. 1976. *Capitalism, Socialism and Democracy*. 5th ed. London: Allen & Unwin.
Schwanitz, F. 1966. *The Origin of Cultivated Plants*. Cambridge, MA: Harvard University Press.
Scourse, N. 1983. *The Victorians and Their Flowers*. Portland, OR: Timber Press.
Scowcroft, W. 2003. "Marquis Wheat: King Wheat Is 100 Years Old." http://www.grainscanada.gc.ca/newsroom/news_tips/2003/marquis-e.htm (accessed September 28, 2006).
Sen, B. 1974. *The Green Revolution in India: A Perspective*. New Delhi: Wiley Eastern Private.
Shakespeare, W. 1623. *The Winter's Tale*. http://shakespeare.mit.edu/winters_tale/full.html (accessed November, 10, 2008).
Shambu Prasad, C. 1999. "Suicide Deaths and the Quality of Indian Cotton: Perspectives from History of Technology and Khadi Movement." *Economic and Political Weekly* 34:5, 12–21.
Shand, R. T. 1973. *Technical Change in Asian Agriculture*. Canberra: Australian National University Press.
Shastry, S. V. S., and S. D. Sharma. 1974. "Rice." In *Evolutionary Studies in World Crops*, ed. J. B. Hutchinson, 55–62. Cambridge: Cambridge University Press.
Shaw, D. 2007. Personal communication. May 11.
Shebir. 2004. Personal communication. October.

Shepherd, J. 1980. "Development of New Wheat Varieties for the Pacific North West." *Agricultural History* 54 (1): 52–63.
Shepherd, R. E. 1978. *History of the Rose*. London: Heyden.
Shipek, F. C. 1989. "An Example of Intensive Plant Husbandry: The Kumeyaay of Southern California." In *Foraging and Farming, The Evolution of Plant Exploitation*, D. R. Harris and G. C. Hillman, 159–70. London: Unwin Hyman.
Shiva, V. 1991. *The Violence of the Green Revolution*. London: Zed Books.
———. 2000. "Poverty and Globalisation." Reith lecture. http://news.bbc.co.uk/hi/english/static/events/reith_2000/lecture5.stm (accessed Aprii 11, 2007).
Shrivastava, A. 2006. "GM Seed Buccaneers Versus the People of India." 7 December. http://www.archivesat.com/Alternative_Medicine_Forum/thread2237722.htm (accessed March 22,2007).
Silvey, V.1994. "Plant Breeding in Improving Crop Yield and Quality in Recent Decades." *Acta Horticulturae* 355:19–34.
Simmonds, N. W. 1976a. "Bananas." In *Evolution of Crop Plants*, ed. N. W. Simmonds, 211–15. Harlow, UK: Longman.
———. 1976b. "Olives." In *Evolution of Crop Plants*, ed. N. W. Simmonds, 219–221. Harlow, UK: Longman.
———. 1976c. "Potatoes." In *Evolution of Crop Plants*, ed. N. W. Simmonds, 279–83. Harlow, UK: Longman.
———. 1976d. "Quinoa and Its Relatives." In *Evolution of Crop Plants*, ed. N. W. Simmonds, 29. Harlow, UK: Longman.
———. 1979. *Principles of Crop Improvement*. Harlow, UK: Longman.
———. 1991. "Bandwagons I Have Known." *Tropical Agriculture Newsletter*, December, 7–10.
Singh, L. B. 1976. "Mango." In *Evolution of Crop Plants*, ed. N. W. Simmonds, 7–10. Harlow, UK: Longman.
Singh, V., and S. Prakash. 1985. "Indian Farmers Rediscover Advantages of Traditional Rice Varieties." http://www.satavic.org/ricevarieties.htm (accessed March 11, 2007).
Slicher van Bath, B. H. 1963. *The Agrarian History of Western Europe A.D. 500–1850*. London: Edward Arnold.
Smil, V. 2000. *Feeding the World: A Challenge for the Twenty-First Century*. Cambridge, MA: MIT Press.
Smith, D. C. 1966. "Plant Breeding, Development and Success." In *Plant Breeding: A Symposium held at Iowa State University*, ed. K. J. Frey, 3–54. Ames: Iowa State University Press.
Smith, J. 2005. "Order 81." *The Ecologist*, February, 41–44.
Smith, P. M. 1976. "Minor Crops." In *Evolution of Crop Plants*, ed. N. W. Simmonds, 301–24. Harlow, UK: Longman.
Sneep, J., and A. J. T. Hendriksen. 1979. *Plant Breeding Perspectives*. Wageningen, The Netherlands: Centre for Agricultural Publishing.

Snooks, G. D. 1996. *The Dynamic Society, Exploring the Sources of Global Change.* London: Routledge Kegan Paul.

Snow, C. P., and S. Collini. 1993. *The Two Cultures.* Cambridge: Cambridge University Press.

Sommer, M. V. A. F. 1966. *The Shaker Garden Seed Industry.* Orono: University of Maine Press.

Spillman, W. J. 1902. "Quantitative Studies on the Transmission of Parental Characteristics to Hybrid Offspring." *Journal of the Royal Horticultural Society* 27:876–93.

Stakman, E. C., R. Bradfield, and P. C. Mangelsdorf. 1967. *Campaigns against Hunger.* Cambridge MA: Belknap Press of Harvard University Press.

Steury, T. 2004. "Full Circle: Perennial Wheat Could Fulfill a Tradition and Transform a Landscape." *Washington State Magazine Online*, Summer. http://washington-state-magazine.wsu.edu/stories/04-summer/wheat/index.html (accessed March 20, 2007).

Storey, W. B.1976. "Figs." In *Evolution of Crop Plants*, ed. N. W. Simmonds, 205–8. Harlow, UK: Longman.

Storey, W. K. 1995. "Small-Scale Sugar Cane Farmers and Biotechnology in Mauritius: The 'Uba' Riots of 1937." *Agricultural History* 69 (2): 163–76.

Strik, B. C., and K. E. Hummer. 2006. "'Ananasnaya' Hardy Kiwifruit." *Journal of the American Pomological Society* 60 (3): 106–22.

Sudaryanto, T., and F. and Kasryno. 1994. "Modern Rice Variety Adoption and Factor-Market Adjustments in Indonesia." In *Modern Rice Technology and Income Distribution in Asia*, ed. C. C. David and K. Otsuka, 107–28. London: Lynne Riener Publications.

Suttons Seed Company. Seed catalogs 1880–1920.

Swaminathan, M. S. 1984. "Rice." *Scientific American* 250 (1): 62–71.

Swingle, W. T., and H. J. Webber. 1897. "Hybrids and Their Utililization in Plant Breeding." In *USDA Yearbook, 1897*. Washington, DC: Government Printing Office.

T. A. F. 1844. Entry under "Miscellaneous." *Gardeners' Chronicle and Agricultural Gazette* 47:788.

Talbot, J. W. 1882. "New Varieties of Pears." *Transactions of the Massachusetts Horticultural Society* 11 (Part 1): 31–43.

Thick, M. 1990a. "Garden Seeds in England before the Late 18th Century. I." *Agricultural History Review* 38 (1): 58–71.

———. 1990b. "Garden Seeds in England before the Late 18th Century. II." *Agricultural History Review* 38 (2): 105–16.

Thirsk, J. 1985. "Agricultural Innovations and their Diffusion." In *The Agrarian History of England and Wales, Volume 5—1640–1750: Agrarian Change*, ed. J. Thirsk, 533–89. Cambridge: Cambridge University Press.

Thomas, G. S. 1986. Foreword to *Daylilies*, by A. B. Stout. New York: Sagapress.

Thornstrom, C.-G., and U. Hossfeld. 2002. "Instant Appropriation-Heinz Brücher and the SS Botanical Collecting Commando to Russia 1943." *PGR newsletter* 129. http://www.bioversityinternational.org/publications/pgrnewsletter/archive.asp (accessed January 15, 2007).

Trehane, J. 1998. *Camellias*. London: Batsford.

———. 2002. "Camellia x williamsii." *The Garden* 127:173–77.

Troyer, A. F. 1999. "Background to US Hybrid Corn." *Crop Science* 39:601–26.

Troyer, A. F., and H. Stoehr. 2003. "Willet M. Hays, Great Benefactor to Plant Breeding and the Founder of Our Association." *Journal of Heredity* 94 (6): 435–41. http://jhered.oxfordjournals.org/cgi/content/full/94/6/435 (accessed November 25, 2007).

Tudge, C. 2000. *Mendel's Footnotes*. London: Jonathan Cape.

Turney, J. 1998. *Frankenstein's Footsteps, Science, Genetics and Popular Culture*. New Haven, CT: Yale University Press.

Van Harten, A. M. 1998. *Mutation Breeding, Theory and Practical Applications*. Cambridge: Cambridge University Press.

Vavilov, N. I. 1951. *The Origin, Variation, Immunity, and Breeding of Cultivated Plants—Selected Writings of N. I. Vavilov*. New York: Ronald Press.

Vilmorin-Andrieux. 1880. *Les meilleurs blés, description et culture des principales variétés de froments d'hiver et de printemps*. Paris: Vilmorin-Andrieux & cie.

VIR. See N. I. Vavilov Research Institute of Plant Industry.

Wallace, H. A., and W. L. Brown. 1956. *Corn and Its Early Fathers*. East Lansing: Michigan State University Press.

Walton, J. R. 1973. "Varietal Innovation and the Competitiveness of the British Cereals Sector, 1760–1930." *Agricultural History Review* 47 (1): 29–57.

Warder, J. A. 1867. *American Pomology—Apples*. New York: Orange Judd.

Warrington, I. J. 1994. "The 'Granny Smith' Apple." *Fruit Varieties Journal* 48 (2): 70–73.

Waters, M. 1988. *The Garden in Victorian Literature*. Aldershot, UK: Scolar Press.

Watkins, R. 1976. "Apple and Pear." In *Evolution of Crop Plants*, ed. N. W. Simmonds, 247–50. Harlow, UK: Longman.

Webber, H. J. 1898. "Improvement of Plants by Selection." In *USDA Yearbook*, 1898. Washington, DC: Government Printing Office.

Webber, H. J., and E. A. Bessey. 1899. "Progress of Plant Breeding in the United States." In *USDA Yearbook, 1899*. Washington, DC: Government Printing Office.

Webber, H. J., and W. T. Swingle. 1904. "New Citrus Creations of the Department of Agriculture." In *USDA Yearbook, 1904*. Washington, DC: Government Printing Office.

Webber, R. 1968. *The Early Horticulturalists*. Newton Abbot, UK: David & Charles.

Wellensike, S. 1939. "The Newest Fad, Colchicine." *Chronica Botanica* 5:15–17.

Wheat Genetics and Genomics Resources Centre. 2001. "Wheat Taxonomy." January 21. https://www.ksu.edu/wgrc/Taxonomy/taxintro.html (accessed August 16, 2006).
Wheatcroft, H. 1959. *My Life with Roses*. London: Odhams.
Wilhelm, S., and J. E. Sagan. 1974. *A History of the Strawberry: From Ancient Gardens to Modern Markets*. Berkeley: University of California.
Wilks, W., ed. 1900. "Report of the First International Conference on Hybridisation and Cross-Breeding." *Journal of the Royal Horticultural Society* 24.
———. 1907 *1906 Third International Conference on Genetics; Hybridisation (The Cross-Breeding of Genera of Species)*. London: The Royal Horticultural Society.
Williams, W. 1985. "Genetic Improvement of Grain Protein Crops—Achievements and Prospects." In *Progress in Plant Breeding*, ed. G. R. Russell. Cambridge: Butterworths.
Winner, L. 1986. *The Whale and the Reactor: A Search for Limits in an Age of High Technology*. Chicago: University of Chicago Press.
Wit, F. 1976. "Clove." In *Evolution of Crop Plants*, ed. N. W. Simmonds, 216–18. Harlow, UK: Longman.
Witcombe, R., and Joshi, A. n.d. "Participatory Breeding and Selection: Decentralization with Greater User Involvement. The Impact of Farmer Participatory Research on Biodiversity of Crops." http://www.idrc.ca/en/ev-85122-201-1-DO_TOPIC.html (accessed March 26, 2007).
Wittmack, L. 1900. "On the Particular Influence of Each Parent in Hybrids." *Journal of the Royal Horticultural Society*1 (24): 252–55.
Wolfe, M. S. 2002. "Organic Plant Breeding." In UK Organic Research 2002: Proceedings of the COR Conference, 26–28 March 2002, Aberystwyth, Wales, ed. J. Powell et al. 303–5.
Wolin, R. 2004. *The Seduction of Unreason, The Intellectual Romance with Fascism, from Nietzsche to postmodernism*. Princeton, NJ: Princeton University Press.
Wolpert, L. 1992. *The Unnatural Nature of Science*. London: Faber and Faber.
Wood, D. 2002. "One Hand Clapping: Organic Farming in India." http://www.cgfi.org/2002/12/12/one-hand-clapping-organic-farming-in-india/ (accessed February 7, 2007).
Wortman, S. 1975. "Agriculture in China." *Scientific American* 232 (6): 13–21.
Wylie, A. 1954. "The History of Garden Roses, Part 1." *Journal of the Royal Horticultural Society* 79:555–71.
Yamanaka, T. 2000. *Japanese Morning Glories II*. Chiba, Japan: Foundation for Museums of Japanese History.
———. 2004. "Morning Glories in Japan." *The Garden* 129:4 (April 2007): 278.
Yates, F., and K. Mather. 1963. "Ronald Aylmer Fisher." *Biographical Memoirs of Fellows of the Royal Society of London* 9:91–120.
Yepsen, R. 1994. *Apples*. New York: Norton.
Zirkle, C. 1935. *The Beginnings of Plant Hybridization*. Philadelphia: University of Pennsylvania Press.

———. 1969. "Plant Hybridization and Plant Breeding in Eighteenth-Century American Agriculture." *Agricultural History* 4 (3): 25–38.
Zohary, D. 1996. "The Mode of Domestication of the Founder Crops of South-West Asian Agriculture." In *The Origins and Spread of Agriculture and Pastoralism in Eurasia*, ed. D. R. Harris, 142–58. Washington, DC: Smithsonian Institution Press.
Zohary, D., and P. Spiegel-Roy. 1975. "Beginnings of Fruit Growing in the Old World." *Science* 187:319–27.
Zuckerman, L. 1998. *The Potato*. London: Macmillan.

Index

Agricultural crops are indexed under their name in American English; ornamental ones, under the scientific name (usually just a genus name). In most instances, under a plant name entry, the names of varieties and types or races of plants are indexed before other references. Only the varieties regarded as most significant are included in the index. Experimental stations, farms, and institutions tend to change their names frequently; they are listed together under a country or U.S. state heading under "experimental stations/farms/institutions."